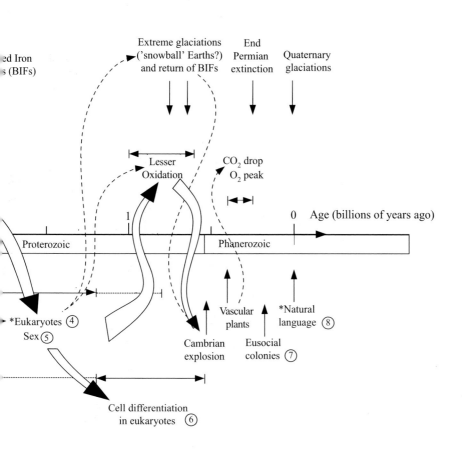

Revolutions that made the Earth

Revolutions that made the Earth

Tim Lenton and Andrew Watson
University of East Anglia,
Norwich, UK

OXFORD

UNIVERSITY PRESS

OXFORD

UNIVERSITY PRESS

Great Clarendon Street, Oxford OX2 6DP

Oxford University Press is a department of the University of Oxford.
It furthers the University's objective of excellence in research, scholarship,
and education by publishing worldwide in

Oxford New York

Auckland Cape Town Dar es Salaam Hong Kong Karachi
Kuala Lumpur Madrid Melbourne Mexico City Nairobi
New Delhi Shanghai Taipei Toronto

With offices in

Argentina Austria Brazil Chile Czech Republic France Greece
Guatemala Hungary Italy Japan Poland Portugal Singapore
South Korea Switzerland Thailand Turkey Ukraine Vietnam

Oxford is a registered trade mark of Oxford University Press
in the UK and in certain other countries

Published in the United States
by Oxford University Press Inc., New York

© Timothy Lenton and Andrew Watson 2011

British Library Cataloguing in Publication Data

Data available

Library of Congress Cataloging in Publication Data

Data available

Typeset by SPI Publisher Services, Pondicherry, India
Printed in Great Britain
on acid-free paper by
CPI Antony Rowe, Chippenham, Wiltshire

ISBN 978–0–19–958704–9

1 3 5 7 9 10 8 6 4 2

For our families

Preface

———◦∞◦———

This book tells a strange and fascinating story—the entangled histories of life and this planet. Our central idea is that a very few profound revolutions have made the Earth as we know it. Each revolution can be traced back to unlikely innovations in the evolution of life, and each involved radical changes in the non-living environment. For the most part, these revolutions are buried in the dark ages of Earth history, known traditionally as the Precambrian, but there is one exception: what we humans are doing now to transform the planet could be the start of a new revolution.

It is hard not to feel a rather urgent concern about how we are altering the Earth. Are the heralds of the climate change apocalypse right to argue that we are dooming ourselves and much of the rest of the biosphere? Rather than dwell on the present to try and answer this, our approach is to look at the past—and examine the course of previous revolutions of the Earth. Life, in every case, was the instigator. Often the revolutions led to near-catastrophe. Ultimately stability was always restored, but these happy conclusions were not inevitable; if it hadn't been that way, we wouldn't be here to remark upon them. Our contention is that the past offers some valuable lessons that can help us navigate the troubled waters that lie ahead. It also reminds us what an utterly remarkable planet we live on.

The revolutions we trace are best viewed as fundamental transformations of the whole system of life coupled to its planetary environment, which we call the 'Earth system'. What exactly is that? Well, we mean the many processes that interact together to set the living conditions at the surface of the planet—atmospheric and ocean composition, temperature and climate, and the life on the land and in the ocean. All-encompassing as this definition is, it nevertheless describes a very coherent system whose properties can only be fully understood by considering it as a whole.

If this sounds very much like James Lovelock's 'Gaia' to you, you're absolutely right. 'Gaia' and the 'Earth system' are for us, close to being synonymous. However, many scientists still react negatively against the idea, or even just the name, of Gaia. Based on a too-literal reading of Lovelock's earlier books and articles, or perhaps instinctively distrustful of such a grandiose theory, they associate it with a teleological viewpoint, an assumption that the biology of the planet somehow knows what's good for it. We disagree with this criticism, but we fear that the name 'Gaia' is so closely tied to Lovelock that it is in a sense defined by his views. 'Earth system science' may be less poetic and resonant, but it is also less personalized and polarized.

Why 'revolutions' that made the Earth? Well, the changes we describe were fundamental ones for life and the planet. They were triggered by new evolutionary innovations, resulting in great and often violent shifts in the environment, which in turn killed off many existing species. Out of this chaos emerged a new system and new life forms. So, in some sense there was a forcible overthrow of incumbents by a new regime. However, unlike human political revolutions there was no foresight or planning on the part of those driving the revolution—at least until now. 'Revolution' also suggests cyclic change, and as we will show, there are theoretical reasons why revolutions have recurred, at long intervals through the planet's history.

We've written for a reader with a science education and a keen interest in how we came to be here, but we have assumed little specialist knowledge. Inevitably we have to cross some complex scientific terrain, and we have put some more technical or non-essential material in boxes alongside the main text, so as not to interrupt the narrative. At the same time we think we have something interesting to say to our colleagues, and we use extensive references to show the links to existing work. The content ranges over the biological and geological sciences, but it is in the synthesis of the two that we think a truly 'Earth system' science emerges. We have attempted such a synthesis, but our belief is that it should still be readable and, we hope, entertaining to the uninitiated.

The idea for this book crystallized in 2003 out of a very special kind of scientific meeting, a Dahlem Conference on 'Earth system analysis for sustainability'. In this week-long retreat in the beautiful outskirts of Berlin, we first began to flesh out the connection between major transitions in evolution and revolutions of the Earth system. Many of those present had an input at the start, especially Eörs Szathmáry, John Schellnhuber, and Wolfgang Lucht.

The prehistory of this book, owes a great deal to Jim Lovelock. Andrew was one of Jim's very few PhD students, and Tim in turn was one of Andrew's PhD students, part supervised by Jim. Lynn Margulis has also been an inspiration to both of us, with her adventurous intellect, undying enthusiasm, and passion for teaching. If there is a family of Gaia-inspired researchers out there then we represent two of its generations. As in all families there have been some disagreements (and what we have written may spark more), but we want to make it clear that whatever you think of Gaia, a book like this could not have been written without Lovelock and Margulis' founding contributions to Earth system science.

At the same time, there are several others who have played key roles in shaping the subject and our thinking. Mike Whitfield has been a great support to both of us, especially in our time in Plymouth where Andrew worked for many years, and Tim started his PhD. Jim Kasting and Lee Kump, both at Penn State University, have been championing Earth system science as a subject for decades. The late Bill Hamilton and the late John Maynard Smith both helped Tim sharpen his thinking about Gaia and natural selection. We'd also like to thank Dick Holland, Peter Liss, John Raven, and David Wilkinson. There are many others we haven't named but you will find their work referenced within.

A number of our students and postdoctoral researchers, past and present, have helped develop the ideas, especially Noam Bergman, Richard Boyle, James Clark, Colin Goldblatt, Ben Mills, and Hywel Williams. Andy Ridgwell (who studied with Andrew alongside Tim) deserves a special mention.

Our attempt to write 'scholarly popular science' has proved a difficult trick to pull off, at least for a couple of first time authors. So, we thank the friends and family who have helped by commenting on earlier drafts, especially Peter Horton, Oliver Morton, John Lenton, Tee Rogers-Hayden, Jackie, Adam and James Watson.

A Reader's Guide

The book covers terrain that ranges in difficulty from easy to strenuous, in the fields that would traditionally be classified as astronomy, geophysics, geochemistry, and biology. Some readers may find the mix of material hard to navigate smoothly, so here we give a quick overview of what to expect.

We have opted for a hierarchical structure of six parts, each containing three or four chapters. Part I is Introductory. We preview the main ideas and material we will be covering, in non-technical mode.

Part II is Theory. We explore two theoretical models, which we subsequently use frequently as 'lenses', through which to view Earth history. In the first of these we reason backwards from our own existence, and identify key rare events in Earth history. In the second we switch to reasoning forwards, exploring the properties of an Earth system dominated by feedbacks. These chapters have some technical content, but they should not be too taxing to those with some general scientific knowledge. Both Parts I and II range over the whole of Earth history.

The remaining four parts proceed chronologically. The technical heart of the book starts in Part III, which focuses on the first great transformation of the Earth system. This hinged on the creators of oxygen gas and its abrupt accumulation in the atmosphere, halfway through the planet's history. We call it the oxygen revolution. Part IV we call the complexity revolution and it tackles the transformation which started with the evolution of a new type of cell, and culminated in the appearance of complex creatures from out of the turmoil of extreme ice ages and rising oxygen, at the 'Cambrian explosion'. Each of these parts begins with a chapter on the biological innovation involved, before moving to chapters on the Earth system response.

Part V covers the interlude between the Cambrian and the present, the part of Earth history that many studies of evolution and geology spend nearly all their time examining. The Earth system transformations during this period are less revolutionary, but we know much more about them and they serve to illustrate important properties of the system.

Finally, Part VI addresses whether we are entering a new revolution. We examine the instability of the recent Earth climate, before summarizing the rise of our own species to prominence as an Earth system phenomenon. Then comes perhaps the most ambitious part of the book, where we attempt to apply what we have learned about the Earth system to the future. The penultimate chapter summarizes the common features of Earth system revolutions. In the last chapter, we examine possible routes ahead.

We believe that if we use our deepening understanding of the Earth system wisely, we can avoid the most painful consequences of past revolutions. It is possible to find a path to a new and dynamically stable state which is secure, both for ourselves and for the majority of our fellow-traveller species on Earth, in an indefinitely sustainable world.

Tim Lenton and Andrew Watson

University of East Anglia, Norwich, UK
May 2010

Contents

———⟳◦⬦◦⟳———

List of Abbreviations

ADP	Adenosine Di-phosphate
AP	Anoxygenic Photosynthesis
ATP	Adenosine Tri-phosphate
BIF	Banded Iron Formation
DNA	Deoxyribonucleic acid
EPICA	European Project for Ice Coring in Antarcica
ESF	European Science Foundation
FOXP2	Forkhead Box Protein P2
HIV	Human Immunodeficiency Virus
LHB	Late Heavy Bombardment
LHC	Light Havesting Complex
LUCA	Last Universal Common Ancestor
MIF of S	Mass independent fractionation of sulphur
NADPH	Nicotinamide Adenine Dinucleotide Phosphate
NASA	National Aeronautical and Space Administration
NPP	Net Primary Productivity
OP	Oxygenic Photosynthesis
PAL	Present Atmospheric Levels
PETM	Paleocene Eocene Thermal Maximum
PDF	Probability Density Function
PSI	Photosystem I
PSII	Photosystem II
RECONS	Research Consortium for Nearby Stars
RNA	Ribonucleic Acid
SETI	Search for Extra-Terrestrial Intelligence
SMS	Szathmáry and Maynard Smith
UV	Ultraviolet radiation
WSC	Water-splitting Complex

PART I
INTRODUCTION

1
Origins

You are slowly awakening from the deepest of sleeps. Your sleep has lasted since the beginning of time. Dreams still echo in your head, myths of creation, voices, music, fragments of beauty and terror. They seem to be remembrances of your past but, as you try to focus on these memories and dreams, they dissolve, leaving you without the explanation for the question in your mind: where do we come from?

Looking up, you see the night sky. Your senses are growing quickly keener and telescopes assist your vision. You see the Earth is not the centre of all things, but that with its sister planets it orbits the Sun. As your vision expands you are shocked to find the Sun is but one tiny dot, quickly lost in the outer swirls of a great galaxy, the galaxy in turn soon a pinpoint among billions of others, stretching out in all directions toward infinity. The distances are unimaginable, and between the lonely galaxies stretch vast deserts of vacuum. As your vision extends into radio frequencies, you see the most distant thing it is possible to see, the faint afterglow of the fireball from which the universe emerged.

Humbled by the enormity of the cosmos, you turn back to the Earth and see it slowly spinning in space. You descend to the surface and begin to look around. All around you the world is alive! Billions of organisms, of every possible variety, inhabit the same world you find yourself in. You are awed by the beauty of this world, animated, full of intricate patterns, colour and sound, perfectly displayed in its planetary setting like the rarest of jewels. It holds you spellbound, and increases your puzzlement still further.

Looking inside yourself, you see your blood circulating, Seizing a microscope you focus on your tissues and discover you are not a unitary thing at all, but a colony of separate cells that cooperate so closely that they become a single organism. Gazing deeper into your inner, cellular and molecular structure and that of the life around you, it is soon obvious that you are related to the other animals, and indeed that all living things form one kin. You see that you are fundamentally the same as them in all respects but one: they seem to be completely fulfilled by simply being, whereas you are driven to ask more and more questions...

1.1 The sleeper wakes

Collectively, we have experienced this awakening into consciousness very recently. In the West we might date the moment we opened our eyes to the Renaissance, or perhaps earlier, to the ancient Greeks. Other cultures might date it differently, but wherever we mark the birth of our collective consciousness, it is just yesterday in human history. So recent is our waking that dreams still fill our heads and we can barely separate them from reality. It is as if we are still in those seconds of confusion we experience after waking, still trying to make sense of who and where we are, in a world that we are clearly a part of and yet somehow different from.

We are ceaseless, inveterate storytellers, and we have made up stories to explain to ourselves how the world was formed and what our place in it is. For the most part these are myths and metaphors, and we recognize them as such. However, in just the last few hundred years, with the help of the 'scientific method', we have begun to construct a story that is different from these. After many revisions we can hope that this new story will converge on what actually happened. This method consists of making narratives of possible cause and effect (dignified by calling them hypotheses, but they are not much different from any other story) and then testing them against the real world to see if they are consistent with what we see. Where they are not (and this is the crucial point, which separates this method from other storytelling) the narrative needs revision. Where they are, we accept them, for now.

In a brief time, the instruments of science have extended our senses immeasurably, with not only telescopes and microscopes, but with hundreds of other inventions. Our eyes have been opened to ever more of both the world outside and the world inside ourselves. Copernicus first suggested that the Earth orbits the Sun, but it was Galileo who truly shattered the illusion of celestial perfection with us at the centre, by taking one of the first microscopes and turning it on the sky as a telescope. Far from a perfect sphere, the Moon was revealed to be pockmarked with overlapping craters, and Jupiter was found to have its own moons. Our (rude) awakening had truly begun. Later, Anton van Leeuwenhoek improved the microscope and turned it on living nature to discover microorganism.

Much more recently we have heard the echo of the big bang through a radio telescope, and seen down almost to the molecular scale with electron microscopes. But perhaps the biggest eye opener of all has come with our ability to leave the Earth and look back at it from space. This has given us a macroscopic view of our home planet and kicked off a revolution in thinking about the Earth and our place in it. At the same time, we are building up a wealth of evidence of how the Earth has changed over its history. And we now know that van Leeuwenhoek's microorganisms are the longest residents of this planet and the most important for its healthy, circulatory functioning.

In this book, we want to weave many strands of science together to present a narrative of Earth's history and how we came to be here. It is a 'systems view' in that it considers the evolution of life and of the non-living environment as one coupled, indivisible process. This process has not been smooth and continuous: a series of just a few revolutions have created us and the world we enjoy today. Each revolution was inherently difficult, and each was built on the previous one. They all had to occur in the sequence they did to allow us—a conscious 'observer' species—to evolve, which could then look back and marvel at this history.

Though each revolution is different in the detail, they share an overall pattern. Each has involved major reorganization in the system, and at each it has moved stepwise towards greater energy utilization, greater recycling efficiency, faster processing of information, and higher degrees of organization. Each revolution has been characterized by near catastrophes and with no certainty of a successful outcome—the revolutions only appear 'successful' because we are looking at them with hindsight, and because had they failed, we would not be here to remark on them.

We can fit our own emergence into this sequence of revolutions—it is the most recent of them, though the one that we represent is far from completed as yet, and may or may not be 'successful'. Humans and the evolution of the Earth system are inextricably intertwined, and there are important lessons to be learned from a study of the past revolutions for our relationship with the planet. These lessons could be vital—they may help us to navigate the dangerous transition that we represent, in safety.

In this first, introductory part of the book, our aim is to sketch out this striking picture of how we came to be here. We will explore a series of perspectives on Earth and its history, and on us and the evolution of life. A variety of instruments will aid our vision, and we will try to explain where some of the most important pieces of evidence come from. First we are going to look back in time from above the Earth. Then we will work our way forwards, starting with the formation of the solar system and the Earth, before descending to the surface of the planet to look for rock-bound evidence of the first fossil life. In Chapter 2 we continue the journey through Earth history, focusing on the key changes in the physical and chemical state of the environment. Then in Chapter 3 we will look inside ourselves to see a hierarchical structure that reflects the major transitions in the evolution of life, before finally in Chapter 4 beginning to connect these to the Earth-scale revolutions, and seeing how we, as a species, fit in.

1.2 The view from above

Picture the Earth in your mind's eye. The blue oceans, the swirling white of the clouds, the mottled greens and browns of the continents, and the icy polar caps. This view of our home planet was revealed to us in just the last few decades, most memorably by the crew (Eugene Cernan, Ronald Evans, Jack Schmitt) aboard Apollo 17—the last manned lunar mission. Their photograph of the full disc of the Earth, the viewer's spacecraft in perfect alignment between the Earth and the Sun, is now the most widely produced image of all time.

Earth has not always looked like this. Suppose you can rewind time, slowly at first—you see the field of lights on the night side of Earth extinguished, these having been lit by industrially inclined humans only in the last few decades. Rewind at a million years a minute, and you can watch ice sheets regularly advancing and retreating across the Northern Hemisphere, every six seconds or so, roughly at the rate that you breathe. Accelerate the rewind to a hundred million years a minute, and you can review the whole history of the planet in three quarters of an hour.

The first thing you'll notice at this speed is the continents skating over the surface. South America and Australia are moving south and join with Antarctica after twenty seconds, and suddenly Antarctica loses its ice sheet and turns green. North America sails back towards Europe, South America towards Africa, and the Atlantic closes up in less than a minute. The continents assemble into the supercontinent Pangaea after two minutes. As you continue into the past, you'd see the white flash of an occasional ice cap, but it's the exception. Usually there is no ice on the planet at all. That is, until something strange happens about six minutes into the rewind. In the blink of an eye, the entire sphere is suddenly encased in white ice, and stays that way for a full ten seconds. It clears to blue again, then another ten seconds or so later repeats the cycle. You've just witnessed two episodes of 'Snowball Earth' and fast rewound through one of the revolutions on which we want to focus.

There are changes too in the vegetation covering the land. The brown grasslands across much of the continents quickly disappear near the beginning of the rewind, but the chlorophyll signature of the forests remains for fully three and a half minutes before giving way in stages to bare rock. After life disappears from the continents, you might strain to see it in the oceans with your unaided eyes, because they have only rather coarse three pigment colour discrimination. But using spectrometers in the visible and infrared, things become much clearer. The chlorophyll in the surface oceans still stands out, and you can see the telltale signature of life in the atmosphere—absorption lines indicating the coexistence of ozone and methane tell you that there is still an active biosphere on the planet. As James Lovelock first realized, this is a 'biosignature'. These gases quickly react with one another and unless there is a source—life—producing them continuously, and using copious energy in the process, they could not coexist at the concentrations you observe. In chemical terms, the Earth's atmosphere is far from thermodynamic equilibrium. You could check this by turning your instruments on Mars or Venus, two planets without life, and with correspondingly 'dead'—close to equilibrium—atmospheres, dominated by carbon dioxide.

Back on Earth, the atmospheric composition changes subtly with time. In particular, once into the period before the Snowball Earths, you see a different mix of trace gases appear, giving a new set of spectral lines, and indications of much lower oxygen concentrations. Much further back in time, more than twenty minutes into your fast rewind, you see a much bigger change: very suddenly, all evidence for oxygen disappears from the atmosphere. Simultaneous with this, there is another sequence of glaciations including at least one Snowball Earth event. This is another of the great revolutions.

Then...the picture goes fuzzy. You have reached one of the limits to what we know. We aren't really sure *what* was in the atmosphere, this far back in time. We know some things it ought to have contained, but we are pushing the limits of our knowledge. We know the early atmosphere contained plenty of 'greenhouse 'gases—carbon dioxide, methane, and probably others too, all of which absorb heat radiation in the infrared, so it will show more absorption lines in your spectrometer than ever. We know this, because it would have been necessary to keep the planet warm given another change which you'll have noticed by now: the Sun is fainter back then. You also notice that the continents are getting smaller and the oceans larger. With the help of spectrometers, you can still see telltale signs of life in the surface of this vast ocean. But the colour you see is shifting, among the red, purple and green, indicating a mix of different pigments and types of photosynthesis. You see the characteristic colours shrink back into ever smaller areas, and then, perhaps forty minutes into the rewind (we're not sure just when) they fade away. It's hard to see from above, but organisms are disappearing from the planet and you're reversing through the origin of life.

Finally, about forty-six minutes into the rewind, you come to the point where the planet begins to disaggregate. Slowly at first, and then in the space of a few seconds, your home planet falls apart in front of your eyes.

Now you've reached the beginning of the story, let's watch the interesting parts again, forwards, and more slowly.

To help you orientate yourself as we cross this vast swathe of time, let us start drawing a timeline (Fig. 1.1). Our rewind, at a hundred million years a minute, covered 4.6 billion years, and our timeline will start 4.6 billion years ago with the formation of the solar system and the Earth, and end at the present. The main divisions are in billions of years (that is every 1,000,000,000 years). We've started by sketching in the

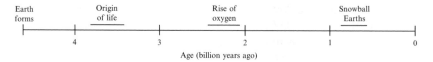

Fig. 1.1 A sketched timeline of the revolutions as seen from above

rough intervals of each of the big revolutions that jumped out of the fast rewind. As we go along we will fill in the details of other key events.

1.3 Setting the scene: origins of a habitable planet

As you begin to play the tape forwards, you see a disk-like bright cloud—the solar nebula* from which the solar system forms. The nebula contains large amounts of hydrogen and helium gas, originally manufactured in the big bang, but also 'dust'—heavier elements which were not made primordially. You see bright spots within the nebula, which are massive stars, and in their nuclear powered hearts, heavier elements up to iron and nickel are being synthesized—including the carbon that is essential for life as we know it. This nuclear synthesis leaves a characteristic signature: The elements generally get less abundant as their mass increases, but iron and nickel buck the trend because they have unusually stable nuclei. Also, elements of even atomic number (such as carbon and oxygen) are more abundant than those of odd atomic number (such as nitrogen and phosphorus)—and this will have a profound influence on the future development of life.

As you watch, there is a blinding flash of light—a supernova explosion—the death throes of a massive and brightly burning, but very short-lived, star. This explosion seeds the dust of the nebula with even heavier elements, including short-lived radioactive ones. The supernova explosion sends a shockwave through the nebula, producing pockets that are denser with material. The picture zooms in on one of these regions and, as you watch, a central blob begins to coalesce—this is the beginning of a new star—our Sun. Around it is a swirling cloud of gas and dust that is destined to become our Solar system.

---

* If you want to know what such a nebula looks like, take a look at the Orion Nebula, which is a region where young stars are forming today. To find it, you'll need to get away from light pollution on a clear night when Orion is in the sky (winter time in the mid-latitudes of either hemisphere). Find Orion's sword, which hangs down (towards the horizon if you're in the Northern Hemisphere) from his belt and looks to be made up of three stars. The middle star looks fuzzy and is in fact a group of many young stars. Through binoculars you can see it is a nebula, though it is too far away to see any disk shapes.

You slow the tape down to see what happens next (*1*). Seemingly in the blink of an eye (10,000 years in real time), the dust grains collide, stick together and grow to the size of boulders and kilometre wide asteroids—in a process called accretion. Then (over the next million years), in the region nearer to the Sun, these larger objects continue to collide and coalesce to from 'planetary embryos'—Moon-to-Mars size objects (*2*, *3*) (Mars in fact, might possibly be a leftover embryo which never grew further). In a final stage (over a period of some tens of millions of years), these baby planets smash together, until there are just a few survivors in well-spaced orbits in the inner Solar System. These are the dense, metal-rich terrestrial planets—you start to count them out from the Sun; Mercury first then Venus and then—to your great surprise—the young Earth has a Mars-sized companion, but it is not Mars because you can see that much further out.

You focus on the Earth and watch as the planet condenses—iron, nickel and other metals sink to the core, above that a less dense molten magma of rocks rich in oxygen, magnesium and silicon forms, and above that the crust solidifies. It is now about 50 million years after the start of the accretion process, and the Earth's smaller companion looks dangerously close. Sure enough, before your eyes, the two objects collide. The energy of the impact is such that it melts the surface of the young Earth and splashes out a huge amount of vaporized rock. This material goes into orbit around the Earth and starts to solidify into rings of debris that quickly coalesce to form the Moon. You have just watched the grand finale of Earth's accretion process. The colliding planet that gave birth to our Moon has been aptly christened 'Theia' (*4*)—in Greek mythology the mother of the moon goddess Selene (see Box 1.1 to find out how we know Theia existed).

Box 1.1 How do we know Theia ever existed?

The moon-forming impact is just a theory, first advanced in the mid-1970s, but it has survived well and looks increasingly probable because it explains some otherwise puzzling observations. The first of these is that Earth is not really a single planet at all, but the bigger partner of a binary planetary system. The Moon is about 25% as big as the Earth, huge compared to the moons of Mars, which seem to be captured asteroids. It is comparable in size to the satellites of the giant planets Jupiter and Saturn, but these have parent planets that are ten times Earth's size and hundreds of times its mass. No other solar system body has a satellite so large compared to itself[†]. It seems that our Moon must have been

---

[†] With the exception of distant Pluto, which also forms a binary system with its moon Charon, but Pluto is tiny and severely errant in its orbit. It was not formed in the same process as the other planets. In fact it is officially no longer a planet at all, but a denizen of the Kuiper belt, in the outer darkness beyond the orbit of Neptune.

formed by some not-particularly-likely event, to explain why we have one but the other inner planets do not. Also, though moon rocks have compositions very similar to Earth, the Moon lacks a heavy core of iron, unlike Earth and all the other rocky planets. These cores contain a good part of the planetary mass, so the Moon overall is much lighter than you would expect by scaling from its size relative to the Earth. This lack of iron means it can't be explained as forming in the same way as the other planets. However, it is neatly explained by the giant impact hypothesis if, by the time of the impact, the Earth's core had already been formed by the melting of rocks and sinking of the iron to the centre. Then the material that was ejected from the Earth and which eventually formed the Moon would have been iron-depleted.

Whilst there is all this exciting action in the inner Solar system, if you look further out from the Sun, you will see things are proceeding a little slower. Although Earth and Mars are nearly full size, the gas giants Jupiter and Saturn are still forming. Their sheer mass is producing a disruptive gravitation that prevents the asteroids between Mars and Jupiter from coalescing, leaving instead a scattered belt of rocks. During the earlier phases of planetary growth, you can watch the presence of gas and small debris acting as a drag on the growing asteroids. They tend to settle towards the central plane of the nebula and their orbits remain nearly circular. However, in the final phase, freed from the damping effects of this envelope, the planetary embryos start to behave rather differently. Collisions and close encounters scatter them into elliptical orbits that can cross from one zone to another. As you watch Jupiter and Saturn grow, some of these rocks are diverted into elliptical orbits that cross the inner Solar system, bombarding the inner planets.

You turn back to the Earth and see that this meteorite bombardment is extremely disruptive—some of the impacts are large enough to melt parts of the crust—but it also brings with it water and other 'volatiles' that are essential for life, including the nitrogen that comes to comprise most of the Earth's atmosphere. It is thus critical for the future habitability of the Earth. These volatiles are missing initially because temperatures in the inner Solar System are too high for them to condense as ice at the orbit of Earth. Instead, the 'snow line' in the solar nebula is much further out, in the region where the giant planets are forming, and from which the meteorites and asteroids come. These embryos carrying volatiles anoint the planets of the inner solar system—all of them, not just the Earth—with water and other life giving substances (5).

Looking back out into the belt between Mars and Jupiter you can see some of these asteroids, including 1-Ceres—the largest example still there today. At nearly 1000 kilometres across, it is an embryonic planet in size, but it is doomed never to grow further because of nearby Jupiter with its huge disruptive gravitational pull. 1-Ceres is no potato shaped asteroid however, but nearly round like a proper planet, indicating that it has begun the process of core formation. It is not very dense, probably because it contains a substantial amount of water ice (6). A few encounters with bodies like 1-Ceres might have been sufficient to deliver all Earth's water.

Thus, water delivery to Earth and the other inner planets was a haphazard affair. How much water a planet ends up with depends on a few chance encounters with water-rich asteroids during its late formation phase. If Earth had been unlucky it might have got nothing, or—equally bad news from the point of view of land creatures like us—it might have ended up with several times as much water and have nothing but oceans at its surface. As a name, 'Earth' is already barely appropriate to a planet that is 70% covered in water. A few more asteroid loads and it would definitely have been 'Ocean'.

1.4 First rocks, first life

So far, about 100 million years after the Solar System first formed, you have followed the building of the planet to the point where it is fully formed. The Moon is in place, as are some essentials for the life we hope it is going to host, like water, nitrogen and carbon dioxide. To follow what happens next you need to leave your viewpoint above the Earth and descend to the newly formed surface—treading warily of course, as it's hot in places. And don't forget your spacesuit, because the atmosphere won't be breathable for another several billion years. You'll probably be keeping a nervous eye on the sky for incoming material, and while looking up you'll be able to admire the Moon, which is much closer and therefore several times larger in the sky than it is now. When first formed it was within a few tens of thousands of miles of the Earth, and it has been receding from us ever since. You'll also notice the days are shorter—because the Earth is spinning faster. As the Moon has drifted away it has gained angular momentum from the planet, slowing the Earth's spin.

You are down here to examine the rocks and to try to discover when life first becomes established. There is actually very little to stand on—only a few volcanic islands protrude from the vast ocean. These are the first pieces of the continental crust, which will steadily accumulate over time. The rocks you stand on are less dense than those at the bottom of the ocean—and hence the young continents 'float' on the ocean crust below. A few pieces of this continental crust stick up above the water line and expose the first dry land. The crust as a whole is, like today, comprised mostly of silicates—rocks that are dominated by a combination of oxygen and silicon (SiO_4). Analyse the rocks you stand on further and you will find aluminium, iron, calcium, sodium, potassium, magnesium and a host of less abundant elements.

The rocks are the record in which virtually all of our information about the Earth's history is written. The problem is that the rocks we have left today are a far from complete record of the past. In particular, reconstructing what you would have seen at any time in the first billion years after Earth's formation is really a challenge, because there is almost nothing left from that time. The reason is that the crust has been repeatedly recycled by the same process of plate tectonics that moves the continents around. Ocean crust is continually being formed at mid-ocean ridges and subducted under the continents and as this happens, pieces of continental crust floating on top are occasionally smashed together, variously churned up, and sometimes dragged down to great depths, where heat and pressure 'metamorphose' the rocks. It is as if we are investigators of a long forgotten crime: there's almost nothing to go on, the crime scene has been heavily disturbed and the intervening billions of years of tectonic activity have trampled all over the evidence. None of the rocks that were at the surface in the first billion years have survived unchanged. In the few cases where they have escaped total destruction, the process of metamorphism has erased most of the clues that they might otherwise contain about the earliest Earth. There are still clues in these rocks, but we need skill, technology, patience and some imagination to reconstruct what was going on back then.

The only objects you can find on the earliest Earth that are still with us today are tiny crystals of a mineral called zircon. The name "zircon" sounds like some exotic substance, but in fact it's a very common mineral—the combination of the metal element zirconium with the abundant silicate (to form $ZrSiO_4$). Zircon grains are typically a few tenths of a millimeter in size and are very hard, so they can survive long after the rock in which they were first formed has been weathered away. Remarkably, single micro-crystals of zircon retain a chemical memory that enables us to date their origin precisely—to within a few tenths of a percent—using highly sensitive instruments called mass spectrometers (see Box 1.2 for an explanation of how).

Box 1.2 How do we know the age of zircons?

When zircon first solidifies, lead is rejected from the crystal structure (leaving it lead free), but impurities of uranium are present. Uranium is radioactive and decays to lead, meaning that as time passes, lead accumulates at a precise rate, set by the half-life of uranium. By measuring the proportions of uranium and lead, we can work out how long it was since the crystal formed. In fact it is even better than that, because there are two isotopes of uranium which decay with different half-lives to two different isotopes of lead, both of which can be used to cross-check the dating. Nature in this case seems to have conspired to help us unlock her secrets (she's not always so helpful!).

Zircons can come from any age, but the oldest ones so far found are in western Australia today, and date from 4.4 billion years ago. Although there are no intact rocks from that time, the zircons provide evidence that there must have been a crust, probably like modern continental crust, even at this early date (7). In addition, they contain evidence that they were formed in water, and therefore the oceans were already in place (8). Apparently, the basic conditions in which life would have been possible existed 4.4 billion years ago. But, at the time of writing, there's no more evidence from this distant time.

To see more you need to move forward a whopping 400 million years to 4.0 billion years ago. Then you will find probably the oldest intact rocks that are still with us today (older candidates have recently been suggested, but their dating has been questioned). Nowadays these rocks are exposed in the northwest territories of Canada (9). They seem to be almost entirely reprocessed igneous rocks—meaning that they formed from the cooling and solidification of molten magma. Not surprisingly, there's no evidence of life in them, but none would be expected given this geological origin. Ideally, we need some sedimentary rocks, laid down in contact with the surface environment, to have any chance of them bearing a signature of life.

To find sedimentary rocks, you need to move forward another 200 million years, which is a welcome leap, because it scoots you through an otherwise unpleasant interval of frequent and large meteorite impacts known as the Late Heavy Bombardment, or LHB for short. This appears to be the final act in the genesis of the modern Solar System. We know about it from dating of the Moon's surface from rocks recovered by the Apollo missions. These yield dates of 3.8 to 4.1 billion years ago, suggesting they crystallized at that time. The Moon's surface was thoroughly reworked during this time, so that nothing earlier remains, apparently due to a sustained bombardment of the inner Solar System by meteorites. The dates cluster at around 3.9 billion years ago, and probably can't be explained simply as the tail of the planetary accretion process, which should have been over and done with long before then. We don't fully understand why (or even whether) the LHB occurred, but a neat theory recently published suggests it may be related to the settling into their final orbits of the outer planets, and in particular to a resonance between the orbits of Jupiter and Saturn that arose some 700 million years after they first formed (10). This disturbed the orbits of Neptune and Uranus, and diverted swarms of asteroids from the outer

Solar System, sending them into highly elliptical orbits that crossed the paths of the inner planets. Whatever the cause, the Apollo rocks seem to bear witness that the surface of the Moon, and by implication the Earth, suffered a period of bombardment by large objects, which occurred long after the initial accretion.

Arriving on the Earth around 3.8 billion years ago you breathe a sigh of relief as the bombardment has been over for about 50 million years. If you take a swim in the shallow waters of the time you can now see sediments being deposited that are still with us today, albeit heavily metamorphosed (so the following interpretation is not certain). These are the next oldest rocks and today they are found in a formation in northern Quebec, known as the Nuvvuagittuq supracrustal group (*11*). Digging into the sediments you find thin, alternating bands of iron-rich and iron-poor material. This is a banded iron formation, or BIF for short. BIFs are an extremely important type of sedimentary rock, not least because they are generally interpreted as evidence for life, and more than that, for a type of photosynthesis occurring in a reducing or low-oxygen environment (see Box 1.3 for an explanation of why). The BIFs you see being deposited could have been formed by an early evolved form of 'anoxygenic' photosynthesis different from the type of photosynthesis we are familiar with today—where oxygen is liberated from the splitting of water. In the case of the very old Nuvvuagittuq rocks, which are heavily metamorphosed, the best guess is that the BIFs were formed in association with some photosynthetic activity but we don't know which type.

Box 1.3 How do Banded Iron Formations (BIFs) form?

This is one of the enduring puzzles in geology. BIFs are not formed in today's world except perhaps in special locations like the Black Sea, where the water below the uppermost layers is anoxic. This is because the iron they are formed from has to have been previously dissolved in water, such that when it is laid down it forms smooth sediments. Iron will not dissolve in water to any extent in the presence of oxygen—it is then in its oxidized form, called ferric iron, which is extremely insoluble (iron may go rusty from years of being out in the rain, but it doesn't actually dissolve). But in its reduced form, called ferrous, iron does dissolve in quantity, and before the rise of oxygen in the atmosphere there would have been a great deal of iron dissolved in the oceans. So BIFs speak of oxygen-free, reducing conditions in the water from which they precipitated. Yet some process must have caused the iron to precipitate out by steadily oxidizing it—because the BIFs are composed mostly of oxides of iron. Many of these deposits in later rocks are very extensive and require huge amounts of such oxidation, occurring over long times. So BIFs are usually interpreted as evidence for photosynthesis, this being the only known process that can cause such extensive oxidation. The BIFs therefore attest to a world in which the ocean is mostly reduced, but in which there is biological activity, driven by some form of photosynthesis, and producing some oxidized products (that need not be oxygen gas), which are responsible for turning the iron from its reduced to its oxidized form.

Sure enough as you continue your swim through the shallow waters of the world 3.8 billion years ago, you think you catch the first tell tale signs of life—because some of the sediments at the bottom of the sea are rich in organic carbon. Nowadays the corresponding rocks are found in western Greenland in a formation called the Isua supracrustal group (*12, 13*). There are BIFs here as well, but more remarkably, from Isua there are specks of reduced elemental carbon in the form of graphite (yes, the same substance used as the humble 'lead' in a pencil may provide the first evidence of life on Earth). These specks of graphite have been found in a variety of environments, but the best preserved today are in a very small section of rock, less than 100 metres long, lying close to the edge of the Greenland ice cap (*14*). They have been described by Minik Rosing of the University of Copenhagen, a native Greenlander and the world's expert on the Isua rocks. Again, these rocks show evidence of being reworked by metamorphism to various degrees. In the least affected, finely laminated slate-like layers containing many carbon granules give them their darker colour, and these are interspersed with BIFs. In rocks subjected to more heating, the layers are progressively obliterated and the carbon granules disappear—an important observation because it shows that the carbon was present in the rock before it was reworked. With the help of an electron microscope you can see the individual granules are a few microns across—a size that might be expected if they were once individual cells.

With a mass spectrometer you can probe even deeper into these granules to find something truly remarkable: the isotopic composition of the carbon is 'light' compared to typical carbonate rocks, by an amount that suggests it was fixed from carbon dioxide by photosynthetic organisms. We are talking here about the ratio of the rarer carbon-13 isotope (^{13}C), in which the atoms of carbon contain 7 neutrons, to the common carbon-12 isotope (^{12}C), which has 6 neutrons per atom. This ratio is about twenty parts per thousand less in the carbon granules than it is in typical carbonates[‡]. Light carbon with this isotopic signature

---

[‡] Geochemists measure the concentrations of isotopes by their difference from a standard, and the carbon granules in the Isua formation have a 'delta-thirteen-C' of about 'minus twenty per mil', written as: $\delta^{13}C = -20\,‰$. The 'per mil' sign looks like a percent sign, but with an extra zero at the bottom, and whereas a percent means one part in a hundred, a per

is usually interpreted as a product of life, specifically, of the capture of carbon from carbon dioxide. Organisms able to do this today all use a process called the Calvin (or more accurately Calvin-Benson-Bassham) cycle, in which a critical enzyme, called Rubisco[§], does the actual binding to carbon dioxide. Rubisco is a key component of the modern biosphere, and is probably the most abundant protein on the face of the Earth. It is thought to be very ancient, and very similar versions of it would have been used by even early carbon-fixers. From our point of view, trying to unravel the history of the planet, it has a particularly useful eccentricity: it has a greater affinity for 'ordinary' CO_2, made with ^{12}C, than for CO_2 made with the heavier isotope ^{13}C, and this is the main reason for the 'light' signature of carbon that has been fixed by organisms. According to Rosing, this signature in the Isua carbon granules is telling us that the carbon was fixed from carbon dioxide, by photosynthetic bacteria.

Because the rocks are heavily reworked, you aren't going to find any true fossils here. These specks are the best you can hope for, and might very well be what we would expect to find in metamorphosed sediments that were originally laid down if a vigorous microbial ecosystem was already flourishing in shallow water. Rosing makes the case that this is evidence not only for such an ecosystem, but for *oxygenic* photosynthesis—the most complex kind, that modern plants use and that totally dominates the Earth's biogeochemical cycles today. The debate is far from settled. Our view (which may be wrong), is that Rosing is most likely correct to infer photosynthesis was taking place, but wrong about it being the oxygenic variety. (We'll return to this and explain the thinking behind it in Part 3).

To find organisms that have left true fossils, you have to go forward another few hundred million years, to the interval 3.5 to 3.2 billion years ago. By this time there are a few examples, found today in South Africa and western Australia, of rocks that are comparatively unaffected by tectonic reprocessing. The word 'fossil' calls to mind dinosaurs, or at least the spiral shells of ammonites, and if you are thinking along these lines you are going to be disappointed. You need the help of a microscope to see them, but all the same, they are more important to our understanding of life on Earth than a museum full of dinosaur bones. At present, the oldest microfossils that are generally agreed to be of bacteria are found in rocks from Swaziland deposited 3.26 billion years ago,

---

mil is one part in a thousand. The value of −20 ‰ means that the ^{13}C content of the granules is about 2 percent less than the standard (set by the international Atomic Energy agency in Vienna), which has a $\delta^{13}C$ of zero by definition. The standard is a carbonate, and carbonate rocks therefore typically tend to have $\delta^{13}C$ values around zero.

[§] Rubisco is a shortened form of its full name, which (you'd probably rather not know) is R̲ibu̲lose-1,5-b̲is̲phosphate c̲arboxylase/o̲xygenase.

and described by Andrew Knoll and Elso Barghoorn in 1977 (*15*). (There are many other descriptions of 'microfossils' from this interval (*16*), including ones from western Australia dated close to 3.5 billion years ago, but serious questions have been raised as to whether some of those structures are truly biological (*17*).)

The Swaziland fossils are rather featureless, round, micron-sized inclusions in the rock. However, this is no more and no less than you should expect—they look very much like modern bacteria, which are more often than not similarly featureless and round when viewed down a microscope, and they also resemble more modern microfossils which are unquestionably biological. Critically, the microfossils include cells caught in various stages of the act of division, as you might see in an active culture in a Petri dish—Fig. 1.2 shows a particularly good example. The original paper (*15*) shows the fossils alongside photo-graphs of modern, living, bacteria in the process of dividing, and the resemblance is so striking that you have to remind yourself that the fossils have been dead for three and a quarter billion years.

Now as you paddle at the seashore 3.3 billion years ago you can see reefs that are being built by these cells, or some of their cousins. They have left layered and banded structures, some of them very large, in the same early, relatively intact rock sequences (*18*). These 'stromatolites', as they are often called, are believed to be formed by the same process as are modern equivalents. They are found today in some extremely salty lagoons on the margins of the sea or lakes, where colonies of photo-synthetic and other bacteria produce similar structures. The bacteria

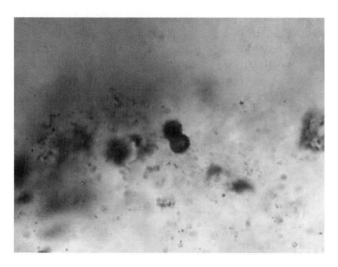

Fig. 1.2 Ancient (3.26 billion year old) microfossils caught in the act of cell division (*15*). [Photograph kindly provided by Andy Knoll of Harvard University.]

excrete mucilage and bind themselves together into a mat, which then acts to trap particles of sediment, building up into pillow-shaped layers.

1.5 Summary

You can now be sure there are thriving microbial ecosystems on the planet. The Earth has gone from a system of physics and chemistry alone to one including biology, and therefore evolution. We won't have a great deal more to say about the origin of life because so little is known about it. But hopefully you will have a flavour of the relationship between the earliest rocks and early life. There are two important points to emphasize: first, for hundreds of millions of years after the origin of the Earth there are no surviving rocks at all. Second, as soon as there *are* rocks of sedimentary origin that might be expected to contain any evidence for life, sure enough it is there. The signs are quite cryptic in the earliest rocks, and it is possible, for example, that the signs of life 3.8 to 3.7 billion years ago are being over interpreted by enthusiasts such as Minik Rosing. However, by the time of the South African and Western Australian rock sequences 3.5 to 3.2 billion years ago, there is no doubt, in our minds at least, that life is thriving on the planet. We can't put a firm date on the origin of life because it most likely occurs in a period where we don't have any rocks, but we can specify latest likely, and latest possible, limits and these will turn out to be very useful in our later discussions. Fig. 1.3 summarizes the main events and key pieces of evidence we have discussed from the first billion years or so of Earth history.

Before we move forward in time, let us take a final look at some of these early microfossils. They include photosynthesisers, there's no doubt, but what kind of photosynthesis? We're not sure. They have often been compared with modern cyanobacteria. Early cyanobacteria were the first organisms to conduct *oxygenic* photosynthesis—the most complex kind that splits water and liberates oxygen gas. They are still responsible for a fair chunk of this essential process in today's world, particularly in the open oceans. However, the argument that the first fossils are cyanobacteria rests on their similar appearance to modern

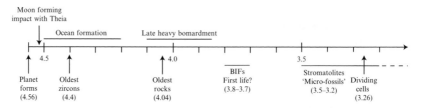

Fig. 1.3 Summary of evidence from the first billion years or so of Earth history

examples—and this can't be fully conclusive, because microbes are hard to tell apart from their appearance alone (even living ones, let alone multi-billion year old fossils). Cyanobacteria can make stromatolites and still do so today, but it does not have to follow that they were part of the first such ecosystems. Instead a good case has been made that some 3.4 billion year old structures, preserved in South Africa, were deposited by earlier 'anoxygenic' types of photosynthesizer (*19*), just as we think Minik Rosing's Isua graphite samples were.

We need to take another large step forward in time before we have convincing evidence for cyanobacteria. Sometime prior to 2.7 billion years ago in a discovery more fundamental to life on Earth than anything that humans have ever achieved, an ancestral cyanobacterium invented oxygenic photosynthesis, and set the Earth on a course that would transform it utterly, with the appearance of toxic, reactive, oxygen gas in the atmosphere.

References

1. A. Morbidelli, *Modern integrations of solar system dynamics. Annual Review of Earth and Planetary Sciences* **30**, 89 (2002).

2. K. Zahnle *et al.*, *Emergence of a habitable planet. Space Science Reviews* **129**, 35 (2007).

3. M. Ogihara, S. Ida, A. Morbidelli, *Accretion of terrestrial planets from oligarchs in a turbulent disk. Icarus* **188**, 522 (2007).

4. A. N. Halliday, *Terrestrial accretion rates and the origin of the Moon. Earth and Planetary Science Letters* **176**, 17 (2000).

5. S. N. Raymond, T. Quinn, J. I. Lunine, *Making other earths: dynamical simulations of terrestrial planet formation and water delivery. Icarus* **168**, 1 (2004).

6. P. C. Thomas *et al.*, *Differentiation of the asteroid Ceresas revealed by its shape. Nature* **437**, 224 (2005).

7. T. M. Harrison *et al.*, *Heterogeneous Hadean hafnium: Evidence of continental crust at 4.4 to 4.5 Ga. Science* **310**, 1947 (2005).

8. S. A. Wilde, J. W. Valley, W. H. Peck, C. M. Graham, *Evidence from detrital zircons for the existence of continental crust and oceans on the Earth 4.4 Gyr ago. Nature* **409**, 175 (2001).

9. S. A. Bowring, I. S. Williams, *Priscoan (4.00–4.03 Ga) orthogneiss from northwestern Canada. Contributions to Mineralogy and Petrology* **134**, 3 (1999).

10. R. Gomes, H. F. Levison, K. Tsiganis, A. Morbidelli, *Origin of the cataclysmic Late Heavy Bombardment period of the terrestrial planets. Nature* **435**, 466 (2005).

11. N. Dauphas, N. L. Cates, S. J. Mojzsis, V. Busigny, *Identification of chemical sedimentary protoliths using iron isotopes in the > 3750 Ma*

Nuvvuagittuq supracrustal belt, Canada. Earth and Planetary Science Letters **254**, 358 (2007).

12. S. Moorbath, R. K. Onions, Pankhurs. Rj, *Early Archaean Age for Isua Iron Formation, West Greenland. Nature* **245**, 138 (1973).

13. S. J. Mojzsis *et al.*, *Evidence for life on Earth before 3,800 million years ago. Nature* **384**, 55 (1996).

14. M. T. Rosing, ^{13}C-*depleted carbon microparticles in > 3700-Ma sea-floor sedimentary rocks from west Greenland. Science* **283**, 674 (1999).

15. A. H. Knoll, E. S. Barghoorn, *Archean Microfossils Showing Cell-Division from Swaziland System of South-Africa. Science* **198**, 396 (1977).

16. J. W. Schopf, *Fossil evidence of Archaean life. Philosophical Transactions of the Royal Society B-Biological Sciences* **361**, 869 (2006).

17. M. D. Brasier *et al.*, *Questioning the evidence for Earth's oldest fossils. Nature* **416**, 76 (2002).

18. J. W. Schopf, A. B. Kudryavtsev, A. D. Czaja, A. B. Tripathi, *Evidence of archean life: Stromatolites and microfossils. Precambrian Research* **158**, 141 (2007).

19. M. M. Tice, D. R. Lowe, *Hydrogen-based carbon fixation in the earliest known photosynthetic organisms. Geology* **34**, 37 (2006).

2
Carbon and oxygen

Let us now jump forward in time to 2.7 billion years ago. In doing so, we pass lightly through a length of time longer than it has taken, just recently, for all the animals on Earth to evolve from their common ancestor—it seems you can't rush this early evolution. This leap forwards in time makes things clearer because rocks from an increasing number of geological settings survive into the modern world. Various lines of evidence are recorded in these rocks, which support the presence of cyanobacteria by 2.7 billion years ago. With the evolution and spread of the cyanobacteria comes an Earth system that in its basic chemistry looks quite similar to what we know today. At this point in Earth history, there's enough evidence to describe how the chemistry of this system is working, and this seems a good moment to take stock of one of the great chemical cycles that keeps the planet habitable. So sit down on a rock—there is now plenty of bare rock to choose from on this still youthful Earth—and let's say a bit more about the most fundamental elemental cycle for life, that of carbon.

2.1 The carbon cycle

The 'primary producers'—the base of the food chain—are photosynthetic. They use light, carbon dioxide and water to get their energy and carbon, producing oxygen as a waste product. The cyanobacteria are playing this role as you look out on the world 2.7 billion years ago, and in the far future they will give rise to all the algae and plants who occupy the planet today. There are still some big differences however, the biggest of which is that there's still no oxygen in the atmosphere, except the odd whiff that leaks out of the shallow waters full of cyanobacteria. For the most part the oxygen is no sooner released by the cyanobacteria than it is eagerly consumed by other microbes that use it for much the same purpose that we use oxygen today. They 'breathe' the oxygen and use it to 'burn' carbon food*. In their case, the food is carbon fixed by the cyanobacteria, which these microbes release back to the atmosphere as carbon dioxide, just as animals recycle the carbon fixed by plants today. Organisms that make their living this way are called heterotrophs. Today they include us humans and all the other animals, but we didn't invent heterotrophy. We are simply carrying on a tradition that we inherited from the deep past. Heterotrophy is a very ancient way of life, and seems to have arisen in the microbes that were among the first inhabitants of the Earth.

Imagine the journey of a single carbon atom. It started life being synthesized in the heart of a giant star and probably came to the young Earth riding on one of the meteorites that also supplied the water of the oceans. On Earth the carbon atom takes the form of a molecule of carbon dioxide (CO_2), initially residing in the atmosphere. Quite often, it crosses the boundary between the atmosphere and the ocean and dissolves there, or goes back in the opposite direction, passing freely between the two. The cyanobacteria now living in the surface ocean, or in stromatolites in shallow lagoons, sometimes take it up and make their bodies from it. When they die, most of the carbon atoms in them are recycled back to CO_2 by heterotrophs. Most that is, but not quite all of them. A very small proportion gets buried in sediments and escapes being recycled. A little of this buried carbon may even survive to this day and form, for example, the microfossils we've been examining in Chapter 1. Over time, the sediments accumulate, burying our carbon atom, and trapping it in a new sedimentary rock, such as in the grains of carbon in the Isua formation. Our particular carbon atom might leave the ocean as organic material like those grains. More likely, it may end up as part of a carbonate rock, which also, as its name suggests, contains carbon. It would then be buried as limestone or a similar rock. Either way, it eventually gets lost into the sediments.

The supply of carbon to, and from, each of the reservoirs on the Earth's surface must be very closely in balance over time. If carbon dioxide is being removed from the atmosphere by one process, it must be replaced by another at the same rate if it is not to run out. Similarly, if life is adding oxygen to the atmosphere, some other process must be taking it out at the same rate, or it will build up and up, to levels that are toxic. The Earth system is so old, and has been going for so long, that if an overall net gain or loss to the surface were sustained over all that time, eventually there would be a crisis of too little, or too much.

---

* In detail the process is far more complex than the word 'burn' implies, but it resembles combustion in that the chemical energy stored in the food is released by combination with oxygen, and the final products are carbon dioxide and water. However, some of the energy released is not wasted as heat, but stored in a chemical form that can be used around the cell wherever it is needed.

The loss of carbon to sediments is a 'sink'—a leak that removes it from the surface where life can use it, and locks it up in the Earth. The leak is very small compared to the amounts that exchange between the air, oceans and living things, but if there were no compensating source, given sufficient time, the leak would drain the surface of all its carbon. For instance, in the modern world, where we can measure the rates of leakage, we can work out that all the carbon in the atmosphere, oceans, soils and plants would be lost into rocks in about a million years, in other words in the blink of an eye in the long history of the Earth.

Carbon however, doesn't remain trapped in sedimentary rocks forever, because the rocks themselves are recycled, thanks to plate tectonics, which began operating within a few hundred million years of the planet's formation. If our carbon atom has landed in the sediments under the deep ocean, it will eventually be carried to a subduction zone by the movement of the plates, where it will be forced down into the upper mantle, under the edge of a continent. There, under great heat and pressure, the rocks are metamorphosed—'cooked', converting the carbon back into CO_2 molecules. This is forced up from the depths of the Earth and eventually released into the atmosphere, perhaps in a volcanic eruption. If instead our carbon atom has landed on the submerged margins of a continent, one of two things can happen. It may also be drawn down to depths, metamorphosed back to CO_2, and released to the atmosphere through a volcano. Or if it is not subducted to great depth, it may remain in a sedimentary rock and instead, in due time, be pushed up to become part of a mountain range.

Eventually, after tens or hundreds of millions of years, the mountains are weathered, by the actions of wind, rain, frost, and (in the modern world, though not on the early Earth) land plants. The mountains are, quite literally, dissolved, ground down, and washed to the sea. Most of the material flows down rivers as sediment, but the carbon reacts to form CO_2 gas, so it escapes again to the atmosphere. Its sojourn in the rocks will typically have lasted for tens or hundreds of millions of years. So the life of a carbon atom on Earth is bleak and mostly subterranean: it spends most of its time locked up in sediments and rocks, and on average is let out into the sunshine for a brief period of a million years, perhaps every hundred million years or so.

Over time—immense and, for us, unimaginable lengths of time, long enough for mountains to be ground to mud and continents to rift and collide—the carbon cycle is nearly closed. The atom of carbon that we followed as it was buried in sediment, eventually reaches the surface again through a volcano, or is weathered out of the rocks of a mountain. Over much shorter times (though still hard to grasp) of a million years or so, the surface budget of carbon must however be balanced: if the volcanic source were to go quiet for a million years—and volcanic activity is by no means constant, so this can happen—the leak into sediments would drain carbon from the atmosphere and ocean. The carbon pool at the surface would shrink, and this would slow down the leak until it balanced the source again. In detail, how this balance is achieved is fascinating and surprising, involving the temperature of the Earth and featuring a starring role for life. We'll go into it in Chapter 7.

2.2 Carbon burial

It is 2.7 billion years ago, and you have been examining cyanobacterial stromatolites on an Archean shore. In due course, the cyanobacteria will cause the appearance of oxygen in the atmosphere, which will transform the planet, but you have to be patient—right now, there's little sign of this happening. In fact it will still be several hundred million years before oxygen becomes a permanent feature of the atmosphere. Having oxygen producers is not enough for this to happen, because almost all the oxygen is being reused by microbes, or reacted with methane they have released to the atmosphere, as quickly as it is produced. There is a small leak of carbon to sediments and a corresponding oxygen source, but there is a counterbalancing sink as well.

Carbon can be leaked from the surface either as carbonate rocks or as 'organic' carbon—the remains of living things, but how much is lost as each kind? This is an important question because the two types of carbon burial have very different effects on atmospheric oxygen. If carbon starts off as CO_2 in the atmosphere and is buried after being fixed by a cyanobacterium, before being recycled back to CO_2, it effectively leaves behind some oxygen. If this process is continued for long enough, it could result in a build-up of oxygen in the atmosphere. On the other hand if it is buried as a carbonate rock, it doesn't affect the amount of oxygen left behind. There is a way to find out what proportion of the carbon is buried as organic carbon, compared to that leaving the surface as carbonates, and we describe it in Box 2.1.

It turns out that the long-term source of oxygen to the atmosphere from the burial of organic carbon was much the same 2.7 billion years ago as it is today (Box 2.1). However, this does not necessarily imply that oxygen accumulated in the atmosphere at the time. Instead there

Box 2.1 The carbon isotope record

Organic carbon is all ultimately fixed by Rubisco from carbon dioxide, and this leaves an isotopic fingerprint. Recall that Rubisco has a preference for light carbon-12 (^{12}C) atoms over heavy carbon-13 (^{13}C), so a record of how that ratio has changed in rocks should tell us about the net amount of carbon being fixed, that is, how much is taken up by photosynthesis but escaped being recycled because it was buried. We have a record‡ of the ¹³C content of rocks, derived from measurements in marine carbonates, covering rocks running from 3.6 billion years ago to the modern day (*1*, *2*), and it is shown in Figure 2.1. Since the record is of the ¹³C in carbonate, it should get heavier (i.e. go up the page) if there is more organic carbon being buried, because it represents what is being left behind by the biology, which selects lighter carbon.

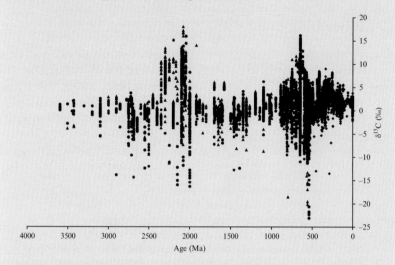

Fig. 2.1 The carbon isotope record of marine carbonates from the databases of Shields and Veizer (*1*) (Precambrian) and Veizer et al. (*2*) (Phanerozoic). Circles indicate analyses from limestones, triangles from dolostones, and diamonds from Phanerozoic calcitic fossils.

Fig. 2.1 covers almost the full sweep of Earth history from 3.6 billion years ago to the modern day, with thousands of individual measurements, though they become sparse in the very old rocks, because there simply aren't many of those. It's plotted in a way that you can see the envelope of the scatter (there is always some scatter, this being real data coming from many individual locations). What you don't see are the thousands of data points all falling on top of one another that characterize especially the more recent measurements. There are two things about this record that are especially significant:

1) **There is no long-term trend.** Over the whole history of the Earth the ¹³C content of carbonates does not drift up or down. What this tells us is that the proportion of carbon leaking from the biosphere as organic carbon has remained about the same over all this time. At first sight that is quite surprising—you might have guessed that the modern world with its

---

‡ We reproduce the record here as if it was generated by magic, but, as with many important results, it did not come for free: it represents decades of careful work by researchers measuring isotopes from thousands of locations around the world, and on whose shoulders—to misquote Newton—we are standing. It is because of such work by our scientific colleagues that we can wander god-like and apparently omniscient through Earth history, so we'll try to pause for a moment's appreciation of their efforts once in a while.

recently evolved big forests and grasslands would bury a larger proportion of organic carbon than the early biosphere, but this doesn't seem to be the case. On reflection it is perhaps not too surprising however. Any drift in this plot actually implies that the carbon cycle is out of steady state, and should be a net source (for an upward drift) or sink of oxygen (for a downward drift). We can use the isotopes to calculate how much oxygen would be released or taken up, and it is potentially large, tens or hundreds of times the amount of oxygen in the atmosphere today. If the carbon cycle were far out of steady state therefore, some other part of the Earth system would have to supply, or take up most of that oxygen—we wouldn't be able to explain it simply as a change in the atmospheric composition, because the atmosphere just isn't big enough.

2) **There are two periods of high variability.** While there is always some scatter on the points, there are two periods where the scatter is very high compared to the rest—the envelope of the data points thickens at these times, like a snake that has swallowed two particularly big meals. The high-scatter periods run from roughly 2.4 to 2.0 billion years ago, and 0.8 to 0.5 billion years ago. During these periods there were times where there were unusually large amounts of carbon being buried, very closely followed by intervals where there was very little—the carbon cycle is out of steady state for (geologically) short times, and varying wildly, whereas over the rest of Earth history it is fairly stable. These two time periods correspond to two great revolutions of the Earth system at the Great Oxidation and the Neoproterozoic, which we will be exploring in more detail in Parts 3 and 4.

was a surplus of other material, chiefly iron, on the surface and dissolved in the ocean, which eagerly mopped up such a reactive gas. Furthermore, metamorphism and weathering of old organic carbon from the crust could also have consumed oxygen. Finally, there's another process for oxygen generation which the photosynthesizers indirectly cause, which in the end may prove the decisive factor: the escape of hydrogen from the top of the atmosphere into space. In Part 3 we'll describe these processes, but for now let's follow through what eventually happens at the surface.

2.3 The Great Oxidation

We have to move forward in time again, three hundred million years or so, to witness the transformation. There may have been brief periods when oxygen appeared for a while in small quantities, only to disappear again, but the big change happens quite suddenly at about 2.4 billion years ago and it's irreversible, the Earth never goes back to its past, oxygen-free atmosphere. No doubt you've been dying to remove the helmet of your space suit and take a few deep breaths, but best to control the urge—oxygen concentrations are still tiny compared to today's atmosphere, perhaps around 1% of the modern values, and you'll suffocate in minutes. However, though the atmosphere remains unsuitable for us, it is still a momentous revolution. It has been christened 'The Great Oxidation' by Dick Holland of Harvard University, who wrote the book (actually several books) on the history of the oceans and atmosphere (3). It is probably the most significant change in the environment of Earth during its entire history from the time of its formation to the present day. As a visitor from another time, what would you see on the surface as signs of this momentous event?

To begin with, the clouds clear. It is likely that the Earth was shrouded in a high altitude haze before the Great Oxidation. It would have been formed in the same way as the clouds that today we see enveloping the Solar system's outer planets and moons. Titan for example, Saturn's largest moon, is often taken as an analogue of the early Earth. It has an atmosphere of nitrogen, like the Earth, but also with trace amounts of methane and ethane such as the early Earth may have had. High in the atmosphere, these react under the influence of ultraviolet radiation to produce an unbroken deck of cloud that obscures its surface from our view. If Earth was like this, the coming of oxygen would have cleared the hydrocarbons from the atmosphere and the cloud would have dissipated, leaving the planet looking much more like its modern blue-white self from space.

Accompanying the appearance of oxygen is an ozone layer in the upper atmosphere. Ozone is three oxygen atoms combined in a single molecule (O_3). It is a kind of supercharged oxygen molecule, fiercely reactive, and generated by the action of ultraviolet radiation (UV) on 'ordinary' oxygen molecules (O_2). An ozone layer begins to form in the upper reaches of the atmosphere as soon as more than a few parts per million of oxygen are present, and once in place it blocks the further passage of UV to the surface. Two billion years later, life on Earth has come to rely on this sunshade, which protects us from UV radiation that is otherwise harmful because it damages DNA molecules. Prior to 2.4 billion years ago, life, especially photosynthetic life which had no choice but to be exposed to sunlight, would have been forced to adopt and adapt to deal with the UV radiation penetrating to the surface. One method might be to live under a natural sunshade, such as a few metres below the surface of water, which would be sufficient to block most of the harmful UV. For shallow lagoon dwellers, the only choice would probably be to evolve efficient mechanisms to repair the damage. However, once oxygen appears and the ozone layer forms, the UV radiation at the surface of the planet is drastically reduced.

In the oceans there is an iron crisis. You'll remember that reduced, ferrous iron dissolves easily in water, but oxidized, ferric iron is resolutely insoluble. Large quantities of iron, which is one of the most abundant elements on Earth, were dissolved in the pre-oxidation oceans. Life, which evolved in those oceans, made liberal use of it for the elaboration of its basic chemistry, and during the long period before the Great Oxidation, iron became a key component of many crucial enzymes. Life is very conservative, and once an effective chemical pathway has evolved it is rarely abandoned, so iron remains an essential element for every living thing on the planet today. But with the appearance of oxygen in the atmosphere, the iron in the surface ocean oxidized to insoluble rust, bringing on a crisis: every living thing at the surface of the ocean was faced with a desperate iron shortage.

In the open ocean far from land, organisms still face this crisis today. After two billion years of evolution, the plankton have learned to cope by curbing their need for iron and conserving what they have, so that they can live with only the most meagre supplies, brought to them for example in dust blown from the land by the wind. However, it is as if the plankton have a memory of the good times past, and a hankering for the iron-rich Archean ocean. The oceans are still chronically short of iron, and the addition of just a few parts per billion of iron to a patch of open sea is sufficient to cause the plankton there to bloom.

You'll notice a change in the colour of the land surface. Once oxygen appeared in the atmosphere, iron exposed in rocks formed the mineral hematite—rust—with its familiar red colour. Geologists recognize surface-formed sediments containing iron as 'red beds', which begin to appear in rocks from this period onwards. Iron is a common constituent of rocks and sediments, and reddish soils and rocks have been a feature of the Earth ever since. Historically, the recognition of the absence of red beds in rocks dating from before about 2.3 billion years ago was one of the first clues recognized as indicating the absence of oxygen in the ancient atmosphere. Another clue, not one you'll notice from a casual inspection, but which you might detect if you have a Geiger counter with you, comes from uranium deposits. Uranium, like iron, is sensitive to oxygen, with two possible states which differ in their solubility. However, whereas iron is soluble before the oxidation but insoluble after it, with uranium it is the other way around. Before the Great Oxidation, uranium would sometimes be deposited in surface-formed soil and sediment, but after that time it is never found there, because it dissolves and is washed away to the sea from such environments.

There are other comparatively subtle indications of the great change that has taken place in the atmosphere, that will nevertheless be measurable by humans in the far distant future. One such is called the 'mass-independent fractionation of sulphur' signal, or 'MIF of S' for short (see Box 2.2 and Fig. 2.2), which in recent years has given us the most precise information on the timing of the rise of oxygen. For you on the surface of the planet however, there would be something rather more

Box 2.2 Mass independent fractionation of sulphur isotopes

This is a relatively new method for dating the appearance of oxygen in the atmosphere (4) Sulphur has four stable isotopes, all of which can be measured by mass spectrometry. All normal chemical and biological reactions fractionate these isotopes in a well ordered way. They may fractionate the isotopes, discriminating strongly or weakly against atoms on the basis of their mass, but they always act only according to the difference in mass, so measuring the ratio of any two of them enables us to predict how much of all the others will be fractionated. However, a few types of reaction do not obey this rule, including those initiated by ultraviolet radiation. In the absence of oxygen in the atmosphere there is no ozone layer, ultra-violet radiation can penetrate into the atmosphere, and reactions of sulphur gases that occur under the action of this UV disrupt the orderly fractionation processes. In the modern world this mass independent fractionation, or MIF, can only be seen above the ozone layer, but in the absence of oxygen it was seen everywhere and is recorded in the rocks.

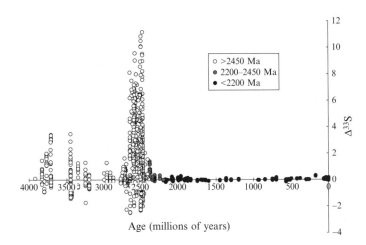

Fig. 2.2 A measure of the "mass-independent fractionation of sulphur isotopes" (MIF of S). Prior to 2.45 billion years (2450 million years) ago there is frequently a large MIF signal. Some rocks between 2.45 and 2.2 billion years ago show a weak MIF of S. In younger rocks there is no MIF signal. [Data compilation and figure kindly provided by David Johnston of Harvard University.]

obvious occurring. The changes in the atmosphere bring on changes in climate. It is getting cold, and the great ice sheets are beginning to advance. The first Snowball Earth is about to occur.

Cast your mind back to the high speed rewind of Earth history in Chapter 1. You saw two periods when these Snowball events occurred, each marking major change and revolution for the Earth. We are now at the oldest of these periods, standing at the transition between the no-oxygen Archean, and the low-oxygen Proterozoic. The Proterozoic Eon, which lasts from this point until a mere half a billion years or so ago, is a long period of mostly stable and warm climates, but it begins and ends with potentially catastrophic glaciations.

The Snowball Earths are unlike anything else experienced by our planet. They last tens of millions of years, during which the entire Earth is encased in ice. It is only in the last ten years or so that it has come to be fairly widely accepted that these catastrophes probably occurred in the distant past. That the Earth should be frozen from poles to equator for tens of millions of years is a bold claim to be made from interpretation of a few old rocks, so before we let you go forward in time into this frozen nightmare world, it's only fair to examine the methods in a little more detail to be sure they actually were as bad as we think they were.

2.4 How to recognize a Snowball

Glaciers and ice sheets leave telltale characteristics that the practiced eye of a field geologist can read. Moving ice carries finely ground rock, and also a random selection of stones, rocks and boulders that ride on the surface, inside or attached to the base of the glacier. When it deposits all this it leaves behind a type of sediment called a tillite, with the larger lumps distributed in a matrix of fine grained material. Typically some of the rocks will be marked with scratches (striations) from having harder stone dragged across them by the movement of the ice. There may also be dropstones, which are rocks released from the bottom of floating ice and icebergs that fall into layered sediment, deforming the layers in a distinctive way. If you see some or all of these effects, you can be reasonably certain that ice was involved in their formation, even if the rock is billions of years old. Thus, glacial periods and episodes are relatively easily discerned.

At around 2.4 to 2.3 billion years ago, a major series of glaciations occurs, essentially simultaneous with the Great Oxidation. While this is not the oldest evidence for ice on Earth (there is an icy episode much earlier, at 2.9 billion years ago), it's the first time that a prolonged series of glaciations had occurred. There are at least three advances of the ice, recorded in rocks in the Huronian system of rocks in the Great Lakes region of North America and spanning 100 million years. Glacial episodes are also recorded in the Transvaal supergroup in South Africa, and these are probably some of the same events as the North American rocks record, though the last may—or may not, geologists disagree—represent a fourth glaciation. This last African glacial deposit, called the Makganyene, is important because the rocks immediately above it and interleaving with it, show evidence that they were lain down close to the equator. The Makganyene is therefore a 'low latitude glacial episode', where apparently ice covered all or nearly all the planet—also known as a Snowball Earth.

Given that the continents have moved around, collided with each other, and broken apart several times since these events, you might justifiably ask: how do we know at what latitude the rock containing the evidence for these ancient glaciations was originally formed? We know because of another of those fortunate and elegant coincidences, where a sensitive measurement works with the grain of nature to produce a profound insight. In this case the sensitive measurement is of the magnetic field of the rock. When sediment is laid down, small grains of magnetic material, such as iron, tend to align themselves with the Earth's magnetic field. Hundreds of millions years later, this magnetism can still be measured, enabling us to reconstruct the direction of the magnetic field, and find the angle it makes to the sedimentary layers. It's a good assumption that these layers were horizontal when they were laid down, so this enables us to get the angle that the Earth's magnetic field made to the horizontal at that time. And this angle, called the inclination, tells us the latitude. You can imagine the Earth's field as being like that of a bar magnet stuck through the planet, nearly along a north-south axis. If you are standing at either north or south magnetic pole, the lines of force come vertically out of the ground at your feet, whereas at the equator the field is horizontal and the inclination is zero. So the inclination of the paleomagnetic field is directly related to the latitude at which the rock was formed.

Joe Kirschvink of Caltech made many of these paleomagnetic measurements, and it was he who, in the early 1990s, coined the name 'Snowball Earth', and began to advocate the idea that the planet had actually experienced such events. An entirely glaciated Earth had been considered as a theoretical possibility by climatologists before that, but up to the time of Kirschvink's work it was thought that such a state could never actually have happened. The reasoning was that it would be irreversible—once entered, the planet would be encased in ice for ever. With the white ice reflecting much of the sunlight away from the planet, the glaciation would be self-sustaining and the Earth would never recover to its warm state again, so all life would perish. Clearly this had not happened on Earth. However, Kirschvink realized that snowballs might not last forever, because a mechanism exists that should eventually, after millions of years, unfreeze the planet. We'll be discussing it in Chapter 7.

So is the Snowball Earth theory agreed upon? If only! Geologists seem to be second only to biologists in the vigour and sharpness that they can bring to a good old-fashioned academic argument. The experts in this field are currently divided into 'hard snowballers' (including Joe Kirschvink), who argue that the whole planet was entirely encased in a deep freeze, and 'slushballers', who declare that, at least at the equator, there must have been some unfrozen ocean. These two camps argue heatedly in the geological journals, shredding the other sides' arguments, reinterpreting their evidence, and getting as close to name-calling as the conventions of academia will allow. Climate modellers have joined the fray (5), but currently aren't able to answer the question definitively, because the models can admit both kinds of solution, depending on the concentrations of greenhouse gases and the arrangement of the continents among other variables (6, 7).

However, for the purposes of the story we want to tell, the snowballers and slushballers don't actually disagree very much. The slushballers do not dispute that there were glaciations at these periods, and that they were deep, long and terrible glaciations, with continental ice spreading to low latitudes, the like of which has not been seen on Earth since. The disagreement is over exactly how bad they were, especially in the oceans near the equator, and on what sequences of events and mechanisms were at work before and between the glaciations. The resolution of the argument is important, particularly for how and where life survived, but the overall pattern of Earth history will be clear whichever side turns out to be right.

2.5 Inside a Snowball

Let us enter then, this snowball world. You will be very glad of the insulation and heating in your space suit. Think of the environment in central Antarctica today: the *average* temperature of Earth if it was covered with ice is likely to have been similar to this, about −50°C. Most of the Earth therefore was like the coldest part of Antarctica today, and it stayed that way for ten million years. The atmosphere would have been exceptionally dry, because at such temperatures it holds almost no water, so there would be little in the way of clouds. Towards the poles wintertime temperatures would have been incredibly low, because the oceans would transport no heat at all to these latitudes, and the dry atmosphere also comparatively little. Average temperatures here might be −70°C and winter temperatures below −100°C. Some carbon dioxide may have precipitated out of the atmosphere as solid CO_2, forming clouds of 'dry ice'.

Are there any signs of life around? It used to be argued that the continued survival of life on the planet for four billion years was proof that no such whole planet glaciation could have taken place. Certainly, if a snowball were to occur today it would mean extinction for the great majority of the multi-cellular life with which we are familiar, but for single cells it is a different matter. Under the right circumstances, it is possible to freeze most prokaryotes and for them to remain viable. It can even be done for small animal cells such as sperm, and they can be kept nearly indefinitely in this way. In the laboratory, the 'right circumstances' generally include suitable antifreeze such as glycerol to surround the cells and help remove water from them, so that they don't burst open when they freeze. However, many organisms make their own antifreeze compounds (called 'cryoprotectants'). Viable bacteria have been cultured from ice cores drilled in glaciers, and though we don't know just how long they might last in their deep frozen state, it is not unlikely that microbes would survive in ice that had have been frozen for millions of years.

There would be other ways of surviving too. By chance you have landed near a rare oasis; a hot volcanic spring on land. These and hydrothermal vents under the sea, would have sustained temperatures suitable for life and provided refuges from the cold. We don't actually need to postulate a role for these during the glaciations that we encounter at the start of the Proterozoic, because at this time, probably, the most advanced form of life on the planet could survive lengthy cold storage. However, if we look ahead to the second series of Snowball Earths at the close of the Proterozoic, things would have been different. Some multi-celled organisms certainly existed by then. It is clear, from the explosion of animal life occurring at the end of those glaciations that their ancestors survived the freeze, and since they would not have survived long-term interment in ice, it seems there must have been unfrozen havens in which they could wait out the snowball. The organisms that survived at this time must also have included multi-celled algae. Since they were photosynthetic, there must have been locations where they could find both sunlight and liquid water. This is not a problem for the slushballers, who envisage open water at the equator. If the 'hard snowballers' are right however, it would seem to point towards locations such as geothermally heated lagoons, found in regions such as Iceland and Yellowstone today, as having been the critical refuges.

2.6 Through the Proterozoic

By about 2.2 billion years ago the glaciations have subsided. As we move on from this turbulent time, it's worth noting how many questions there are about it. What caused the glaciations, and was it just a coincidence that they happened at the same time that the first oxygen appeared in the atmosphere? Why did the oxygen concentrations begin to rise, but then seemingly stall at very low levels? These will be tackled in Part 3. For now let us summarize the main events surrounding the Great Oxidation, by expanding the relevant part our timeline (Figure 2.3).

We are now into the Proterozoic Eon, which by current convention started 2.5 billion years ago (although it would be more natural to treat the Great Oxidation as the divide between the Archean and the Proterozoic). The Proterozoic will last until the Cambrian ushers in the

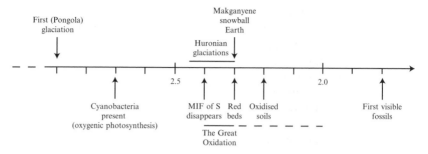

Fig. 2.3 Summary of events around the Great Oxidation

modern world of complex life, dated at 0.54 billion years ago. The climate was warm for a long period in the mid-Proterozoic, with no evidence of ice on the planet. A variety of pointers accumulating over the last decade supports the view that oxygen concentrations remained low, perhaps a hundredth of present levels at the beginning, possibly (or possibly not) rising slowly through time, but to no more than a tenth of today's level. You can't take your helmet off as you would still suffocate at such levels, but from a geochemical point of view this is a fully oxidized atmosphere, one in which all the chemical indicators we described that mark the transition through the Great Oxidation are satisfied.

Geologically, the mid-Proterozoic is a peaceful time—it has been christened 'the boring billion' by some Earth scientists, impatient for a bit more excitement in their rocks. Evolution was quietly elaborating its designs for living things, with the first fossil evidence for multi-celled creatures. Still, it seems that the rate at which structures were evolving remained very slow compared to the frenetic pace at which things would begin to move once into the Cambrian. As we pass through you'll notice the first signs of multi-cellular life, the first living things big enough to see with the naked eye, around 1.8 billion years ago. These take the form of spiral coils a couple of centimetres across that may be photo-synthetic algae. But you have to wait until 1.4 billion years ago to see something that is definitely red seaweed.

Peaceful and warm perhaps, but the mid-Proterozoic would still seem a very alien world to you. For example, based on analysis of the sulphur budget, Don Canfield of Odense University in Denmark pro-poses that the deep oceans were oxygen-free but sulphide rich (8). If he is right, you would find the seaside in the Proterozoic to be a pretty unpleasant place. The ocean would emanate sulphurous gases including hydrogen sulphide, which gives rotten eggs their overpoweringly awful smell and is toxic at parts per million concentrations. Canfield's analy-sis is supported by the inferred presence of types of bacteria near the surface of the ocean which can't tolerate oxygen but need hydrogen sulphide (9), and arguments based on the availability of molybdenum (10) and sulphate (11). However, not everyone is convinced by the sulphurous 'Canfield Ocean': Dick Holland, for one, is sceptical, but he too agrees that atmospheric oxygen remained low at that time (12). We will examine this controversy in Part 4.

How and when did atmospheric oxygen reach modern concentrations? Most probably, a second comparatively rapid increase occurs towards the end of the Proterozoic (Figure 2.4). At this time there are also renewed swings of climate, including at least two snowball glaciations, around 0.71 and 0.64 billion years ago. This also coincides with a second bulge in carbon isotope scatter (see Box 2.1 and Figure 2.1) from 0.8 to 0.5 billion years ago, indicating a carbon cycle that is swinging between extremes. Finally the Proterozoic gives way to the explosion of complex life at the start of the Cambrian, by which time we know there must have been concentrations of oxygen sufficient to support animals. At last, you can take that helmet off!

In Part 4, we'll describe the mechanisms that we believe gave rise to the extraordinary series of events surrounding the Proterozoic-Cambrian boundary, and which finally set the stage for the modern Earth, with its dominance of large multi-cellular plants, animals, and fungi. We'll argue there for a second step rise in oxygen. We are among the majority who accept this idea, but the evidence for it is not as yet conclusive. We can't rule out a slow rise occurring over a billion years, for example. The popularity of the 'second step' hypothesis rests just as much on the fact that it makes a great story, helping to fit other evidence into place in the satisfying way of a well plotted detective story. In this case, the 'crime' to be solved is: what caused the Cambrian explosion, when complex animals suddenly burst on the scene? This is one of the great puzzles in the history of the Earth, speculated upon for 200 years. All animals need substantial amounts of oxygen in the atmosphere to breathe, and a rise in oxygen has long been a prime suspect to pull the trigger on the starting

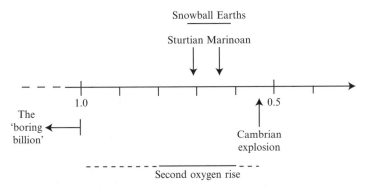

Fig. 2.4 Summary of events approaching the Proterozoic-Cambrian boundary

gun for the Cambrian explosion. We like this idea, but we readily admit that at present it is not fully established by the evidence alone, and might turn out to be wrong: it fits the evidence, but its attraction is due just as much to its power at explaining the subsequent events[†].

We've reached the Cambrian, the place where most discussions of life on Earth begin. Animals are suddenly scuttling through the shallow seas, large algae are there for them to eat (as well as each other) and the atmosphere is, very broadly, the same as today. The really fundamental Earth system development has mostly occurred by this time, though we still have the invasion of the land by plants, fungi and animals to come. We'll stop our travel through time there, and instead take another, radically different journey that gives another perspective on Earth system evolution. Life and the planet are one system, and we can learn as much about that system by looking inside living things, as we can by looking at rocks.

References

1. G. Shields, J. Veizer, *Precambrian marine carbonate isotope database: Version 1.1. Geochemistry Geophysics Geosystems* **3**, DOI 10.1029/2001GC000266 (2002).

2. J. Veizer et al., $^{87}Sr/^{86}Sr$, $\delta^{13}C$ and $\delta^{18}O$ evolution of Phanerozoic seawater. *Chemical Geology* **161**, 59 (1999).

3. H. D. Holland, *The Chemical Evolution of the Atmosphere and Oceans.* Princeton Series in Geochemistry (Princeton University Press, Princeton, 1984).

4. J. Farquhar, H. M. Bao, M. Thiemens, *Atmospheric influence of Earth's earliest sulfur cycle. Science* **289**, 756 (2000).

5. W. T. Hyde, T. J. Crowley, S. K. Baum, W. R. Peltier, *Neoproterozoic' 'snowball Earth' simulations with a coupled climate/ice-sheet model. Nature* **405**, 425 (2000).

[†] The debate about the history of oxygen in the atmosphere illustrates how the scientific method works in areas which are short on hard evidence, but rich in the ingredients of a good story. There is an ideal in which scientists are completely dispassionate, weighing the evidence with open minds unfettered by favouritism for any particular theory, keeping alternate hypotheses in play until the evidence is overwhelming in favour of one over the other. But what actually happens is more haphazard—and also more fun. We divide into camps, backing one idea or another, and argue, sometimes for decades or even centuries, until one view prevails over the other—often we have to wait for the retirement of all the senior academics on the losing side until an idea is finally abandoned. And it is not just raw evidence, but the attractiveness of the story they support that is important in deciding which theory eventually succeeds. So for example, the evidence for the Great Oxidation is now very strong, but there is still a minority camp that is not convinced. On the other hand the case for a second step rise in oxygen is not so strong, but here too the majority opinion is that such a rise did occur.

6. J. P. Lewis, A. J. Weaver, M. Eby, *Snowball versus slushball Earth: Dynamic versus nondynamic sea ice? Journal of Geophysical Research-Oceans* **112**, (2007).

7. A. Micheels, M. Montenari, *A snowball Earth versus a slushball Earth: Results from Neoproterozoic climate modeling sensitivity experiments. Geosphere* **4**, 401 (2008).

8. D. E. Canfield, *A new model for Proterozoic ocean chemistry. Nature* **396**, 450 (1998).

9. J. J. Brocks et al., *Biomarker evidence for green and purple sulphur bacteria in a stratified Palaeoproterozoic sea. Nature* **437**, 866 (2005).

10. G. L. Arnold, A. D. Anbar, J. Barling, T. W. Lyons, *Molybdenum isotope evidence for widespread anoxia in mid-Proterozoic oceans. Science* **304**, 87 (2004).

11. L. C. Kah, T. W. Lyons, T. D. Frank, *Low marine sulphate and protracted oxygenation of the proterozoic biosphere. Nature* **431**, 834 (2004).

12. H. D. Holland, *The oxygenation of the atmosphere and oceans. Philosophical Transactions of the Royal Society B-Biological Sciences* **361**, 903 (2006).

3
Russian dolls

Looking inside a human, or any other complex organism, reveals nested structures, one layer inside another, like Russian dolls. As we go deeper into this structure we also go further back in time.

3.1 The multi-cellular community

The first layer of organization can be seen with a light microscope at quite low magnification. All animals, and also plants and many fungi, are multi-cellular, made up of trillions of cells (about 10 trillion in us humans, for example) which act in supremely close cooperation, mostly using chemical signalling mechanisms to coordinate their actions. The adoption of the multi-cellular habit has enabled a huge increase in the complexity of size and form of living things. The three kingdoms of plants, animals and fungi, which have raised multi-cellularity to a high art form, and which contain virtually all the living things that you can actually see with the unaided eye, all appear in force in the fossil record after the most recent Snowball Earth event. This is not to say that nothing before that time was multi-cellular or that there weren't some fungi and seaweeds, at least, that predate the revolution, but there was a great increase in their number and their size afterwards.

It is worth pausing for a moment to consider what a wonderful thing it is to be multi-cellular. Those ten trillion or so cells in your body can be any of about two hundred distinct cell types, each with its own role in the body. All of them 'know' precisely where and when they can and must grow and divide, where and when they cannot and must not, and when they must commit suicide to end their lives. On the—thankfully extremely rare—occasions when they forget these abilities and begin to divide inappropriately, they form a tumour, sometimes a fatal, cancerous one. Animal bodies are totalitarian states, with their citizen cells slavishly and absolutely dedicated to the greater good of their society. Without this iron and absolute discipline not even the simplest nematode worm could exist.

3.2 The intra-cellular community

If you turn up the magnification to look inside a single one of your cells, you'll see a second layer of organization. The cell itself is also a community. A typical animal, plant or fungal cell contains many 'organelles', different kinds of internal structures bounded by membranes that separate them from the cytoplasm in which they reside. One of these organelles is the nucleus, which contains most of the DNA in the cell and can be thought of as the central library and copy centre. It gives our type of complex cell its scientific name—'eukaryote' (from good or true nut). Within nearly all eukaryote cells, there will also be at least one, and potentially thousands, of organelles called mitochondria. These are critical in producing the energy that the cell requires, by the carefully controlled burning with oxygen of molecules derived from the organism's food.

One of the great recent advances in our understanding of evolution has been the realization, championed by Lynn Margulis, that ancestrally, some of these organelles were derived from free-living bacteria. This is certainly true of the mitochondria. At some time in the past the ancestor of the parent cell engulfed the ancestor of the mitochondrion, but did not destroy it. Instead, the mitochondrion lived on as an 'endosymbiont', a partner organism living inside another. But despite having taken up residence permanently inside their hosts over a billion years ago, mitochondria still look and behave in many ways like bacteria. At a few microns in size, they have about the same dimensions as bacteria, and they divide and reproduce inside your cells like their free-living cousins outside. Some of their genes are contained in their own genome, which is a single loop of DNA, again just like bacteria, and they also retain their own machinery for reading, copying and utilizing this genetic material.

If you turn the microscope on a plant cell, you will find a different kind of organelle, called a chloroplast, which carries out photosynthesis and the fixation of carbon dioxide. The chloropolasts are also endosymbionts, descended from a free-living ancestor, which in this case was one of the oxygen-producing cyanobacteria that we have already met.

Though they retain some of their own genes, both mitochondria and chloroplasts have transferred many important genes to the cell nucleus. They now rely on the cell to supply most of their needs, so they can no longer survive outside the host cell. Depending on your outlook, you might describe the host as enslaving the endosymbiont, tricking it into giving up sufficient of its genes that it can no longer break free and is shackled to the larger cell forever. Or perhaps the last laugh is with the organelles, who have found a feather-bedded niche, provided with food and protected from the outside world in their own miniature environment. They can even get many of their genes copied for free at the central library. It's amusing to think of them like this, but of course these are just interpretations in terms of human relationships which don't capture what is really happening. The organisms themselves are value-free and really don't care who is dominant. What matters is that the partnership worked, and worked well, and the resulting organisms have survived and prospered. In evolutionary terms, both host and symbiont are winners.

The free-living cousins of mitochondria and chloroplasts are members of the 'prokaryotes'—cells without a nucleus, which vastly outnumber the eukaryotes on Earth. Free-living prokaryotes are also part of the community that makes up our bodies. They line our guts, help us digest our food and perform a host of other useful functions. They outnumber our own cells by about tenfold, but because each of the prokaryote cells is so much smaller than our own, we barely recognize their existence.

The prokaryotes used to be all classed simply as bacteria, and up to now we have been using this word for them, but today they are divided into two fundamentally different domains of life—the archaea and the bacteria (sometimes called the archaebacteria and the eubacteria)—and we can find examples of each in our guts. Under the microscope you can't see the difference—it was only recognized in the 1970s and is most strikingly apparent at the molecular level. Most of the microorganisms in our guts are bacteria, but there are also some archaea, and it is the latter that you may be most unwittingly familiar with, as they produce the methane in farts.

In the modern description therefore, life is usually divided into three domains: Eukarya, Archaea, and Bacteria (see Box 3.1 and Fig. 3.1). The Eukarya include all animals, plants and fungi, but also a large and diverse group called the protists. These last are an unsatisfactory grouping, something like the 'miscellaneous' bin in a supermarket, and there are various ways of classifying them. To a first approximation, they consist of all the eukaryotes (e.g. having one or more complex cells

---

Box 3.1 The tree of life or the tangled bush?

The consensus view of the tree of life (Fig. 3.1a), around 10 years ago, was constructed from analysis of small subunit ribosomal RNA, a technique pioneered by Carl Woese. It recognized the symbiotic origin of mitochondria (from proteobacteria) and chloroplasts (from cyanobacteria). It also suggested that Bacteria branched first after the last universal common ancestor (LUCA), before the Eukarya and Archaea split. The revised tangled bush or net of life (Fig. 3.1b) recognizes that lateral gene transfer was extensive before the three domains of life emerged (1). Therefore the universal ancestor is best thought of as a community exchanging genetic material rather than as an individual (2). We now think the latter picture is closer to the truth. A sub-tree of life may still be discerned for the Eukarya and to some degree for the Archaea and Bacteria but a universal 'tree' is no longer a useful metaphor.

(a)

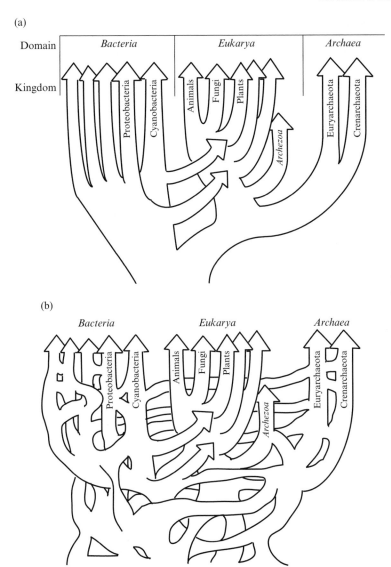

Fig. 3.1 (a) The tree of life, or (b) the tangled bush of life. [Redrawn from Doolittle (*1*)]

with a nucleus) that can't be fitted into a description as plants, animals or fungi. The majority of them are in fact single-celled. Some organisms such as seaweeds may be classified as plants in one context, and protists in another. (We will give a more up-to-date view of the subdivision of eukaryotes in Part 4.)

When did the modern eukaryotic cell, evolve? This is a subject for heated debate. The extremes of the views range from as early as 2.7 billion years ago to as late as 0.8 billion years ago. Presumably, eukaryotes evolved in stages, adding new organelles sequentially, so there is no one moment at which they could be said to come into existence. We'll discuss the subject in more detail in Part 4, but for now we state our preference, on the basis of fossil evidence, for the fully functioning eukaryotes evolving roughly 2 billion years ago. Their appearance or at least their increase in abundance, sometime after atmospheric oxygen was established would be consistent with the observation that almost all eukaryotes need oxygen.

3.3 The genetic code

A third level of organization is apparent at still higher magnification, in the genetic material in your cells (or their symbionts). This is the population of genes which resides in the molecules of DNA. The genetic information is written in a universal code, shared by all living things on the planet today, so you would find the same code inside the bacteria and archaea in your gut, for instance. The genetic code evolved before the archaea, bacteria and eukaryotes split from one another, back at the very origin of life.

The DNA molecule is made up of four different sub-molecules, called bases and conventionally labelled C, G, A and T. These combine to form long strands, wound around another complementary strand in the famous double helix. Each group of three adjacent bases, called a codon, specifies a single amino acid, which are the building blocks of proteins. For example, GAC specifies aspartic acid and CAG specifies glutamine. There are twenty such amino acids, each specified by different codons consisting of such three-base combinations. Each protein manufactured by the cell is written in the DNA as a string of codons whose order specifies the sequence in which to put the amino acids together. Each string specifying a protein is a gene, and it is terminated by codons that signify a full stop.

The 'genetic code' is the name given to the table of instructions that translates each triplet of bases to a given amino acid. This code is arbitrary, but it is also universal. It would be possible for a different code to have evolved where, for example, GAC specified glutamine and CAG aspartic acid, rather than the other way around. It would have worked (though perhaps not quite as well), but at some point in the distant past there was an organism which was the ancestor of everything alive today, and that organism had the present code, so as a consequence all life today also has that code*. The origin of the code lies back at the very

---

* Well, almost. It is an unwritten rule of biology that for every grand and general statement that a biologist can make, somewhere there is an organism that is an exception to it. Actually there are minor variations in the genetic code—vertebrate mitochondria for example have several codons that differ from the standard version, but these seem to be due to subsequent mutations. They are sufficiently rare that they make the generally conservative nature of the code all the more obvious.

origin of the cell—indeed it was the first cell that really deserved the name, and most likely it existed before 3.8 billion years ago.

To summarize then, three levels of organization—the genetic code, the eukaryotic cell, and the 'multi-cellular community'—representing increasing complexity and progressively later evolution—are nested like Russian dolls in our bodies. They date from respectively, the earliest Earth, the period after the Great Oxidation, and the period around and after the late Snowball events. The biological evolution of these levels is intertwined with the evolution of the Earth system in a network of cause and effect, both contributing to causing the crises in the Earth system, and being shaped by them, as we'll show in later chapters.

References

1. W. F. Doolittle, *Phylogenetic Classification and the Universal Tree. Science* **284**, 2124 (1999).
2. C. Woese, *The universal ancestor. Proceedings of the National Academy of Sciences USA* **95**, 6854 (1998).

4
The revolutions

In 2002, the European Geophysical Society held its science congrega-
tion in the city of Nice on the Cote d'Azur. This annual meeting is huge,
the biggest gathering of Earth scientists in Europe. Every discipline
related to the study of the Earth and the planets is represented. The subjects range from the origins of the Solar System through to pollution in
the Rhine, from the dynamics of the Earth's core to the ecology of
microbes in soil. There was the usual bedlam in such meetings, with
dozens of sessions going on in parallel and delegates rushing purpose-
fully everywhere like termites in their mound. At lunchtime and in the
evenings the city was invaded by thousands of earnest looking scientists, instantly distinguishable from the well dressed and tanned locals,
the delegates by comparison shabby, pallid, and frequently still wearing
nametags. We were both there and one lunch time we were sitting outside a small café, eating bruschetta, and talking enthusiastically about
the revolutions in the Earth system and how they meshed with the evolution of life. We were particularly excited about a book (*1*) and paper
(*2*) published by the evolutionary biologists Eörs Szathmáry and John
Maynard Smith, entitled 'The major transitions in evolution', which we
had found fascinating because it was reaching similar conclusions to us,
but starting from a completely different discipline. The authors were
evolutionary biologists, whereas we were geochemists interested in the
history of the Earth, but the same sequences and transitions were apparent in both disciplines. The man at the next table seemed interested in
our conversation, and after a while he leaned across and introduced
himself as Eörs Szathmáry. We spent the next hour or so in excited discussion of the convergence of these ideas, on the kind of intellectual
high which is one of the best rewards of science. Eörs nearly missed
giving his own talk, about life on Mars, as a result.

4.1 The major transitions in evolution

The work we found so much to talk about was a description by
Szathmáry and Maynard Smith of the entire sweep of evolution, going

from primitive molecules to human society and language, as a series of eight transitions (see Table 4.1). The transitions—we'll call them the SMS transitions after their authors' initials—are defined as steps in which simpler structures combine to make more complex ones such that 'entities which were previously able to reproduce independently before the transition can only replicate as part of a larger unit after it'. We could hardly ask for more authoritative evolutionary biologists to write on this topic: John Maynard Smith, who died in 2004 but whose career in biology spanned fifty years (and was his second, his first being as an aeronautical engineer), was perhaps the most influential evolutionary theorist of his generation. Eörs, who is much younger, is an intellectually adventurous Hungarian who has made particular contributions to perhaps the hardest theoretical challenge of all, understanding the origin of life.

What excited us so much that day in Nice was the fact that, though these transitions are defined quite differently to the changes in the Earth system we were interested in, they are often in effect the same transitions. Looking at the SMS transitions, we can see that the nested levels of organization that we have identified within ourselves in the last chapter are part of this longer series. The origin of the genetic code and the first cells are at step 3, eukaryote cells form at step 4, and multi-cellular plants, animals and fungi at step 6 (and step 5, the origin of sex, might arguably be lumped with 4, the origin of eukaryotes). The SMS transitions are however more comprehensive than our series, extending both inwards in space and backwards in time, down to the molecular origins of life, and outwards and forwards, up through animal societies to human language. For example, extending outwards, the nesting of levels of organization does not stop with us as individuals. We in turn form elements in our animal society (step 7). Within human society in particular, you are able to read this book because our ancestors developed a sophisticated language which can be used to convey complex ideas—step 8.

Table 4.1 Major transitions in evolution, after Szathmáry and Maynard Smith (2)

No.	Major transition
1	Replicating molecules to populations of molecules in compartments
2	Unlinked replicators to chromosomes (linked genes)
3	RNA as gene and enzyme to DNA Genetic code
4	Prokaryotes to eukaryotes
5	Asexual clones to sexual populations
6	Protists to animals, plants and fungi (cell differentiation/metazoans)
7	Solitary individuals to colonies (non-reproductive castes)
8	Primate societies to human societies (origin of symbolic language)

A key aspect of the SMS transitions is that they are hierarchical. Each transition is contingent on the previous ones. For us humans with grammatical language to arise required all the previous transitions to have occurred in sequence. Furthermore all the units except the ones preceding the genetic code are still with us. The asexual prokaryotes for example, remain the most indispensable part of the modern biosphere, which is layered with each subsequent level of evolutionary unit.

The SMS transitions vary considerably in their difficulty. Multi-cellularity has emerged independently many (perhaps 20) times—overwhelmingly in the eukaryotes, though with some interesting forebears in the prokaryote world. This indicates that, once other conditions are right (for eukaryotes, we guess enough oxygen in the atmosphere to be one of them) it is relatively easy for evolution to create a multi-cellular structure that is naturally selected for over its component parts. Social organization has evolved frequently and repeatedly in animals (insects, fishes, birds and mammals) so apparently is also a relatively easy transition, once the appropriate preconditions are met. Other transitions are thought to be extremely difficult. Of the three that precede prokaryotic cells, the third—the origin of the genetic code—is thought by Eörs to have been particularly difficult. The origin of eukaryotic cells was probably also very difficult, and does not seem to have occurred multiple times. Finally, the origin of grammatical language in humans has been argued to be difficult on theoretical grounds. It certainly appears to have happened only once, though animals as diverse as insects, amphibians, reptiles, birds and mammals communicate using sound. But our insistence on its importance may in part reflect a very human desire for us to retain for ourselves a special place in the evolutionary pantheon, as it is unique to us. The SMS transitions describe, from a biological standpoint, the ascending ladder of complexity that marks the broad structure of evolution. This has had profound consequences for the whole Earth system, and we'll have cause to come back to their work at several points in the rest of the book.

The revolutions of Earth are marked therefore not only by physical and chemical changes to the Earth environment, but also by increases in the complexity of the organisms living on it. They are also marked by increasing use of energy, increased efficiency in recycling materials, and increased information processing, as we'll now discuss.

4.2 The energy revolutions

Today the Sun, of course, is the source of energy for almost all living things on the planet. Our food energy derives from the one-part-in-a-thousand of the total sunlight falling on the Earth that is captured by green plants through the mechanism of oxygenic photosynthesis. We are so used to this process, which provides everything that we are made of and everything that we eat, that it can be a shock to realize what a star-tlingly difficult chemical trick it is. It involves splitting water using the energy from sunlight, and obtaining electrons and hydrogen ions (pro-tons) from it, while the oxygen is released as a 'waste' product. The electrons are then further boosted in energy before being transferred to carbon dioxide, where they provide the energy to make sugars and all the other molecules of life. Using water as a source of electrons was an evo-lutionary 'stroke of genius' by the first cyanobacterium, because water was already an absolute requirement for life and so was always going to be available in any situation where photosynthesis was possible. But water is a very tightly bound molecule, and the machinery required to split it, and then further boost the energy of the electrons, is formidably complicated. In practice two different biochemical 'machines', called photosystems I and II, are coupled in tandem to achieve it. We will describe these systems and their evolution in more detail in Chapter 8.

Remarkably, the two photosystems used in oxygenic photosynthesis appear to have separate evolutionary heritages. We have already men-tioned that there are other, earlier evolved, types of photosynthesis, which are 'anoxygenic': They use substrates other than water to provide the electrons, and do not produce free oxygen. Purple and green bacte-ria are still flourishing today in environments such as volcanic springs that would have been more common on the early Earth. The purple bacteria photosynthesize using a system that is related to photosystem II, and the greens have a system related to photosystem I. Some of them use sulphur in place of oxygen, starting with hydrogen sulphide instead of water (H_2S instead of H_2O). And some purple bacteria can get the electrons they need for photosynthesis by directly oxidizing iron (*3*). In all cases, the amount of energy that has to be captured from sunlight to fix a molecule of carbon dioxide is less than in oxygenic photosynthe-sis, because the starting materials used are more willing donors of their electrons than is water.

Oxygenic photosynthesis probably evolved when an ancestral pro-to-cyanobacterium obtained the genes for both photosystems and then succeeded in coupling them together. This is a good candidate for an extremely difficult evolutionary transition, which has happened only once in the Earth's history. It was difficult firstly because of the com-plexity of the machinery, involving the two separate photosystems and about 100 carefully ordered proteins. Secondly, once the machinery was in place, oxygenic photosynthesis would only be naturally selected over earlier, anoxygenic forms of photosynthesis when their substrates were in short supply. This is because splitting water requires the organ-ism to capture more energy to fix each molecule of carbon dioxide. We don't know exactly when oxygenic photosynthesis evolved (more on this debate in Chapter 9), but as already mentioned, it was probably before 2.7 billion years ago in the Archean, and definitely before the start of the Great Oxidation 2.4 billion years ago. Before it evolved, the anoxygenic photosynthesisers would have been the sole converters of the Sun's energy for the rest of the biosphere.

There was, we think, a yet earlier biosphere, powered by an altogether different source of energy. At deep ocean vents today, there are whole communities which do not directly rely on the Sun for energy, but are powered by chemical energy. The water coming from the vents is reducing, containing no oxygen, having lost it by reaction at depth with hot rocks. Instead it contains among other constituents sulphur gases, hydrogen and also many metals such as iron and manganese. Some prokaryotes can exploit the energy available from the reaction of these reduced chemicals with the more oxidized environment of the deep sea, and some of this energy at least can be traced ultimately to the geothermal heat emerging from the Earth's interior. The organisms that can do this are called 'chemo-litho-autotrophs', not a pretty word, but the 'chemo-litho' indicates that they derive their energy from simple chemicals coming from the Earth, and not from light or by consuming organics. Their genes indicate that they are very anciently evolved organisms. They are often heat tolerant, not surprisingly for bacteria that make their livings close to volcanic vents, but a property that would also have been useful on the fiery, earliest Earth. These are the modern representatives of the earliest biosphere, dating from a time before photosynthesis of any kind had evolved.

This earliest chemically driven biosphere could only access a tiny fraction of the energy of the modern solar powered biosphere, and the interim, anoxic photosynthetic biosphere would also have had less energy available than the oxygen dominated world, because the substrates it used to supply electrons were less abundant than water. Once oxygenic photosynthesis evolved in cyanobacteria they could have found a niche wherever the substrates for anoxygenic photosynthesis were in short supply. They probably colonized parts of the surface ocean (where, incidentally, they are still often the dominant primary producers), and also lakes and other damp places on the land surface. Around and after the second oxygen rise and Snowball interval, life really took off on the land surface, and was able to capture yet more energy. Thus, there has been an increasing trend in the energy flow that life has been able to access, the increases apparently occurring in steps, as new means of capturing energy evolved.

With all our modern measurements on land and sea, and with satellites to show us the whole Earth, we know pretty well the total productivity of life on the planet today. After accounting for their own respiration, the plants of Earth currently extract about 3 thousand tonnes of carbon from atmospheric carbon dioxide every second, half of it on the land and half in the ocean. They fix energy from the sun at the rate of about 100 000 Gigawatts as they do so—to give you some perspective on this number, it is about ten times the rate at which the entire human race is currently using energy through burning fossil fuels, so the energy flow through the biosphere still comfortably exceeds our own power hungry habits.

How did the earlier biospheres compare for energy use? It is much more difficult to get accurate figures for the past, but we can make some guesstimates. Before life gained a foothold on the land, we might estimate a throughput equivalent to today's ocean biota, which is half the total—probably an overestimate because, as we'll discuss in Part 4, the algae in the modern ocean benefit from their land cousins providing nutrients for them. Going back before the invention of oxygenic photosynthesis, there have been a number of attempts to put a figure on global productivity. A recent analysis suggests that photosynthetic bacteria using reduced iron in the oceans as their electron source would be the main carbon fixers, and might be at maximum about one tenth as productive as the oceans today (4).

Finally there would be the biosphere before any sort of photosynthesis, which relied on the energy that could be extracted from chemicals entering the oceans and atmosphere from the interior of the Earth. We can use estimates of these fluxes today to work out how much biological productivity could be sustained in this way on the modern Earth (5). Even if we then multiply by ten to account for more active volcanism at that time, the pre-photosynthetic biosphere would have been more than a thousand times less productive than the modern one.

To summarize then, our best guess is that starting from this low level, the energy available to life on Earth increased by a factor of about 100 when anoxygenic photosynthesis was invented, another factor of ten with the advent of oxygen-producing photosynthesis, and a further factor of two or more with the colonization of the land, for a total increase over time of at least a thousand-fold.

4.3 The recycling revolutions

To achieve these massive increases in productivity, the biosphere had to do more than just evolve new means of capturing energy—it also had to find new ways of supplying the materials needed by life. All organisms build their bodies by transforming materials taken in from their environment ('food' in the broadest sense) and excreting waste products. For example, imagine that you, like Tim, weigh 70 kg, then 43 kg of you is oxygen, 16 kg carbon, 7 kg nitrogen, 1 kg calcium, 780 g phosphorus, 140 g sulphur, 140 g potassium, 100 g sodium, 95 g chlorine, 19 g magnesium, 4.2 g iron, 2.6 g fluorine, 2.3 g zinc, 1.0 g silicon, and so on, through most of the periodic table (6). Many chemical elements are essential for all life forms, the 'big six' being carbon (C), hydrogen (H), nitrogen (N), oxygen (O), phosphorus (P), and sulphur (S), but there are a host of other 'micro-nutrients' that life requires in at least trace quantities. All these elements, as well as energy, have to be taken up by 'autotrophs'—the primary producers at the base of the food chain. These are then eaten by 'heterotrophs'—secondary consumers, like us—that get almost all their nutrients directly from their food.

The Earth as a whole contains large amounts of all the elements required by life but they are mostly locked up in the core, magma and crust, and their proportions are rather different to life's requirements. Of the big six essential elements for life, all but oxygen are relatively scarce in the Earth's crust and interior. A significant mass of the lighter elements, including those essential to life, had to make a late arrival in the meteorite bombardment that also brought water to the Earth. Recall that within this mixture, the process of nuclear synthesis had left an abundance of even atomic numbers (such as carbon, atomic number 6, and oxygen, atomic number 8) over odd atomic numbers (such as nitrogen, 7, and phosphorus, 15). This is the fundamental reason why nitrogen and phosphorus are potentially limiting nutrients for growth.

The young atmosphere is thought to have been more massive than today's and to have been dominated by carbon (mostly in the form of carbon dioxide, with some carbon monoxide and methane), with a similar amount of nitrogen to today (mostly as nitrogen gas, plus some ammonia), and small amounts of sulphur (as hydrogen sulphide, and other trace gases). The young ocean would have contained much carbon as dissolved carbon dioxide, and sulphur as sulphate, along with some nitrogen as ammonium, and phosphorus as phosphate. However, the volume of the atmosphere and ocean—the thin surface layer of the Earth that life inhabits—is tiny compared to the planet as a whole. If the biosphere had consisted only of autotrophs, taking their resources from the chemical pools in the atmosphere and oceans, they would have run out of raw materials in a geologically short space of time. Furthermore, the essential elements needed were not always present in a readily available form to biology.

Once photosynthesis was available to supply energy, life as a whole had to find ways to increase the availability of the materials it needed. There were two different means it could adopt: increase the inputs and conversion to biologically available forms, or reduce the outputs and recycle what it already had. In fact both strategies have been used. The inputs come partly from volcanic activity, which was out of life's control, but also from the weathering of rocks exposed on the continents. Before life became well established on the land, this process may also have been largely non-biological so the only option for life was to learn to recycle efficiently. Eventually, the land was colonized, and weathering too became a biologically controlled process. One element which is in short supply and which is particularly difficult to recycle because it has no gaseous form is phosphorus, and land life eventually succeeded in finding ways to coax this out of the rocks, as we'll discuss later (in Part 4).

The other five of the big six elements for life—C, H, N, O and S—all have gaseous, as well as dissolved, forms and hence can be cycled between the oceans, atmosphere and land surface. The elements hydrogen and oxygen have never been in short supply as both are available in water. The other three are however, recycled energetically by living processes alone. Carbon for example, is supplied to the Earth's surface by volcanoes, metamorphism and weathering, at a rate of around 100 million metric tonnes per year. Today's biosphere however fixes around 100 *billion* metric tonnes of carbon per year from carbon dioxide. Consequently to maintain the current productivity of life on the planet, the biosphere as a whole has to recycle 999 carbon atoms for every 1 that exits the system into sedimentary rocks. Only one part in a thousand of the carbon taken up by primary producers can be lost if present productivity is to be maintained. The rest must be recycled.

The main reservoir of nitrogen is not the crust but the atmosphere. You might therefore think it would always be readily available to life, wherever organisms are on the surface, but that's not so. Gaining enough nitrogen has posed a problem for life, because the two N atoms in gaseous nitrogen (N_2) are extremely tightly bound by a triple bond. To split this bond requires even more energy than splitting water in oxygenic photosynthesis. Once again it is those great chemists the prokaryotes, and they alone, that have mastered this reaction, the 'fixing' of nitrogen. Microbes that can do this are called 'diazotrophs'. Gardeners and farmers know that some plants, mostly members of the legume family such as clover and soybeans, can fix nitrogen. But in reality it is not they, but symbiotic *rhizobia* bacteria in their roots, that are performing the task. Some cyanobacteria can also do it, and these organisms are therefore able to perform not one, but two of the most vital and difficult chemical transformations needed for life—oxygenic photosynthesis and nitrogen fixation. There are also other, free-living bacteria and a few archaea that can fix nitrogen, including some of those that produce methane.

Both energy requirement and nutrient availability can limit the productivity of life, and the earliest biosphere, before photosynthesis, was very probably energy limited. However, once anoxygenic photosynthesis emerged, the supply of energy itself could become a function of the efficiency of recycling and hence the supply of essential elements to the primary producers. This created a 'bootstrapping' biosphere where more efficient recycling of essential materials would have allowed more primary productivity (energy capture). With early forms of photosynthesis that used hydrogen (H_2) or hydrogen sulphide (H_2S) in place of water, it seems likely that the supply of such substrates and the efficiency of their recycling was the ultimate limit. With the type of anoxygenic photosynthesis that uses iron (Fe^{2+}) it may have been the physical recycling (upwelling) of iron-rich water from the deep ocean that limited productivity. With the innovation of oxygenic photosynthesis using water as a substrate, this limiting factor was removed (at least in the oceans), and it is likely that the supply of an essential element (probably either phosphorus or nitrogen) would have become limiting.

Over time then, the biosphere has become more efficient in obtaining and recycling the essential elements, which in turn has enabled more efficient use of energy. Much of the basic chemistry required for efficient recycling was 'worked out' by bacteria early in the planet's history, but there have also been important developments in more recent times. In today's world, the biota are highly productive and efficient, but they are still ultimately limited by a combination of nutrient and energy availability. In much of the tropics and sub-tropics nutrient limitation is the norm, but towards the poles and in winter, light supply (and low temperature) become limiting.

The key point is that for the overall energy processing of the biosphere to increase in a way that is long-term sustainable, requires more than just innovations in energy capture, such as the origin of a new form of photosynthesis, it also requires a corresponding increase in the recycling of essential materials. The mark of each successful revolution of the Earth system has been a coupled increase in both energy processing and material recycling by the biosphere. The two must go hand in hand. Initially in a revolution things can go awry, often with biological innovations leading to an increase in carbon burial and the removal of essential materials from the surface system, followed by wild oscillations (see Fig. 2.1). But such a state of affairs cannot persist and it appears to force new innovations that lead to a recovery of recycling and a reduction in organic carbon burial back to the original flux.

4.4 The information revolutions

Life depends on energy and matter transformations, but something else is essential to it as well. This is heredity—the passing on of information to new generations. Over time the increasing complexity of the biosphere has gone hand in hand with the ability to pass on progressively more information. The first replicating molecules must have been very simple compared to even the least complex living cell today. Rather small amounts of information needed to be passed on for their type to survive, and the copying of information was very error prone. In the modern day some such simple replicating systems exist as viruses, though modern viruses are not directly descended from early life but rather have evolved secondarily. They nevertheless have some features in common with early life. For example, the human HIV virus uses the simpler RNA rather than DNA as its genetic molecule, passes on only a few genes (nine in the particular case of HIV) and does so in a way that is highly error prone—making copying errors about a hundred thousand times more frequently than a modern cell. This high mutation rate enables it to evolve very rapidly and makes it difficult to target with drugs, but strictly limits how complex it can become.

A bacterium by contrast has typically 500 genes contained on its DNA. Some of these genes are devoted to making enzymes which proofread and correct errors that occur during the copying. This is essential: it means that the copying fidelity is greatly increased so that it is possible to make a copy of its 500 genes without them being too garbled to assemble a new working bacterium.

When eukaryotes evolved, information transfer was moved up to a new level which enabled much greater complexity still. Typically a eukaryote cell has a thousand times as much DNA as a bacterium, and ten to a hundred times as many genes. To enable this expansion of the information content of cells, entirely new structures and mechanisms of gene copying were required, not just simple tweaking and tinkering with the bacterial mechanics. Though the fundamental molecular chemistry of the genes did not change, the reproduction process was revolutionized. We regard this change as perhaps the most fundamental, and difficult to achieve, revolution in the history of life, and we'll go into exactly why it was so important for the modern world in Part 4.

There has been one other major change in the information transfer capacity of living things, invented by the animals. That is culture—the passing on of information to new generations by direct communication between animals, rather than through their genes. While many, probably most, social animals have some form of culture, humans are of course pre-eminent here, because with our language we can pass on information about a huge range of subjects. And with the invention of writing, printing, computers and the internet this capacity is expanding explosively. We view this as part of the most recent, and as yet incomplete, revolution of the Earth system, the one driven entirely by our species. The last part of this book will discuss this revolution in the context of Earth history, but let's now briefly set the scene by discussing how our human transformations mirror the earlier revolutions of Earth.

4.5 The human predicament

The lesson from the past is that the successful revolutions of the Earth system have involved a sustained increase in energy use and information transfer. In order to maintain these, increased recycling of material was necessary. It's hard not to see the parallels with human activity today, but there is also a crucial difference: the past revolutions had to be successful in order for us to be here at all, but the same is not true for the present. If humans represent another such revolution, there is no guarantee it will be successful, and at present the omens don't seem very promising that it will be. Currently we are faced with a dual problem—most of the energy fuelling our societies comes from finite, unsustainable sources, as do many of the materials we consume or use for construction.

Historically, as a species, we have followed the pattern of past biospheres and progressively increased the amount of energy we access and transform. This started with traditional agriculture and progressed through the replacement of human with animal power, the invention and refinement of water wheels and windmills, peat burning and the use of biomass to produce charcoal for metallurgy. Then we began in earnest to dig or drill into the crust to extract, and then combust, ancient organic carbon—fossil fuels. The fossil fuel age took off with exponential, global increases in coal, then oil, then gas production. Total human fossil fuel use has now reached about a tenth of the energy being captured by the photosynthetic biosphere. This makes us comparable with the earlier biosphere based on anoxygenic photosynthesis—a remarkable and unprecedented feat for a single heterotroph species to achieve in just a few thousand years (7).

However, fossil fuels are clearly finite. They represent the energy of ancient sunlight in concentrated form, but as we've seen, only a tiny fraction of primary production is buried as organic carbon. Each year we burn an amount of organic carbon in fossil fuels that it would take current global net primary productivity at least 400 years to deposit. Add to this that fossil fuels are only formed in particular geological circumstances and it is clear that 'modern civilization is withdrawing accumulated solar capital at rates that will exhaust it in a tiny fraction of the time needed to create it' (7). Best estimates of current, commercially extractable, 'reserves' are around 1000 billion metric tonnes of carbon, and total 'resources' (not currently economically viable to extract) are around 4000 to 5000 billion tonnes of carbon, mostly as coal. With annual fossil fuel combustion currently 8 billion tonnes of carbon and rising, it is clear that the fossil fuel age can last at most a few centuries.

To fuel our material consumption, construction and food production, we extract and refine materials from the Earth's crust, ocean or atmosphere, the reservoirs of which are finite—albeit in some cases, large. This also uses energy. Take the example of food production: To produce fertilizer we expend large amounts of (mostly fossil fuel) energy industrially fixing atmospheric nitrogen at high temperature and pressure in the Haber-Bosch process (something which nitrogen fixing organisms can achieve at room temperature and pressure). We also mine the crust for phosphorus and potassium and then expend more energy refining them—particularly phosphates which are treated with acid. The resulting boost of land based primary productivity from fossil fuel produced fertilizers effectively feeds about 40% of the world's population (7).

As food production illustrates, we are collectively dreadful at recycling. Other, older methods of boosting food production were better in this regard. Our home county of Norfolk is famous for the development of a four crop rotation system which uses natural nitrogen fixers to boost productivity—but the short-term productivity gains are not as great as with synthetic fertilizer. So, we add nitrogen and phosphorus into the system and of course, this is washed into surface and ground waters, boosting productivity in other ecosystems on the land, in freshwaters, coastal seas and ultimately, the open ocean. This might sound like a good thing, but such 'eutrophication' often involves rapid shifts in eco-systems leading to the dominance of a few species.

Our waste products are rapidly accumulating in the non-living environment, visibly in landfills and floating as flotsam and jetsam on the water, and less visibly, but of more concern for the cycles of elements, in the soils, the atmosphere and dissolved in the ocean. We are strongly boosting the inputs to the natural global biogeochemical cycles of the essential elements C, N, P and S. Among the consequences, the greenhouse gases emitted from our activities are measurably altering the climate. Increasing carbon dioxide in the atmosphere is the main culprit, primarily from fossil fuel use, but with a further roughly 2 billion tonnes of carbon per year from 'land use change'—mostly deforestation in the tropics. Although about half of this CO_2 we add each year is removed into the natural carbon cycle, the other half accumulates in the atmosphere. Methane (CH_4) is the next most important anthropogenic greenhouse gas, produced mostly from agriculture and from its extraction and use as a fossil fuel. Much of the fertilizer nitrogen we add to the natural cycle ends up being denitrified by bacteria and returned to the atmosphere, but in so doing a third greenhouse gas, nitrous oxide (N_2O), is released. The warming effect of the increasing burden of greenhouse gases is amplified, by roughly a factor of two, by an equivalent contribution from increased water vapour in the atmosphere. Most of this change has become apparent in less than one hundred years, geologically an instantaneous moment in time.

If we hope to achieve long-term sustainability, and not be just a striking 'blip' in the future rock record, then we must learn and follow the lessons of the past. The past revolutions of the Earth system were ultimately successful because the sources of energy tapped were sustainable, and because the system 'learned' to recycle the waste products of the innovators. However, it took tens of millions of years for evolution to come up with solutions to these challenges. During that time there were major swings of climate that potentially could have destroyed the emerging system, and possibly even have left the Earth inhabited only by extremophile bacteria. Fortunately, we are equipped with brains and imagination, and we *ought* to be able to foresee and understand these dangers. Thereby we could avoid them, and also the millions of years of climatic upheaval that the system has previously had to endure before 'the blind watchmaker' of evolution (8) was able to restore a smoothly working global system.

4.6 Summary: the whole system view

The sleeper with whom we started Chapter 1 is now very much awake. Looking back at our origins, the fog that surrounds much of Earth history is beginning to clear. We human, conscious beings are only here because of a remarkable series of revolutions in Earth history. We have awoken to find ourselves in a system built on a series of previous ones. The transitions between these systems each involved qualitative changes in energy processing, material recycling and information transfer by the biosphere. These were reflected in a changed overall state of the non-living environment and a diversification or differentiation of the Earth system into a wider range of environments (or niches). Earlier biospheres are still with us just as in the major transitions of evolution earlier units of evolution are still with us. In ecological terms, the addition of a new layer at each revolution of the biosphere represents a form of niche construction on a grand scale.

For the Earth system then, we can list the following revolutions:

1) **Inception**: This is the revolution about which we know the least, both in terms of the mechanisms involved and their timing. It represents the establishment of a global biosphere, beginning with the origin of life. We are not even sure that life originated on Earth, but it was probably present by 3.8 billion years ago, and certainly by 3.5 billion years ago. As life diversified, recycling loops must have soon established. At first the energy for the biosphere came from chemical gradients, but at some point an early form of photosynthesis got going, greatly increasing the energy captured by the young biosphere. This enabled early life to sequester large amounts of carbon in the rocks, though with lower primary productivity and less efficient recycling than today. This would have altered the composition of the atmosphere, lowering the level of carbon dioxide, adding a host of new compounds and tending to remove others. By 3.3 billion years ago, if not before, there were recognizable prokaryotes forming ecosystems and mediating the key biogeochemical transformations.

2) **Oxygen**: This revolution began in the Archean with the origin of water-splitting oxygenic photosynthesis, probably sometime before 2.7 billion years ago. This generated an order of magnitude increase in energy supply to the biosphere, demanding a corresponding increase

in the recycling of carbon and other essential elements for life. Oxygen gas was released into the micro-environments around the photosynthesizers and some leaked into the atmosphere, but initially it was all used up and held at a low concentration by reactions oxidizing the abundant supply of reduced material entering the system. Then around 2.4 billion years ago, the scales tipped and oxygen rose dramatically in the Great Oxidation. At least one extreme Snowball Earth glaciation at the time suggests the levels of greenhouse gases dropped through this transition before recovering. By around 2.2 billion years ago in the early Proterozoic, a new oxygen-tolerant surface system had emerged.

3) **Complexity:** The presence of oxygen in the atmosphere and surface ocean provided an energy-rich environment in which eukaryotes began to flourish and diversify, developing much larger and more complex cells and genomes, and the ability to transmit more information to their offspring. A long period of slow change marks the period in which the first multi-cellular organisms evolved and elaborated. Meanwhile the anaerobic members of the older prokaryotic biosphere, still vital for the recycling of materials, were excluded by oxygen from the surface, but flourished in the deeper ocean and in sediments. We think the third great revolution really got under way around 1 billion years ago, as eukaryotes gradually came to ecological prominence and—we postulate—began to colonize the land surface. In so doing they unlocked a much greater availability of phosphorus, a nutrient that had previously limited the Earth system. Eventually they caused a second step rise of oxygen, and the Snowball Earth glaciations that accompanied it, during the interval 0.8 to 0.6 billion years ago. In the aftermath of the great glaciations, with a high oxygen concentration now in the atmosphere, the first multi-cellular animals flourished. Once again energy use by the biosphere increased, and the advent of animals allowed complex food webs with multiple trophic levels to develop. The arms race that ensued between the eaters and the eaten left its most striking mark in the fossil record with the advent of hard shells as protection, recorded in the Cambrian explosion around 540 million years ago. The familiar ecology that characterizes the modern world, the Phanerozoic Eon, had been born.

4) **Us?:** The time since the Cambrian is the period with which most studies of life on Earth are concerned, but from our perspective it is no more or less eventful than the previous periods—it's simply that we have more fossil evidence to go on. From 400 to 350 million years ago, the rise of vascular plants on the land surface doubled energy capture by the biosphere. This led to a reduction in atmospheric carbon dioxide, planetary cooling, and further (but compared to the snowballs, minor) glacial episodes. Fungi evolved to consume and break down woody plant material and recycling was boosted again.

Animals followed plants onto the land surface and under the high oxygen atmosphere they became larger and more mobile, promoting the evolution of nervous systems and the brain. Through a period of climate cooling, culminating in periodic glaciations, a range of bipedal ape species evolved and one of them developed a surprising language faculty, enabling them to transmit much more information to their offspring. Because of this they were able to develop technical skills to a high degree, including the use of fire, tools and weapons. On emerging from the last glaciation these *Homo sapiens* spread worldwide, causing widespread extinction of other species, inventing farming and writing, and congregating in larger numbers in towns and cities. The rest, as they say, is history. We are the end of the story so far. Whether this marks a successful or unsuccessful revolution, is yet to be written.

In the next part of the book we will explore two contrasting (but not incompatible) theories for how the Earth came to support intelligent life. We start by treating the question of whether life is common in the universe, and how much we can say about this problem from our particular, very peculiar vantage point, as products of four billion years of evolution on this planet. This is the problem of 'anthropic bias', and we will introduce a very simple model which can account for it. We then examine whether the model fits the data, and show that each of the revolutions of the Earth system that we have just identified, are reasonable candidates for evolutionarily difficult, 'critical' steps. Then we turn to reasoning forwards from first principles, starting with an uninhabited Earth and an Ark full of life, we examine how easy or difficult it is to make a thriving, self-regulating Earth system, or 'Gaia' (9), and ask whether there any theoretical reasons to expect revolutions to occur? The results suggest that whilst 'microbial Gaias' like the Archean Earth may be commonplace wherever life gets started, a 'complex Gaia' like we have on Earth today is an extremely rare occurrence.

References

1. J. Maynard Smith, E. Szathmáry, *The major transitions in evolution.* (WH Freeman, Salt Lake City, 1995).

2. E. Szathmáry, J. Maynard Smith, *The Major Evolutionary Transitions.* Nature **374**, 227 (1995).

3. F. Widdel et al., *Ferrous Iron Oxidation by Anoxygenic Phototrophic Bacteria. Nature* **362**, 834 (1993).

4. D. E. Canfield, M. T. Rosing, C. Bjerrum, *Early anaerobic metabolisms. Philosophical Transactions of the Royal Society B-Biological Sciences* **361**, 1819 (2006).

5. D. J. Des Marais, *Evolution—When did photosynthesis emerge on earth? Science* **289**, 1703 (2000).

6. J. Emsley, *The Elements*. (Clarendon Press, Oxford, ed. 3rd, 1998).

7. V. Smil, *Energy in nature and society: general energetics of complex systems*. (MIT Press, Cambridge, 2008).

8. R. Dawkins, *The blind watchmaker*. (Harlow, Longman Scientific, 1986).

9. J. E. Lovelock, *Gaia—A New Look at Life on Earth*. (Oxford University Press, Oxford, 1979).

PART II
THEORY

5
The anthropic Earth

Ever since the Copernican revolution, when it was first realized that the Earth was not the centre of the Universe, we have been asking; is there anybody out there? In most cases the answer has been 'Yes, of course'. For example in 1698, Christiaan Huygens, one of the most famous scientists of his day, published posthumously *'The Celestial Worlds Discover'd: or, Conjectures Concerning the Inhabitants, Plants and Productions of the Worlds in the Planets'*. Huygens was convinced that all planets must have inhabitants, not just life but intelligent life '...not men perhaps like ours, but some Creatures or other endued with reason'. Huygens, and most other commentators of the time, based their arguments largely on theology. Why should God create the other planets of the solar system, if there were no beings to view them? And what of the other stars, which it was realized by the late seventeenth century, were in fact distant suns? Must they not also have planets, with inhabitants? Some of the theological arguments were much more explicit than Huygens. For example, in 1715 the Reverend William Derham published a book with the crisp title 'Astro-theology'—which sounds a bit like the modern research area of 'Astrobiology' and was indeed, more or less the eighteenth century equivalent, in that it was carefully argued from the generally held assumptions of the time and was supported by the major institutions and funding agencies of the day (today, space agencies such as ESF and NASA, then, the Church).

In the modern era, the basis for the belief in extra-terrestrial life may have changed somewhat, but the popular conviction that there are a great many aliens out there waiting to be discovered seems just as strong. Starting with the classics of the science fiction genre by Jules Verne and H. G. Wells, through *2001*, *Star Trek* and *Star Wars*, countless fantasies have habituated us to the idea that there are other beings pretty much like us waiting for us to contact them. We are sociable, playful apes, we spend most of our time chatting to, playing and competing with others like ourselves, and the idea of discovering human-like creatures from other planets we can do all this with is deeply attractive. We want to find father- and mother-figures, who'll teach us at last to be wise, generous and loving, or friends with whom we can share the wonder of the Cosmos, or enemies we can fight (and of course win against)—basically we want company. The thought that we might be alone, the only creatures that are self-aware in a vast universe, is unsettling and alarming. It also seems to defy reason. If complex, intelligent life can arise here, surely it must be able to arise elsewhere too?

The idea that other planets will most likely be just like the Earth, (and by extension might very probably have life, and intelligent life, on them), is called the 'Principle of Mediocrity'. It states that there is nothing special about the Earth, and the same processes that are going on here must be going on everywhere. On the face of it there is nothing very special about the Sun: it is a type-G main sequence star, brighter than most stars, but not spectacularly so—we can see millions of other such stars in our part of this galaxy. However, it is possible that there *is* something special about the Earth, which is related to our presence. The Earth has life on it, which has persisted for a very long time. As we discussed in Chapter 1, life is almost as old as the planet itself, having been around for at the very least 3.5 billion years and probably more than 4 billion years, with its presence marked in the first sedimentary rocks that could show such evidence (*1, 2*). We are part of that life, and as very lately-evolved, complex organisms, we owe our existence to the very long history of life on the planet. Possibly, the chances that life can arise and persist for so long could be quite small, and the likelihood that self-aware creatures like ourselves evolve could be vanishingly small. In this and the next chapter we want to address the question of how likely the sequence of events that has led to our evolution might be, and thus put some bounds on the likelihood that we may one day find other Earth like planets. To do this we need a framework that takes into account the fact that our own existence is bound up with the singular history of the Earth.

5.1 The anthropic principle

The notion that we need to take account of our own existence in thinking about certain problems lies behind a set of ideas that go under the name of the anthropic principle. There are several versions of the anthropic principle to be found in the literature: for example, an authoritative book on the subject by John Barrow and Frank Tipler (*3*) defines 'weak', 'strong' and 'final' versions of the Cosmological Anthropic Principle. Here we will invoke exclusively 'weak' anthropic arguments, which are really just an uncontroversial piece of common sense (in their definition at least—not necessarily in the uses to which they are put!)

The weak anthropic principle states simply that the universe as we see it must be compatible with our own existence. We are complex, carbon-based life-forms that require the existence of carbon, oxygen, nitrogen and hydrogen and many other elements in the Periodic Table. We could not exist in a universe in which these elements were not present. The hydrogen and helium in the universe were mostly formed at the time of the Big Bang, while all the heavier elements were forged in the thermonuclear hearts of stars, and dispersed by their explosions as supernovae before the Sun and the Solar system were formed. It turns out that for this sequence of events to have occurred requires that some of the fundamental properties of the universe are 'fine tuned' within quite narrow bounds. We can imagine universes where these properties take different values, and some cosmologists suggest they may really exist as part of the 'multiverse'. For instance, the ratio of electromagnetic to gravitational forces, the charge-to-mass ratio of the electron, or the starting mix of gravitational and kinetic energy, might be substantially different from those in our universe. We can imagine these universes, and we may in time even develop our theories of physics and cosmology sufficiently to become confident that they exist, but we know that no creatures like us will live inside them and be able to observe them, because they could never evolve in such universes. There would be no carbon, or no oxygen, or the universe would not exist for long enough, to allow them to arise.

There is a large, and growing, literature on the anthropic principle as applied to cosmology, which reflects the increasing importance that anthropic arguments have in that field (for an authoritative popular description by one of the foremost exponents, see *Just six numbers*, by Martin Rees (4)). By contrast, there has been rather little written on the anthropic principle applied to the study of biological evolution and Earth history. However, it is just as important to correctly factor in 'anthropic bias' to this subject as it is in cosmology, and that will be our aim in this chapter.

We could not arise in a universe that was too short-lived, not only because it takes time for galaxies, stars and heavy elements to be made, but also because, after all that is done, a further long time is needed for life to evolve from simple molecules to beings that are complex enough to ask questions about their own existence. You are able to read this book because of the persistence, over a very long time, of a favourable environment at the surface of the Earth. On this planet, all the elements necessary for life have been available, and the temperature has almost always remained in the range where water remains liquid—between 0°C and, roughly, 100°C—for several billion years. Liquid water has specific properties that make it an essential solvent for the chemistry of living things and our kind of carbon-based life absolutely requires it. Furthermore, this carbon-water chemistry is the only type we know of that has the enormous flexibility required to produce the complex molecular structures needed for life. The maintenance of a habitable environment over such a length of time requires a stable star burning fuel at a nearly constant rate, and a planet with a particular set of properties.

5.2 The habitable period

Let us start by considering the set of properties which determine the length of time that a planet remains habitable. We assume that life, evolution and the origins of intelligence happen on rocky planets which orbit stars, and on which the surface temperature is such that liquid water exists at the surface. Such planets are not indefinitely inhabitable. We define the *habitable period* as the time for which a planet can actually support life.

A planet which is going to be suitable for life must be in the right position in its solar system, called the 'habitable zone' (5), which is the region around the star where the surface temperature of the planet allows water to be liquid. You can think of this as a comparatively narrow ring surrounding the star. There is a complication however, because the star gets brighter over time and so the habitable zone doesn't stay in the same position. Stars such as the Sun are on the 'main sequence', which is the most stable and longest lived part of their lives. They are powered by thermonuclear reactions which fuse hydrogen nuclei into helium in their cores. But helium is denser than hydrogen, and slowly, as the helium is formed, this causes the Sun to increase its output of radiation. The Sun has become about 30% brighter since the Earth was first formed, and this process is slowly causing the habitable ring to expand outwards. Fig. 5.1 shows an estimate of the habitable zone around the Sun, and how it has changed with time, taken from (5). When the inner edge of the habitable zone expands beyond the orbit of the Earth, it will be too hot for life to continue here and time will be up for life on Earth.

A planet's habitable period is not just a property of its star however. It depends on other factors too, in particular what kind of atmosphere it has. The thickness of the atmosphere depends in turn on the size of the planet. Smaller planets and satellites tend to lose gases from their atmospheres, and this can mean that they are either never habitable at all (their habitable period is zero) or have only short habitable periods. Lighter elements are lost the most quickly. The Earth tends to lose from its atmosphere the two lightest elements, hydrogen and helium, but nothing heavier, with the result that it has held on to its atmosphere of nitrogen and, more recently, oxygen for billions of years. Earth is big enough too, that it has a molten core of iron, which generates a magnetic field. The magnetic field of the Earth deflects the stream of charged particles blowing away from the Sun—the Solar Wind—which otherwise would slowly strip even these heavier molecules from the atmosphere. Mars

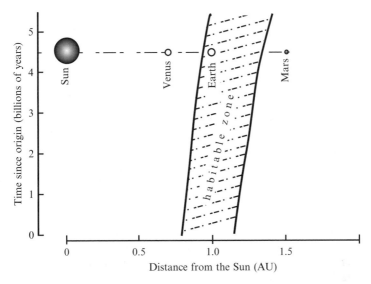

Fig. 5.1 An estimate of the habitable zone around the Sun, as a function of time since the origin of the Solar system. The horizontal dashed line represents the present day, 4.6 billion years after the origin. [AU-Astronomical Unit. Data from the 'most conservative' case of Kasting *et al.* (5).]

on the other hand is smaller. Its gravitational grip is weaker, and it has no molten core and no magnetic field, so it tends to lose even the heavy gases. Mars today has a much thinner atmosphere than Earth, mostly consisting of carbon dioxide, but very early in its history, and perhaps at intervals since, it has had liquid water at its surface. This suggests that its atmosphere was thicker then (with a stronger greenhouse effect). The progressive loss of its atmosphere means a Mars-sized planet, even if it were at Earth's position in the Solar system, would have had a shorter habitable period than Earth.

Hydrogen is essential to life, and a planet that lost all its hydrogen would be uninhabitable. The Earth does lose hydrogen, but today this process is fortunately very slow, because on Earth most of the hydrogen is in the form of water in the oceans. Gases containing hydrogen can penetrate into the upper atmosphere, where the harsh ultraviolet from the Sun splits them to release hydrogen atoms, and these can escape. However, such gases are at very low abundance in the Earth's atmosphere—methane has the highest concentration, and that is only at about a part per million. Water vapour is abundant in the lowest part of the atmosphere, but is confined there by a 'cold trap' effect. Air that ascends in the atmosphere is cooled, causing the water vapour to condense out as clouds, and fall back to the surface as rain or snow. If you have flown in a passenger jet with a display of the outside temperature, you'll have noticed it is pretty chilly outside at the cruising altitude, typically minus fifty or sixty Celsius, and this is cold enough to freeze out virtually all the water in the air before it gets high enough to be split by solar UV. Without this atmospheric 'cold trap', Earth would by now have lost much of its hydrogen to space and there might be no water left on the planet. This is what is thought to have happened to Venus, which is just as large as the Earth and is able therefore to retain the heavier gases, but which is too close to the Sun to have an efficient cold trap. All the water on Venus is thought to have dissociated and the hydrogen then escaped the planet, in a process called a 'runaway greenhouse'—of which more in Chapter 7.

A third factor that influences the habitable period is also determined by a planet's size. In order to retain volatiles like carbon dioxide and water at its surface, rather than locked up as components of rocks in its interior, the planet must have a supply of internal heat. On the modern Earth, water and carbon dioxide tend to combine and react with rocky materials at the surface of the Earth. They form carbonate and hydrated minerals in sediments, and these are removed from the atmosphere and oceans as these sediments are washed to the sea and laid down as rock. Over an immensely long time (hundreds of millions of years), these rocks are subducted into the interior of the Earth, around the margins of the continents where the tectonic plates carrying the ocean floor are pushed underneath the continents.

Over time then, this process removes water and carbon from the surface of the Earth. If this were the end of the story, Earth would have absorbed these vital materials for life into the interior by now and would be dry and lifeless. This has not happened because another part of the cycle returns them again to the surface. As the rocks are taken down into the Earth they are heated, and the heat 'cooks' the volatiles out of the rocks, reversing the reactions that occurred at the surface. The carbon dioxide and water are released at depth in the Earth, but eventually they work their way upwards and are released to the surface again by volcanoes. The process driving plate tectonics, that keeps the continents moving restlessly across the surface of the Earth, is also therefore responsible for returning to the surface the water and carbon needed for life.

The geothermal heat to drive this process comes from two sources: A small part of it today comes from the residual heat that was generated when the Earth first accreted—from the huge amount of energy released when that great mass of material smashed together to form the planet. Initially, the Earth was molten from this release of energy, but a solid surface formed within a few million years, and as the oldest zircons have told us, oceans probably existed by 200 million years after formation began. Since that time most of this primordial heat has escaped. A much longer-lasting source of energy has come from the decay of radioactive elements in the Earth interior, and it is this radioactive source that mostly accounts for the heat inside the Earth today. However, most of the heat of these radiogenic elements has now also dissipated, as they decay away to stable products.

A plate tectonic cycle and an adequate supply of interior heating, such as the Earth currently possesses, are important therefore to sustain the habitability of a planet. Once the planet cools, there is nothing to cook the rocks so as to dehydrate them and de-carbonize them, and the carbon dioxide and water will be lost from the surface into the interior. Something like this has probably happened to Mars. There is plenty of evidence, from channels cut on the surface, from the remains of flood plains, and from the Mars Rover missions that have sampled sulphate rocks on the surface, that Mars once had flowing water at its surface. Also, Mars has volcanoes—many and varied, and including Olympus Mons, three times higher than Everest, the largest volcano in the Solar System. However, the volcanoes are all now extinct and there is no longer any water flowing at the surface, though quite large amounts may exist as hydrated minerals not too far underground. Likewise the atmosphere of carbon dioxide is very thin—less than 1% of the pressure at the surface of the Earth. Mars is cold and dead inside, while the Earth is still hot. Mars has cooled more rapidly than the Earth, because it is so much smaller. Like a big cake and a small cake left to cool after baking, the small planet has lost its heat already while the big one is still warm inside. But the Earth too is cooling, and this sets a limit to how long it will be habitable. It may be a close-run thing whether it will be this process, or the increasingly brighter Sun, which will finally spell the end of life on Earth (see Box 18.1 for further discussion).

To summarize, there are several processes that limit the habitable lifetime of a planet such as the Earth: its star becomes brighter, its interior cools and eventually re-absorbs its atmosphere and oceans, and if it is too small it may lose much of its atmosphere to space. To quote George Harrison, all things must pass, and eventually the Earth will die. But it is important to our story to establish whether life on Earth is young—say with the expectation of a longer time ahead than has so far passed since the origin of life, or whether it is old—most of the time that the planet will be habitable has already passed.

Much of our discussion in later chapters will concern the habitability of the Earth, but for now let us give one of the conclusions, which comes from an understanding of the processes setting the surface temperature and composition of the atmosphere which we develop in Chapter 7. To find out how long the Earth will be habitable in the future, we run such models through time, with the Sun getting brighter. This exercise has been carried out a number of times (*6–9*), with models of increasing sophistication, since James Lovelock and Mike Whitfield first looked at the problem (*10*). We've already seen the results of one such model, used by Kasting and colleagues to construct the evolution of the habitable zone (Fig. 5.1). The results suggest that Earth was well positioned for a long sojourn in the Sun's habitable zone, being comfortably inside it at the birth of the Solar System. The zone has moved outwards over time as the Sun has become brighter, and the planet is now positioned towards the inner edge, as shown in Fig. 5.1. Our best guess is that for Earth, the habitable period is about 5–6 billion years, of which 4.5 have already passed and only 0.5–1.5 remains. Earth's biosphere is therefore in its old age, somewhere between 75% and 90% through its expected lifetime. If she were a human, Gaia would be collecting her old age pension.

5.3 The expectation time to observers

If it were very easy for complex and intelligent life to evolve, we might expect that it would have done so while the Earth was still young. Instead, it has actually evolved towards the end of the Earth's habitable period. This observation seems to carry some information about how easy or difficult the process of evolution is. Can we use it to say something about how long in general, on *any* habitable planet, we should expect to wait before intelligent observers evolve? This sort of question was first posed by Brandon Carter, the astronomer and theoretical physicist who first named the 'anthropic principle', in a 1983 paper—one of only a few ever written on its biological ramifications (*11*). We lay out his arguments below, but first we need to define some terminology.

Suppose an event occurs during evolution. The *expectation time* of the event is the period after the origin of life that, on average, you would expect to pass before it occurred. It is not the period that actually elapses before the event occurred in any one realization, such as that which occurred on the Earth, but the *average* that you would find if you were able to observe the process being repeated many different times—if, in Stephen Jay Gould's phrase, you could replay the tape of evolution many times over, or if you could have a God's-eye view of the whole universe, and see evolution occurring on many different planets. If you were such a cosmic experimenter, you could ask what the expectation time is to reach any particular level of complexity in evolution. But let's start, as Carter did, by asking what the expectation time would be for an 'intelligent observer species' such as humans, to arise? This immediately begs several issues: how do we define intelligence? Does a chimpanzee qualify for example, or perhaps a dolphin? These are important questions, but they would need another book to fully explore them. For the present, we'd like to defer this argument, so we define an intelligent observer species as one whose members—some of them anyway—are able to ask such questions about their evolution, and have an interest in the answers. (If you've read this far, you qualify!)

We are assuming in our definition of expectation time, that when we replay the tape of evolution, the environment remains constant, and favourable to the development of life, so there is nothing to stop us from allowing it to play for as long as we like. If it takes say, 100 billion years on average for observers to evolve, then their expectation time would equal 100 billion years, despite the fact that actually, the Earth and other similar rocky planets aren't going to be around for that long. Defined in this way, the expectation time for observers depends only on biological processes, and is independent of astronomical or geological events that affect the habitability of real planets.

Conversely, the habitable period of a planet, which we have just discussed, is dependent mostly on astronomical and geological variables, and only secondarily (or not at all) on biological ones. For comparison with the average expectation time to observers, we should consider the average habitable period for all potentially inhabitable planets. Armed with these two independent variables, Carter (*11*) asks us to consider three possibilities:

Case 1: The average expectation time to observers is a much shorter period than the average habitable period.

Case 2: The average expectation time to observers is of the same order as the average habitable period.

Case 3: The average expectation time to observers is a much longer period than the average habitable period.

In the first case, intelligent observers would normally be expected to arise in a much shorter time than the habitable period of a typical Earth-like planet. On most such planets therefore, intelligent life would be expected to arise before the planet became uninhabitable, and intelligent life should be common in the universe. From the evidence of the Earth, this would seem to be an unlikely scenario. We know that here, it has taken several billion years for intelligence to evolve, a time of the same order as the habitable period for the Earth. Unless life has been unusually slow in evolving here, we can rule out Case 1.

Another piece of evidence that argues against Case 1 is that we see no evidence for intelligent life anywhere else in the universe. In the last 40-odd years we have been searching seriously, with various SETI ('Search for Extra-Terrestrial Intelligence') programmes, of which the California based SETI Institute is the most famous and well organized. So far, there is nothing to report. Of course, the number of potential stars that are close enough and visible enough to be targets for SETI lies in the millions, and only a proportion of them have so far been surveyed, so we should not immediately conclude that there are no other intelligent civilizations out there. Still, the lack of success is thought-provoking. If life is common and intelligent civilizations sometimes arise, why do we not see some evidence for them?

This puzzle is called the 'Fermi Paradox'. Many books have been written, and hypotheses put forward, to explain it. For instance, a good one, by Stephen Webb, has the title *If the Universe is teeming with aliens, where is everybody? Fifty solutions to Fermi's Paradox'* (12). The reason why there are so many possible solutions is that we can't tell what aliens may be thinking or how they may behave, so it is hard to argue that 'The Great Silence', as it has been called, tells us much about the actual frequency of alien civilizations. It may be, for example, that they have formed a Galactic Club with a strict non-intervention policy towards primitive civilizations, similar to the one that the crew of the Starship Enterprise were supposed to observe (but never did of course—otherwise most of their storylines would have been finished in the first five minutes). Speculating about how aliens will behave is great fun, but there are too many possibilities to make this a promising line of scientific inquiry. Proposed solutions to Fermi's Paradox are usually not falsifiable even in principle—there is no way to prove them wrong, unless of course, SETI were to be successful in detecting an alien civilization, in which case they would be redundant. So while we think that the lack of evidence for intelligence elsewhere argues against Case 1, it's not a very strong argument. Nevertheless, the evidence about evolution *here on Earth* allows us to rule out Case 1, unless life on Earth has been exceptionally slow to evolve.

Case 2 is obviously compatible with our observation that for Earth, the actual values we observe for the time it takes observers to evolve and the habitable period are both a few billion years. However, Case 3 is *also* entirely compatible with what we observe. In this case, Earth must be very unusual—on most habitable planets, intelligence does not evolve because it takes too long. However, the Earth, in common with other rocky planets, is only inhabitable for a few billion years. If intelligence is going to arise here, it must do so within that time period. So, on the rare occasions when intelligent beings like us beat the odds and evolve, they do so in a period of the order of the average habitable period and much shorter than the average expectation time to observers. It's then no surprise that *we* have evolved in such a time period.

We cannot tell however, just by looking at the time of evolution here on Earth, which of Case 2 or Case 3 is correct. Both cases would be compatible with intelligence evolving relatively late in the habitable period of the planet. On the other hand, Case 2 predicts that intelligent life should be common in the Galaxy, as a good fraction of planets would evolve intelligence (about half for instance, if the average expectation time to observers actually equals the average habitable period and if the probability distributions are symmetric). Case 3 however, would predict that intelligent life was rare. So a solution of the Fermi Paradox falls naturally out of Case 3. However, if we accept Case 3, we have violated the Principle of Mediocrity. Earth is now special, precisely because intelligent life has evolved on it, and because this is an unusual event.

This is the essence of the anthropic principle as applied to Earth history. In terms of probability and statistics, the effect can be described as 'observer self-selection'. Suppose a researcher from the Galactic Federation wants to establish how common life is in the Galaxy. If he or she (or it) can persuade the Federation to devote a starship to the project, to voyage through space at will and study a representative subset of all the planets in the Galaxy, the estimate will be unbiased. But we humans are confined to the planet on which we evolved, and at present can observe only the Earth, so our sample of just one planet is biased by 'self-selection'.

Carter argued that Case 2 was intrinsically unlikely, and that therefore Case 3 was to be preferred. His argument was as follows: The processes that determine the average expectation time to observers are those important in biochemistry and evolution. They are entirely different from the processes determining the average habitable period which are those important in astronomy and geology. Time scales defined by astronomical processes span tens of orders of magnitude, and time scales defined by biological and evolutionary processes similarly range over many orders of magnitude. Why then should the seemingly unrelated time scales turn out to be so similar? *A priori*, it is more likely that one is much larger than the other, i.e. Case 1 or Case 3. Having already ruled out Case 1, we are then left with Case 3.

You may, or may not, be convinced by Carter's argument for preferring Case 3 over Case 2—we're ambivalent about it. For example, it would be invalidated if there were some deeper reason, one that we don't as yet know about, relating habitability lifetimes to evolutionary time scales. We can conceive that such 'deeper' reasons might exist, for example if both are ultimately traced back to the quantum physics of the atom. After developing the ideas a bit more, we will return, at the end of the next chapter, to the question of predicting how common complex life in general, and observers in particular, might be.

5.4 The 'critical steps' model

In his 1983 paper, Carter went on to propose a conceptual model for the evolution of complex life and 'intelligent observer species' (who, remember, are defined as species who are interested to ask this kind of question about their evolution). This is a 'toy model'—deliberately highly simplified so that it can be analysed with fairly simple maths, but still capturing some essentials of the process. The model supposes that, over billions of years, the pace of evolution towards complex organisms and observers is governed by the necessity to pass a number of difficult transitions. By 'difficult', we mean that the probability that any one of them will happen in a period of several billion years is low, so they are unlikely to occur in the habitable lifetime of a planet. Initially, we do not know how many of these critical steps there are in the sequence that leads to observers: we use *n* to stand for this unknown number. All other evolutionary steps are assumed to be easy, meaning they occur quickly by comparison with the difficult steps, so they do not slow down evolution. The events are pictured as occurring on planets with a fixed habitable period, starting from the mix of simple chemicals to be found on a newly-accreted planet. The steps are *sequential*, meaning that each must be passed before the next can occur, but otherwise they are assumed to occur randomly in time. We then ask the questions: How often will observers evolve on a planet? On the occasions they do evolve, how far through the lifetime of the planet do they appear?

If the probability of even one event occurring during the habitable period is small, the probability that two events will occur, one after the other, is much smaller still, and the probability that all *n* events occur is miniscule. Imagine again that you have a God's-eye view of the universe. You can see all the billions upon billions of planets on which intelligence might have evolved, and you can see that on most of them not even the first stage was passed. Putting these unsuccessful planets aside, you are still left with a huge number that did pass that step. Of these, you see that on most of them, the second step was not passed. You go on, through all *n* steps, winnowing out the unsuccessful planets at each stage. By the end you have only a miniscule fraction of the planets you started with, on all of which observers did evolve. Now you ask, *when*, on average, in these planets' habitable lifetimes, did the intelligent creatures evolve? Was it late in the planet's life, or early, or somewhere in between?

The mathematics gives a precise answer to this question (see Box 5.1). Remarkably, it turns out not to depend on the probabilities of the individual steps, providing these are small enough to qualify them as 'difficult'. The steps might all be very, very difficult, or all just quite difficult, or a mixture of both—it doesn't matter to the question of *when* the observers evolve on the successful planets, though it would of course affect *how many* successful planets there were. The 'when' question however, turns out to depend only on how many steps there are—the key result is that the more steps there are, the later the observers are expected to evolve.

A formula can be derived for the average time at which the intelligent species is expected to arise (see Box 5.1). This is only an estimate of the actual time that it will happen in any one instance, so when we look at the particular instance of our own evolution on Earth, we have to bear in mind that we are unlikely to arise at exactly the expected time, though we ought not to be a very long way from that time either. For the case of humans on Earth therefore, if we know how many difficult steps had to be passed for our evolution, we could use this value of *n* to obtain a 'best guess' of when in the habitable lifetime we would be most likely to evolve. Alternatively, we can run the calculation backwards: knowing when in the habitable lifetime of Earth we have evolved, we can use it to get a best guess of the number difficult steps involved in our evolution.

When Carter first derived this formula (Box 5.1) and applied it to the Earth, he took the habitable period for Earth to be 10 billion years, which is roughly the length of time the Sun will spend on the main sequence. At the end of this time, the star runs out of easily burnable nuclear fuel. Eventually it will burn out entirely, but it does not simply fade away to a dignified death. Rather, like geriatric tearaways, on leaving the main sequence, stars like the Sun begin a violent old age of expansions, contractions and explosions that will certainly destroy any planets that orbit them. Since we have appeared 4.5 billion years after the Earth formed, we have evolved about half way through the stable period. Carter therefore concluded that the best fit to the model was *n* = 1—just one unlikely event was needed for intelligence to evolve.

However, as we have seen already, this estimate for Earth's habitable lifetime, reasonable enough when Carter was writing in 1983, turns out to be too long. Present estimates place us about four-fifths through the habitable period, which would suggest a best estimate for *n* of 4. Certainly we appear to be more than ¾ but less than 9/10 through the habitable period, so *n* probably lies between 3 and 9.

Carter's original paper suggests he was surprised to find that a single step was the best fit to his model. Had he been a biologist, he probably would have given up before publishing, because from an evolutionary standpoint a value of *n* = 1 is very hard to accept. Thinking back to the Szathmáry-Maynard Smith major transitions that we described in the

---

Box 5.1 The key results of the critical steps model

We use the symbol $<t_o>$ to stand for the expectation time for observers to arise (the angular brackets denote an average, and the subscript o reminds us that we are talking about the evolution of, not just life, but of intelligent observers). We use t_h to stand for the habitable period of a planet, and n is the number of critical steps in the sequence that leads to observers. On those planets on which observers do evolve, the expected time is:

$$\langle t_o \rangle = \frac{n}{n+o} t_h$$

This formula was first derived by Carter (*11*), and a derivation is also given in Andrew's paper on the subject (*9*). Even if maths is not your strong point, this equation is fairly straightforward to understand. Imagine first that *n* is equal to 1—evolution of an intelligent species requires just one, very unlikely, evolutionary event, which occurs randomly in time. Normally it does not occur in the habitable lifetime of a planet. On those worlds where it does occur, it can happen at any time, so the *average* time that it occurs is at the half way mark, i.e. at $1/2t_h$. If you substitute 1 for *n* in the equation above you get just this answer. Now, suppose instead that there are two rare events that must occur in sequence for observers to evolve. Observers now appear on many fewer planets. Considering only that subset on which they do appear, this tends to happen later than in the case where *n* is one, closer to the end of the period available, in fact on average at $2/3t_h$, which is what you get if you replace *n* in the equation with 2. As the number of steps increases, the actual number of planets on which observers evolve gets smaller and smaller. On the remaining cases where they do evolve, they appear closer and closer to the end of the habitable period.

The formula above refers only to the expectation time of the last step, the emergence of observers. However, we can derive a more general expression that covers for all the earlier steps as well (*9*). If we number the steps from *1* to *n*, then the expression for the expected time of the mth step (where *m* can take any value from *1* to *n*) is

$$\langle t_m \rangle = \frac{m}{n+o} t_h$$

What this means is that, having winnowed the set of planets down to only those on which all steps occur and observers do evolve, we will find that on average, the earlier steps should tend to be evenly spaced through the history of the planet. For example if *n = 3*, the first step (with *m = 1*) would tend to occur around ¼ of the way through the habitable period, the second (*m = 2*) half way through, and the observers would be expected to appear at around ¾ of the way through.

The fuller analysis also derives probability distribution functions (PDFs) which show how the probability of occurrence of each step changes as a function of time (*9*). These functions are shown in Fig. 5.2, for the cases where *n* = 1, 2, 3 and 4. In the case that there is a just a single step required to produce observers, the PDF is constant independent of time—the step is equally likely to occur at any point in the habitable period. However for larger values of *n*, the first step becomes more likely to happen nearer the beginning and the last step more likely towards the end, with intermediate steps having their maximum probabilities between these. It is notable however, that the PDFs, especially of the intermediate steps, are quite broad. This means that in any one realization, the steps could depart from their expected even spacing by quite a wide margin.

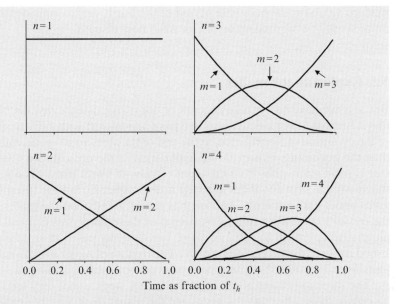

Fig. 5.2 Probability distribution functions as a function of time (shown as a fraction of t_h, the habitable period) calculated for the critical steps model, for cases in which 1, 2, 3 or 4 steps are required for the evolution of observers (9). With only one step in the model, it is equally likely to occur at any point in the habitable period. As more steps are added, the first becomes more likely to happen towards the start, and the last towards the end, with intermediate probability steps being most probable in between

last chapter, for this to be correct requires that the only really difficult step in the evolution of human intelligence from simple inorganic chemicals was the last transition, the one that occurred in our own very recent past and is connected with the origin of our language, and which distinguishes us from our close relatives who are not 'intelligent observers'. The earlier steps, origins of the genetic code, of bacteria, eukaryotes, sex, multi-cellularity etc, must all by comparison, be 'easy', compared to the origin of language and intelligence.

Most biologists would reject this idea. One would think that a long and difficult road was travelled in evolutionary terms between amino acids and chimpanzees. By comparison, the further step to humans seems comparatively small. It is perhaps for this reason, that his conclusion seemed biologically unsound, that Carter's paper was largely ignored by biologists. (Or more likely the explanation is simply that biologists don't read papers published in volumes devoted to physics and cosmology, which is where the paper was published—the next one in the volume was by Stephen Hawking.) However, a larger value of 4 or so for the number of critical steps, such as we get when we use a modern value for the habitable period of the Earth, makes considerably more sense in evolutionary terms, as we'll now discuss.

5.5 Exploring the model

Because we have had to take account of our own existence and the bias that it imparts to our calculations, we have had to start with a discussion of the evolution of intelligence, at the end of the story so to speak, rather than the beginning. However, the application of the critical step model that is most interesting from the point of view of Earth history is not so much what it says about the last step in the sequence, but the insight it gives about earlier events. Andrew has recently extended the maths of the model to provide a fuller description of these properties (9), and an alternative approach to some of these was first suggested in an unpublished paper ('Must early life be easy? The rhythm of major evolutionary transitions') posted on the internet by the economist Robin Hanson.

The most important property of the earlier steps is that they tend to be *evenly spaced* through the history of the planet (Box 5.1). For example, for a two-step sequence, the most likely times for the steps are at 1/3 and 2/3 of the habitable period. For our 'best guess' of the case for the Earth, of four steps, their expectation times would be 1/5, 2/5, 3/5 and 4/5 of the habitable period, or roughly at one-billion year intervals up to now. This tendency for even spacing holds no matter what the probabilities of the individual events provided only that they are sufficiently improbable to fulfil the assumption of the model, that a priori they are unlikely to occur during the habitable period. It does not matter that some may be only just unlikely enough to meet this definition, while others may be wildly improbable.

A second important property is that the range of times over which a given step can occur is quite wide. What this means is that although on average steps should be evenly spaced in time, if we examine any particular sequence there may be a fair amount of variation from this rule. This is not too much of a problem however, since, having a complete mathematical description of the model, we can calculate the probability that *any* observed sequence is consistent with it. We can use such calculations to test hypotheses in the usual way that statisticians do (9), rejecting scenarios that have say, less than a 5% or 10% chance of occurring. Steps that are too irregularly spaced—nothing happening for 4 billion years say, and then four steps in a fraction of a billion years, are unlikely to occur and can be rejected using these tests.

If we accept, for the moment at least, that this model may describe something important about evolution on Earth, the properties that are outlined above give us a sort of recipe for recognizing these critical steps. We should look for a small number of events (more than two, less than ten, best estimate of four), the last one of which is the development of ourselves, big-brained, grammatical-language-enabled human observers. The others should be rather evenly spaced through the history of life on Earth. Each step should be needed before more complex organisms were possible in a sequence traced from simple chemicals to humans (so, to take a random example, the origin of dinosaurs would not be such a step, since it was not essential to our development). Each would have occurred only once (because the steps are improbable: while one occurrence of each is necessary for us to be here, the chance of two occurrences is improbable squared, that is to say, negligible). By following such a recipe, it seems we ought to be able to identify candidates for these critical steps.

There are complicating factors however, that make the task unexpectedly difficult. These stem from the key simplifying assumption of the model, that evolutionary change comes in just two sorts—very easy and very difficult. This is a pretty drastic simplification. Imagine again that you have a God's-eye view, this time of evolution on Earth, and know everything there is to know about it. You could rank all the evolutionary steps that have occurred or might occur in order, from the most likely to the least likely. We know of no reason to think that if you could do this, you would find they fell neatly and naturally into two distinct groups, one easy, one difficult. More likely you would find that there were representatives of all different likelihoods. Taking account of intermediate difficulty events tends to slow the process of evolution between the hard steps, leading to the conclusion that the simple model probably over-estimates the number of hard steps in order to compensate for the fact that those of intermediate difficulty are ignored.

The critical steps model is too simple therefore to be a fully accurate guide to the evolution of the Earth system, but that doesn't mean it is useless. As we'll see in the next chapter, there are some candidates for unlikely transitions in Earth's past that are critical to understanding how we arrived at our present state, and they are well spaced through the history of the planet, so the model has some correspondence with reality. However, the transitions are not always definable as single point events, and they sometimes set off a lengthy train of consequences, which may all be considered part of the transition. With these due cautions about the simplicity of the model, let's now turn to address the question, what are the critical steps in our own history?

References

1. M. T. Rosing, *13C-depleted carbon microparticles in > 3700-Ma sea-floor sedimentary rocks from west Greenland. Science* **283**, 674 (1999).

2. J. W. Schopf, *The first billion years: When did life emerge? Elements* **2**, 229 (2006).

3. J. D. Barrow, F. J. Tipler, *The Anthropic Cosmological Principle*. (Oxford University Press, Oxford, 1986), 706 pp.

4. M. J. Rees, *Just six numbers: the deep forces that shape the universe*. (Weidenfield and Nicholson, London, 1999), 183 pp.

5. J. F. Kasting, D. P. Whitmire, R. T. Reynolds, *Habitable Zones around Main-Sequence Stars*. *Icarus* **101**, 108 (1993).

6. K. Caldeira, J. F. Kasting, *The Life-Span of the Biosphere Revisited*. *Nature* **360**, 721 (1992).

7. T. M. Lenton, W. von Bloh, *Biotic feedback extends the life span of the biosphere*. *Geophysical Research Letters* **28**, 1715 (2001).

8. N. M. Bergman, *COPSE: a new biogeochemical model for the Phanerozoic*. PhD thesis, University of East Anglia, Norwich (2003).

9. A. J. Watson, *Implications of an anthropic model for the evolution of complex life and intelligence*. *Astrobiology* **8**, 175 (2008).

10. J. E. Lovelock, M. Whitfield, *Life-Span of the Biosphere*. *Nature* **296**, 561 (1982).

11. B. Carter, *The anthropic principle and its implications for biological evolution*. *Phil. Trans. R. Soc. Lond., Series A* **310**, 347 (1983).

12. S. Webb, *If the Universe is teeming with Aliens...where is everybody?: fifty solutions to Fermi's paradox and the problem of extraterrestrial life*. Springer, New York (2002).

6
The critical steps

From the billions of steps on the road of our evolution, our task in this chapter is to identify the most likely candidates for critical steps—truly difficult events that may have determined the pace of evolution through Earth history. A good place to start is the list of 'SMS' transitions introduced in Chapter 4: these are critically important events in the increasing complexity of life, at least according to the two eminent evolutionary biologists Eörs Szathmáry and John Maynard Smith. However, these transitions are defined differently from our critical steps, and we can't simply take their list as our definitive answer. SMS are interested in how the unit of natural selection changed through time, and they define their transitions as occurring when a new type of reproductive unit evolves, such as the gene, the cell, or the colony. There is a good theoretical reason why such transitions from a lower to a higher unit of reproduction can be very difficult: selection at the lower levels usually acts to destabilise higher level units. So, the SMS transitions are a reasonable place to start looking, but they are only a guide, and some of them may turn out not to be sufficiently difficult to qualify.

Conversely, a hard step does not necessarily have to involve the emergence of a new unit of reproduction, in which case it would not be on the SMS list at all. In fact there is one such event in the history of the Earth which seems to us to be an excellent candidate for a critical step. It really must be included on our short-list, but it is not an SMS transition. This is the invention of photosynthesis—the oxygen-producing variety, which is chemically and energetically the most difficult. This invention seems to have occurred just once, and we would not be here had it not occurred, even though the ancestral cyanobacterium that first evolved the process is not a direct ancestor of humans. Adding this to the SMS transitions—we number it 3.5, since it occurs between the third and fourth in the original list—we obtain the following shortlist:

(1) Replicating molecules in compartments,
(2) Genes,
(3) Genetic code/prokaryotes,
(3.5) Oxygenic photosynthesis,

(4) Eukaryotes,

(5) Sex,

(6) Cell differentiation,

(7) Societies,

(8) Language/observers.

We'll now go through this list, asking what evidence there is for including each transition in a final list of critical steps. To evaluate this we use two criteria; the number of occurrences of the transition, and their timing. First we ask; is there evidence that the transition occurred more than once? If so, it is almost certainly not very difficult. Second we ask; do the steps meet the expectation that we have from the critical step model that they should tend to be evenly spaced through time? To answer the latter we have to date them, and Fig. 6.1 summarizes the evidence we will discuss for the timing of some candidate critical steps, and for the start and end of the Earth's habitable period.

6.1 The early life problem

We start with the controversial topic of dating the origin of life. Recall that the mathematics suggests that critical transitions should be approximately evenly spaced through time. We can see immediately from Fig. 6.1 that there is a problem with the first three SMS transitions, which must all occur before the origin of the first prokaryote cell. We show two 'latest' dates for the first cell, following on from our discussion of the evidence for early life in Chapter 1—an absolutely 'latest possible', and a 'likely latest'. The absolutely latest possible date is

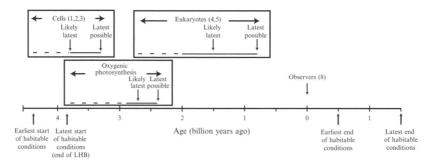

Fig. 6.1 A timeline of Earth history and future, indicating the habitable period and some candidate critical steps with ranges of ages within which they occurred. Numbers in brackets refer to the corresponding SMS major transitions. Dashed lines indicate that the relevant step might have occurred whereas once the line becomes solid the evidence is more convincing that it has occurred. Likely latest and latest possible dates are also given

3.3 billion years ago, and is that of micro fossils which have stood the test of time, and are taken to be relics of living microbes, including cells caught in the act of dividing into two. At the latest, the first three SMS transitions had been passed by this time. Our 'likely latest' date of 3.8 billion years ago comes from the Isua formation, and assumes that Minik Rosing's specks of carbon-13-depleted graphite are the remnants of a microbial ecosystem. Also on the timeline are marked possible earliest times after which the Earth was continuously habitable. Again we give two, the earlier of which is at 4.4 billion years and is the age of the oldest zircon crystals, dating a solid crust and the presence of oceans. The later one is at 3.85 billion years and marks the end of the 'late heavy bombardment', or LHB.

If, as our first assumption, we suppose that life could not have survived the LHB and therefore originated on Earth after it, the first three of our candidates for critical steps would have to all occur after 3.85 billion years in a time of 350 million years (at most, if we ignore Isua). If we accept that the Isua formation as evidence for prokaryotes, then the events must all occur in just 50 million years between 3.85 billion and 3.80 billion years ago. Neither is compatible with the assumption that they all are critically difficult steps—we would expect them to be more evenly spaced through the habitable period of Earth. The critical step model can in fact give a precise answer for just how unlikely this would be, suggesting that the probability is less than 3% if we ignore Isua, and less than 0.01% if we accept it. In science, hypotheses are commonly rejected if they have such low probabilities of being consistent with observations. Either the theory is wrong, or some of these steps are not critical.

Before revising our list of critical steps, let us try to see if there is a way around this problem, which preserves the view that the origin of life involves at least three very unlikely events. For example, could life have originated before or during the LHB and survived it? This would ease our problem by allowing more time to accomplish the necessary evolution. Judging by the battered state of the Moon, much of the surface of Earth was probably molten at one time or another during the LHB. However, since we have zircons that predate the LHB, we also know that not everything melted, and probably for most of the time, most of the surface remained solid. Life would have been a good deal less robust than zircon crystals however. To kill all known organisms for example, it is sufficient to bake them for half an hour in an atmosphere of superheated steam, in a type of oven called an autoclave, and this is regularly done in hospitals and research labs to sterilize equipment. An autoclave would however have no effect on the dates of zircon crystals. Large impacts during the LHB would probably have boiled off much of the early oceans and would have turned the atmosphere into something resembling the inside of an autoclave. But provided these were transient events, and that there was never a time that *all* the water on the planet was turned to vapour, life could have survived deep in the oceans.

There is one valuable piece of evidence that suggests this might have happened, though like virtually everything relating to this distant period, it is open to other interpretations as well. It emerges from the science of molecular phylogenetics and attempts to reconstruct the tree (or tangled bush) of life that we touched on in Chapter 3. One group of organisms that clearly comes near, and possibly at, the root of the tree is called the hyperthermophiles. As the name suggests, they live at very high temperatures and are found today at home in thermal springs, or at hot deep ocean vents. The temperatures at which they can survive and metabolize are above 80°C (and sometimes even above 100°C at ocean vents where the pressure keeps the water from boiling). This is no mean achievement, since most of the chemistry that lies at the heart of life can be expected to be completely dysfunctional at these temperatures: proteins and nucleic acids denature, unwinding, unfolding, dissociating and generally neglecting their allotted tasks. To cope with this, the hyperthermophiles have had to evolve a new type of cell membrane and many new heat-resistant enzymes and processes, just to survive in their chosen niche.

Some scientists take the deep rooting of the hyperthermophiles as an indication for a 'hot' origin for life, for example at deep ocean vents, while others argue for a cool beginning—like most facts about the origin of life, this one is not agreed upon. But there is another possible reason for finding the hyperthermophiles at the root of the tree. Perhaps this is the 'memory' that life has of the LHB. Perhaps life, having originated in a relatively quiet period after the accretion of the Earth, was then subjected to several hundred million years of heavy bombardment and near-misses where the planet was almost heat-sterilized. Nothing survived except the hyperthermophiles.

Armed with this possibility, let's return to our shortlist of candidates for critical steps. Even if we assume that life originated before the LHB, around 4.4 billion years ago, the probability would remain low (less than 10%) that three critically difficult steps could have occurred by 3.5 billion years ago. There just does not seem enough time therefore for three critical transitions to occur before prokaryotes arrive on the scene. As a last resort to fit the theory, we might allow that life could have arisen elsewhere in the Solar System even before the Earth formed, and sheltered on some friendly asteroid until Earth was cool enough to be habitable. Even this doesn't help much however, as the process of Solar System formation is thought to be pretty fast: the curtain rises on the Solar System at 4.57 billion years ago, and the Earth has an ocean and stable crust by 4.4 billion years ago, so, unless life came from outside the Solar System, which really seems unlikely to us, we can move back less than another 0.2 billion years. We are therefore forced to the conclusion that not all the first three SMS transitions, which mark the evolution from molecules to microbes, can be critically difficult.

Suppose instead there is only one difficult-to-pass bottleneck in this evolution, not three separate and independent ones. This is the opinion of Eörs Szathmáry, who believes that of the three SMS transitions that precede prokaryotic cells, just one—the origin of the genetic code—was a really difficult one. This fits much better. We would still have a problem only if we insist that the Isua formation contains evidence for microbial life, *and* that evolution could not begin before the end of the LHB. With only around 50 million years between these events, the chances of even one critical step occurring in such a time interval are only a few percent. However, if we assume that life evolved before the LHB and survived it, or if we reject Isua as evidence for life having reached the stage of cells, then the model can be reconciled with the observations.

Alternatively, given the right conditions, perhaps the time required to go from replicating molecules to the establishment of prokaryote cells is not so long at all, but is more or less inevitable in say, 100 million years. For this to be the case, none of the three steps on this road could be difficult in the sense of critical steps. It is a measure of the denseness of the fog that still surrounds our understanding of the molecular origins of life that we simply do not have the information to decide between these scenarios. There is no agreement on just how difficult was the evolution from replicating molecules to bacterial cells. In the modern scientific literature two polar opposite assumptions are sometimes held: on the one hand that life is virtually certain to occur, and quickly, once conditions are right or, on the other hand, that it is an extremely unlikely event. The evidence is not strong—it is pretty nearly non-existent—for either of these positions.

To summarize, at least two of these first three SMS transitions must be dropped from our final list. We are persuaded by Minik Rosing's evidence for the significance of the Isua rocks as evidence for life, but also by Szathmáry's expert advice on the difficulty of evolving the genetic code, so we propose to keep this as a critical step. We can fit in one such step before Isua, provided that life arose before or during the LHB, which was never sufficiently intense to sterilize the entire planet. (Alternatively we might speculate that the Isua graphite could have been produced by early life forms that did not yet possess the genetic code, and that it did not arise until around 3.5 billion years ago.) We must admit that, given that we have only 'latest' dates for prokaryotes, we can't really calibrate just *how* difficult this step is and whether it really is hard enough to qualify as 'critical'. All we can say is that it could be such a step, and it will turn out to be the least secure of the steps we propose.

6.2 Transpermia and deep biospheres

The possibility that life was already advanced to the bacterial stage before the end of the LHB leads naturally to a slightly startling conclusion, that microbes could easily have been spread around the other bodies of the Solar System as well as Earth—and that life could in fact have started just as well on any of them and been spread to the Earth, as it could have made the journey in the opposite direction. This is because the Earth was not isolated during this time. In fact it is wrong to think of Earth as being isolated even today—about 40 of the meteorites that have been found on Earth are actually pieces of Mars—and Mars presumably has meteorites on it that are pieces of Earth. So the planets are still exchanging material that life might conceivably hitch a ride on. At the time that the Solar system formed, there would have been much more of this exchange going on. Today the large impact necessary to knock a piece out of the gravitational pull of one of the established planets is very rare, but as they were forming, there would have been a ready to-and-fro of matter in the form of asteroids passing into elliptical orbits that crossed the formation zones of the different planets.

In 1996, one of the rare Martian meteorites, ALH84001, found in the Allan Hills of Antarctica, made the headlines when NASA scientists announced that it contained evidence for bacterial life from Mars. Subsequent debate has failed to convince most experts of this claim, but it remains tantalizing, and certainly not definitively disproved. Several lines of evidence suggest the possibility of Martian biological activity having occurred inside the meteorite. First, though it is mostly igneous rock, there are globules of carbonate formed by interaction with water at low temperatures. Within these, crystals of magnetite (of a kind almost always on Earth associated with bacteria), relatively complex organic molecules, and structures that could be (very small, it is true) fossilized bacteria are all found. But sceptics have advanced non-biological or Earth-contaminant explanations for all of these, so we are left with the present 'not proven' judgement. However, the Allan Hills meteorite, though knocked off Mars less than 20 million years ago, is made of very old rock indeed—at 4.5 billion years old, it is much older than any intact rock on Earth, and formed less than 100 million years after the origin of the Solar System. So, if it contains evidence for life, it could be some of the very first life (which might perhaps explain why the fossil cells, if that's what they are, are so small). Staring at the pictures of the ALH84001 'bacteria', the thought occurs that they could be our ancestors, and if so, we are all, ultimately, Martians.

If anything about the ALH84001 was ever alive, it was certainly dead by the time it began its journey through space. But in principle, in the past it would have been possible not just for the dead remains of bacteria to make this journey, but for viable ones to do it too, and by this means life could have spread to all the rocky planets and moons of the inner Solar System. To survive such a journey, the bacteria would have to be entombed sufficiently deep in rock that they would be shielded from the extreme heat that the outside of a meteorite would be subjected to, both when first ejected by an impact from its old home planet and when falling through the atmosphere of its soon-to-be-new-home planet.

Since today we mostly think of life as a phenomenon that resides at the surface, it might come as a surprise that there would be any living things actually inside rocks. But there is no doubt that this 'deep biosphere' does exist. It has been found, here on Earth. Drilling into sediments at the bottom of the ocean, bacteria are found living more than 500 metres down below the sea bed (*1*). Viable bacteria have been recovered from 3 kilometre deep drill holes, and there doesn't seem to be an obvious reason why the biosphere should not extend substantially deeper still. The pressure of overlying rock is enormous of course, but this does not seem to bother bacteria. It has been claimed that the biomass of this hidden deep biosphere may be greater than that of all the biology with which we are familiar that lives at the surface. In this sense life may be mostly a sub-surface bacterial affair even on the Earth, where living at the surface is possible, indeed pleasant.

What are bacteria doing to make a living inside rocks? Of course they are not photosynthetic, there being no light down there, but rather chemosynthetic—utilizing chemical gradients of one sort or another as an energy source. But the flux of energy down where they live is today extremely low. Therefore, the bacteria at extreme depth also have extremely slow metabolisms. Doubling times, which in an energetic environment such as the human gut can be less than an hour, are thousands of years in the deep biosphere of the Earth (*2*). They tick over at the slowest possible rate, and in all probability can remain viable for millions of years. Controversial claims have been made that bacteria have been cultured after recovery from inside a crystal of halite which was laid down 350 million years ago, and bacterial spores recovered from the inside of a mummified bee, preserved in amber 25 million years ago, have been cultured (*3*). So bacteria can survive in suspended animation, for long enough to enable them to make the journey between two planets in good shape.

This being the case, once we allow that life was probably already started during the bombardment period, it is quite plausible, in fact it is hard to avoid concluding, that it would be spread quickly onto all of the rocky bodies of the Solar System. Since that time of course, all those planets and moons have evolved and on none of them except Earth can life survive today at the surface. On some of them (Venus and Mercury for example, which are hot and dry from surface to core) it is a fair bet that they are sterile today. However, some of them still have internal zones in which liquid water may exist. Mars is an obvious example— colder inside than Earth, but not that cold. Some of Jupiter's Moons, especially Europa, which may have an ocean, tidally heated sufficiently to keep it liquid, beneath a thick covering of ice, are also possible candidates. If so, life may have simply migrated inside these bodies as the surface cooled, and could still be there, still viable, still metabolizing very slowly. Not a very interesting sort of life by our standards perhaps, but a heck of a long one. This view of life, as a low energy, subsurface phenomenon, is very different to our surface-dwelling perspective, but it may be closer to the original. Life on Earth became supercharged, a high-energy and fast-living surface dweller, only after a later step, the invention of photosynthesis, which supplies orders of magnitude more energy than is available to the deep biosphere.

6.3 Oxygenic photosynthesis

Oxygenic photosynthesis is our own insertion into the SMS list of transitions—number 3.5 on our shortlist—put there because we think the case for its being a critical step is very strong. Photosynthesis powers the planet, and we'll describe in Part 3 the evolution of this remarkable mechanism, and just how profoundly it has transformed the Earth system. Here we simply want to establish reasons for believing it may have been one of our critical steps, using the two criteria of uniqueness and timing by which we can hope to recognize such transitions.

The bacterium that first produced oxygen by splitting water with sunlight, coupled together two earlier-evolved photosynthetic pathways, called photosystems I and II. In addition it possessed a unique enzyme containing four manganese atoms, called the water splitting complex. The biochemistry of oxygenic photosynthesis is staggeringly beautiful and complex, involving the absorption of a total of eight photons in the process of splitting two water molecules to release one molecule of oxygen. Among all the prokaryotes, only the cyanobacteria evolved this ability, and later, they became the ancestors of all the chloroplasts in all the eukaryote algae and plants.

The evidence is consistent with the original invention occurring just once, with no sign that it evolved separately ever again, and the water-splitting complex has remained essentially unchanged over billions of years (4). This is significant, because oxygenic photosynthesis is an amazingly useful trick for any organism to possess, since it enables the uptake of carbon *and* the production of energy using just light, carbon dioxide and water—three of the most abundant resources on the surface of the planet. An organism that can do it is well equipped to make a living in most habitats on Earth and, therefore, you might suppose that if it were easy in evolutionary terms, it might have happened more than once. Since it did not, we can assume that it was no everyday event: this was a red-letter day for life on Earth.

Such is the case for the uniqueness of oxygenic photosynthesis. What about timing—is it well-spaced compared to other candidate critical steps, as the model would lead us to expect? Of course, it absolutely must have occurred before free oxygen appeared in the atmosphere at the 'Great Oxidation', about 2.4 billion years ago. This gives a latest possible origin of around 2.45 billion years ago (Fig. 6.1) that has some vocal supporters (5). Most probably however, it was invented before that time: there is evidence of a whiff of oxygen entering the atmosphere as early as 2.7 billion years ago. Euan Nisbet and colleagues argue that the carbon-13 content of the oldest carbonate reefs at about 2.9 billion years ago betray the presence of cyanobacteria then. All these inferences have their weaknesses and we will discuss them in more depth in Part 3. However, the chemical composition of the remains of a microbial ecosystem laid down in the South African Buck Reef Chert, 3.4 billion years ago, shows a complete lack of oxygen even locally. Apparently, there were photosynthetic bacteria here, but they were of the pre-oxygenic variety, so perhaps the invention occurred between 3.4 and 2.7 billion years ago.

If the origin of oxygen-producing photosynthesis occurred within this interval, it would fit our model rather well, because it would be in the range 0.5–1.5 billion years after the origin of prokaryotes, which is our previous step, with a best guess of around 1 billion years after. Before congratulating ourselves too warmly however, we should be honest about the reliability of the inference that oxygen production evolved after the date of the Buck Reef Chert. The evidence is that oxygen was locally absent within the formation, at this location. However, just because there were no ancestral cyanobacteria at Buck Reef, it does not necessarily follow that oxygen evolvers did not exist at that time. You may recall for example that Minik Rosing has argued, based on carbon-13 measurements, that oxygenic photosynthesis is implicated in the Isua rocks at 3.8 billion years of age. We think, or rather we predict from the critical steps model, that he should be wrong: a case can be made that the invention of oxygenic photosynthesis is really difficult. It requires the pre-existence of prokaryotes, so the theory leads us to expect some good fraction of a billion years should elapse between the appearance of prokaryotes and photosynthesis. If Rosing turns out to be right (and who knows what future research will show), it will imply that photosynthesis is not so hard after all.

6.4 Eukaryotes and sex

Next on our list we would like to put a date to the origin of the eukaryote cell (number 4 of the SMS transitions). We will discuss this in more detail in Part 4 of the book. For now, we note that eukaryotes have a much more complex structure than prokaryotes, and this gives good reason for thinking that this was a difficult transition.

According to one view, the key innovation that enabled eukaryotes to evolve was the ability to flow around, engulf and 'eat' other cells, a process called phagocytosis. Before this could happen, the forerunner of eukaryotes had first to develop an internal macromolecular structure and scaffolding, called a cytoskeleton, that allowed the interior of the cell to flow and move, and also to lose the rigid outer wall that bacteria have. There is no agreed pathway or time by which these changes took place, but they required the accumulation of adaptations in a long series, such that the divergence from prokaryotes may have begun early, and modern eukaryotes have many unique genes with no prokaryote equivalent. The acquisition of phagocytosis may therefore have been intrinsically unlikely, even though once accomplished many benefits would flow from the innovation.

In this view, once the capacity for eating things was in place, the early eukaryotes began to acquire internal components. Phagocytosis involves folding into the cell part of its external membrane, and this helped create the several kinds of internal membranes which characterize eukaryotes, and organelles such as the nucleus. They also acquired new organelles by the process of endosymbiosis, which we introduced in Chapter 3—by engulfing other organisms but failing to digest them and instead incorporating them as internal symbionts. By this means they obtained their mitochondrial power houses, and one of their ancestors obtained a photosynthetic chloroplast by engulfing a cyanobacterium.

Endosymbiosis occurred at least twice (one event for mitochondria and one for chloroplasts) but there are strong clues that suggest it has probably occurred several other times as well. For example, a large group of algae called the heterokonts, which include seaweeds such as giant kelp, have chloroplasts with an unusual four-layered outer membrane. It seems likely these arose after one of them ingested an already photosynthesising eukaryote, which had in turn evolved from the original chloroplast-forming event—the Russian doll structure again. There is also one known organism, a photosynthetic amoeba living in fresh water called *Paulinella chromatophora*, which appears to have picked up its chloroplast from a cyanobacterium recently (6). This event was completely separate from the one which, probably more than 2 billion years ago, gave rise to every other known chloroplast. Therefore, endosymbiosis seems comparatively common, and by itself is probably not a 'critical' event. However, the fact that eukaryotes seem to be so good at hosting endosymbionts, but prokaryotes are not, points again at the evolution of the ability to engulf food as possibly involving such a critical step.

After the origin of eukaryotes the next transition suggested by SMS is the origin of sex. Sexual reproduction, the sharing of genes between two individuals to produce offspring, is normally viewed as a primary character trait of all eukaryotes. There are many eukaryote species that *can* reproduce asexually, but there are only a few that *must* reproduce asexually, and in those cases it is usually clear that they stem from sexual ancestors—they have given up sex, rather than never having had it in the first place. Eukaryotes and sex seem to be inextricably associated, and sex seems to be important for the long-term survival of eukaryote species. With one exception*, no group of organisms that has forsworn sex entirely seems to have prospered in evolutionary terms. Most celibate eukaryotes have probably given up sex relatively recently, and are destined for extinction in the fairly near future. We males can take some comfort from this: it seems we must be good for something after all, though it's not quite clear just what!

We would like to know if the evolution of sex is really separate from the evolution of eukaryotes, or whether the two are so closely related that sex co-evolved with the eukaryotic cell. To answer this, it would help if we knew precisely why organisms bother with sex, but we don't. The advantage of sex remains a major problem in biology which is unresolved today. Sex, as John Maynard Smith pointed out, is costly, and a naïve Darwinian analysis would lead us to believe it shouldn't happen (7). It comes at a 'twofold' cost, because only half of a parent's genes are passed to any given descendant. By comparison, in an asexual population, everyone can reproduce and can pass on all their genes. Under favourable conditions, when populations are not limited by resources for example, an asexual population should in principle multiply twice as fast as a sexual one and should quickly come to dominate. There *must* be counter-advantages to sexual reproduction which become apparent over longer time frames, when conditions turn less favourable for instance, to account for the near-universality of sexual reproduction in eukaryotes. It is usually assumed that these elusive advantages stem from the additional genetic diversity that sex promotes in a population. Sexual populations are more varied genetically because new genotypes are created with each generation, and the genes from the entire breeding population are mixed together and potentially available to many descendants, whereas asexually produced offspring are clones, genetically identical to their mother-parent except for mutations. This means that a sexual species ought to be able, in some circumstances at least, to adapt more rapidly to changing conditions than can an asexual one.

---

* The bdelloid rotifers, whose existence was described as 'an evolutionary scandal' by John Maynard Smith. Recent research has established that they gave up sex about 80 million years ago. They are a class of microscopic animals that live in puddles but have the ability to survive complete desiccation for long periods—which perhaps also serves to free them of parasites, removing one powerful reason for the persistence of sex.

For example, one prominent theory, the 'Red Queen'[†] hypothesis, holds that sex persists because of an unending arms race, especially between parasites and their hosts but also between predators and their prey. Rapid evolution of parasites to exploit their hosts for example, leads to equally rapid co-evolution in the hosts to defend against the parasites, with both sides having to evolve continuously, and therefore benefiting from the genetic diversity that sexual reproduction allows. The Red Queen effect is only one of several possible benefits that help to explain why sex persists, but it is particularly relevant to the question we are asking, of whether sex is essential to eukaryotes, because they would have become particularly susceptible to parasitism when they evolved the ability to ingest food wholesale. With the ability to eat by engulfing foreign cells, came the danger of ingesting potential pathogens. So if sex is part of the protection by which species fight their parasites, eukaryotes would be expected to need this protection more than bacteria.

Thus, we think that the origin of eukaryotes can be seen as a candidate for a really difficult transition and that the origin of sex is part of the same story, arising essentially at the same time as the modern eukaryotes appeared. But when was that?

Dating the first eukaryotes is difficult, and not just because, unlike the simplistic assumption in our mathematical toy model, it is really impossible to define them as coming into existence at a single point in time. We go into the dating issue in more detail in Chapter 12, but here give our main conclusion from that discussion: our best guess is that the key event, the evolution of phagocytosis, followed relatively quickly by the acquisition of symbionts, occurred roughly 2 billion years ago, give or take 0.5 billion years. This is about a billion years after our best estimate for the origin of oxygenic photosynthesis, in good agreement with the hypothesis that they are both critical steps. (However, again we need to acknowledge the uncertainties: as Fig. 6.1 shows, the latest possible date for the origin of oxygenic photosynthesis postdates the earliest possible date for the origin of eukaryotes, which certainly wouldn't be consistent with the choice of these two as critical steps.)

6.5 The last three SMS steps

As we cautioned above, while the SMS transitions are a reasonable place to start looking for critical steps, there is no guarantee that they are truly

---

[†] Named for the Red Queen in Lewis Carroll's *Alice through the looking* glass, who tells Alice: 'Now, here, you see, it takes all the running you can do, to keep in the same place. If you want to get somewhere else, you must run at least twice as fast as that!'

difficult. Transition six, cell differentiation and multi-cellularity, and transition seven, the formation of colonies and societies, are cases in point. With both of these transitions we have evidence that they have happened independently a number of times in different classes of organisms. Multi-cellularity and cell differentiation, which we'll discuss in more detail in Chapters 12 and 14, occurs independently in plant, animal, fungal and some protistan lineages of eukaryotes, and it also occurs to a limited extent in some prokaryotes. Social organization has arisen more than ten times in insects alone, as well as in mammals, birds, and fishes, to name a few. This could not happen if the steps were in themselves so difficult that they would not normally be expected to occur for billions of years, so they cannot be critical in the sense of the critical step model.

Nevertheless there is more to be said, particularly for the role of the multi-cellularity and cell differentiation. Eukaryotes repeatedly took this step, sometime in the period leading up to the Cambrian explosion, and they have continually differentiated since. Animals are the differentiators par excellence, so that today there are of order one hundred different cell types in higher animals, and this kind of complexity would seem to be a requirement for self awareness and 'observer status'—being able to ask the question 'where did we come from?' So for example, of the two hundred-odd cell types in humans, at least a dozen are types of nerve cell, which is what we use to think. By contrast, though bacteria frequently live together in associations that are loosely multi-cellular, they exhibit only very muted cell differentiation, with the number of different types of cells rarely exceeding two (for an example, see Box 6.1).

Something changed over the billion years from the first signs of cell differentiation in eukaryotes to the Cambrian explosion, which let the brakes off and allowed hugely greater cell differentiation in several lineages of eukaryotes, leading to structurally much more complex organisms. In fact several major changes, including much faster information processing, a rise in atmospheric oxygen, and catastrophic snowball glaciations, were needed to set the stage for the Cambrian explosion. We will tell this story in Part 4, but the important point for our current discussion is that we believe these all to be consequences of the rise of the eukaryotes, albeit that they took more than a billion years to really take effect. Just now what we are concerned with is the fit, or lack of it, to a critical step model of evolution, and it is clear that multi-cellularity, cell differentiation, and social organization can't themselves be interpreted as critical steps. Rather they are delayed consequences of the *previous* step—eukaryote evolution. It is one of the shortcomings of our toy model that it can't deal with such delays, tending instead to overestimate the number of critical steps.

We are now at last up to the point in Earth history where most paleobiologists begin their story—the Phanerozoic, the eon of plants and animals, everything from mosses through to dinosaurs. We have

> **Box 6.1 *Trichodesmium***
>
> An example of cell differentiation in prokaryotes is *Trichodesmium,* a cyano-bacterium that is responsible for fixing much of the nitrogen from the atmosphere into the oceans today. It is found in the warm subtropical oceans, where it forms golden brown clumps of filaments a centimetre or so across, which float at the surface. Inside these filaments, differentiated 'interior' cells are responsible for the nitrogen fixation, protected by their surrounding 'normal' colleagues from atmospheric oxygen which interferes with the process. *Trichodesmium* is hugely important for the modern Earth system, responsible as it is for splitting the strong N-N bond in atmospheric N_2 gas, to provide much of the fixed nitrogen needed by life in the oceans. It is a chemical marvel, performing at room temperatures and pressures a reaction that we humans can only emulate in an industrial plant at temperatures of hundreds of degrees and pressures of hundreds of bars. However, while *Trichodesmium's* two-fold cell differentiation may be sophisticated by prokaryote standards, it is frankly primitive compared to animals. We eukaryotes with our hundreds of cell types can afford to be smug: structurally, we are much more sophisticated than the humble, hard-working *Trike.*

proper fossils to study at last, ones that you don't need an electron microscope to see. However, by the very exacting standards of this chapter, where we are looking for really rare, unusual and critical steps, nothing interesting happens in most of the Phanerozoic, so we skip straight to the modern day, and (a roll on the drums please) the final step, the emergence of *Homo sapiens*.

6.6 The emergence of intelligent observers

It is an important assumption of the critical step model that the emergence of humans, a representative 'intelligent, observer' species, is a critical step. If this is right, even on a planet on which complex animals had evolved, we would not expect a self-aware, language and technology equipped species such as ourselves to evolve except rarely. Can we find evidence that bears on that assumption?

We are digging here in an intellectual garden, or perhaps it is a minefield, that has already been thoroughly dug over by well-qualified experts who know their fossils. But (and somehow, this isn't a surprise), these authorities don't agree with one another, in fact they can reach diametrically opposite conclusions by considering the same set of events. Famously, Stephen Jay Gould, surely the most articulate and engaging paleobiologist ever to put pen to paper, argued that evolution was so highly contingent—his word for it—that it would never replay the same way twice. He imagined life as a tape recording: 'You press the rewind button and, making sure you thoroughly erase everything that actually happened, go back to any time and place in the past. . . . Then let the tape run again and see if the repetition looks at all like the original' (8). His answer was that it would not. The chance that, starting again at say, the Cambrian explosion, we would end a half-billion years later with a sentient and self-aware approximation to a human being was miniscule.

But more recently, Simon Conway Morris has championed the opposite view. Readers of Gould's best-selling *Wonderful Life,* in which he put the contingency argument most forcefully, may be forgiven for being confused by this. They will remember the youthful Conway Morris as the hero of that story, and wonder how he came to be one of its sharpest critics. He was the brilliant student who re-interpreted the Burgess Shale, the most important collection of fossils from the early Cambrian, to show the amazing variety of different animal forms that arose shortly after the Cambrian explosion. The fact that the great majority of these weird and wonderful creatures then went extinct, in such a hurry that they didn't even make it to the next geological period, demonstrated for Gould the essential contingency of evolution, dominated by chance events so that the outcome was entirely unpredictable.

However, in *The Crucible of Creation: The Burgess Shale and the rise of animals,* an older Simon Conway Morris argued that his own youthful exuberance for classifying the Burgess Shale animals as new phyla had been, at least sometimes, mistaken, and that many of them were indeed ancestors of modern animals (9). Gould had therefore misinterpreted the Burgess Shale to support his own preconceived idea of rampant contingency, but in fact it did no such thing. Then in a more recent book, *Life's Solution*, Conway Morris argues that the dominant paradigm for evolution should not be contingency, but *convergence* (10). He describes dozens of elegant examples to illustrate how life takes similar evolutionary paths in response to similar needs and environmental conditions, so that similar structures and functional forms evolve repeatedly. We have already remarked how multi-cellularity and social cooperation evolved many times, but Conway Morris gives examples of such convergence in almost every branch of biology. To mention a few of the more famous examples of convergence, the wings of bats and birds look remarkably similar but have evolved separately. The grey wolf (a placental mammal) and the recently extinct Tasmanian wolf (a marsupial) both look unmistakably dog-like, but only one is closely related to the domestic dog, and the camera-like eyes of vertebrates such as ourselves, and cephalopods like the octopus are nearly identical in design but independently evolved. After reading Conway

Morris's book, no one can doubt that that evolutionary convergence is commonplace—perhaps more the rule than the exception.

But Conway Morris goes further. He subtitles the book *Inevitable humans in a lonely universe* and proposes that essentially all evolution is convergent, including intelligence, which should eventually evolve many times. He suggests that several animals besides humans, including crows, octopuses and dolphins, have most of its elements. He implies that given time, the emergence of human-like intelligence is inevitable and if it had not been us or our cousin apes, then one of the descendants of these animals would have become language enabled and big-brained enough to be recognizably intelligent at some time in the, not perhaps too distant, future.

Evolution is often convergent, perhaps more often than not. But is it *always*? If so, then there is no place for critical steps in evolutionary theory, because they are by definition not convergent—they are rare and happen only once. In fact in the limit of total convergence, there is no place for any kind of contingency or deviation from a single pathway. Everything would be effectively pre-ordained as soon as the first bacterium emerged. Like a train rolling down a hill, once started, the course of evolution would be fixed on an invisible track until intelligence evolved, around four billion years later. Unless the origin of life is spectacularly difficult so that the process almost never actually begins, this would imply that complex, intelligent life is probably comparatively common in the universe.

We are not very convinced by this idea, that convergence is so powerful a tendency that evolution is pre-ordained to follow a given path. Imagine for example, that the dinosaurs had not been wiped out by a meteorite impact 65 million years ago. Presumably they would still be here, and the descendants of our mammalian ancestors would still be scurrying around their ankles, because all the ecological niches for larger animals were filled by big-bodied, but very small-brained, reptiles. So a chance event, at least this one time in the shape of the meteorite, has been involved in evolution up to the present day. If it had missed, mammals might have had to wait a few hundred million years longer—and recall that the Earth is in its old age and that, on the geological scale, time is running out for life on Earth.

Neither does it seem to us very likely that crows, octopuses or even dolphins will prove ancestral to 'observer species' in the foreseeable future—though we'd have to agree that the course of evolution can be hard to predict (it's contingent!) and a visitor to the planet a few million years ago might not instantly have spotted the potential of our ancestors, who at the time were one among many species of apes in Africa. In the 500 or so million years since the Cambrian explosion, only one animal species has become sentient, linguistic and technological. There does not seem any very fundamental reason why this had to happen at just this point in time, rather than hundreds of millions of years sooner or later, so we think that it is at least a defensible view that this is an unlikely, and randomly timed, step, of the kind that the critical step model envisages. The fact that we may represent such a step does not of course mean that any big divide in biological or genetic terms exists between us and our nearest relatives—we know that there is no large division. The divide in this case is not biological but cultural, in our language, libraries, art and science.

We've now completed consideration of our short-list of possible critical steps, and just four have passed muster. They are: the origin of prokaryotes before 3.8 billion years ago, oxygenic photosynthesis between 3.4 and 2.5 billion years ago, eukaryotes, which evolved in stages but to which we assign a critical step at roughly 2.0 billion years ago, and humans, 0 billion years ago. These are not particularly evenly spaced through Earth history, since three of four events occurred before two billion years ago, only just over halfway from the formation of the Earth to the present day. However, we can test this spacing using the full maths, and it turns out it is regular enough to be consistent with the model—recall that perfectly regular spacing is only the expected *average* outcome if we were able to sample many planets on which observers evolve. To be tediously exact about the statistics, we find that in a four-step model where the habitable period of the Earth begins at 4.4 billion years before the present and extends to 1 billion years in the future, the first three steps would have occurred by 2 billion years ago in 24% of cases. This figure would need to be 10% or less for us to consider rejecting the model according to normal statistical practice. We've had to strip out uncertainties and caveats, and somewhat arbitrarily plump for exact dates for the steps, to be able to make the comparison at all, but the important thing we have learned is that Earth history is not obviously incompatible with the assumptions underlying this model.

6.7 Predictions regarding life elsewhere

In this section we make some predictions from the assumptions of the anthropic model, concerning unknowns about life elsewhere. These are interesting in themselves, and since they are open to test in the future as our knowledge increases, they could help establish or disprove the anthropic view of Earth history which we have been exploring.

So let us now review those fundamental assumptions and what they imply. The main one is that the pace of evolution on Earth to ourselves, complex, intelligent observers, has been constrained by the necessity to pass a small number of intrinsically very unlikely events. These are sufficiently improbable that, a priori, they would not have been expected to all occur during the limited time that the Earth will be inhabitable. However, on Earth, by lucky chance, they occurred considerably more quickly than was to be expected, which is how we came to be here and are able to ask these questions.

This idea has implications for how much life, of what kind, and around what kind of stars, we ought to expect to find in our part of the universe. These implications have some hope of being tested, because over the next few decades, astronomers and the space agencies will be putting huge effort into discovering and characterizing extra-solar planets, using telescopes both at the Earth surface and in space. Ultimately, the goal will be to analyse spectroscopically the atmospheres of any planets that are found, looking for the raw materials we know to be necessary for life such as water and carbon dioxide. But the investigators will also be hoping to detect 'bio-signatures'—in particular, ozone. This is thought to be diagnostic for oxygen, which itself is difficult to detect in a planetary atmosphere since it does not exhibit visible or infrared absorption lines. Ozone however is comparatively easy to detect, and is expected to be present only when there is also significant oxygen.

The planet-finding missions of this century will build on ideas going back to the middle of the last one. Jim Lovelock was the first to suggest that analysis of planetary atmospheres could be used to diagnose the presence of life, an idea he developed with the philosopher Dian Hitchcock (*11*). They pointed out that the best bio-signature is not just one gas, but the presence of two that are in strong chemical disequilibrium with one another. They suggested that if you trained an infrared telescope on the Earth, you would be able to detect the simultaneous presence of ozone (hence oxygen) and methane in the atmosphere. Since methane and oxygen react with one another rapidly to produce carbon dioxide and water, you would be able to deduce that something must be producing them from their reaction products with equal rapidity, and this something must be life. (Actually, it's not just life, but photosynthesis, that you would have diagnosed from the measurements.) Some 25 years later, in one of his last papers, Carl Sagan and colleagues (*12*) demonstrated that the technique worked for the Earth, using the instruments on the Galileo spacecraft looking back to Earth while on its way to Jupiter.

One way or another, in the first half of the twenty-first century we are going to get lots of evidence that bears on the habitability of nearby solar systems. There are some 250 star systems within 30 light years of Earth. Let's be optimistic and assume we find that many of them have planets in their habitable zones. What does the anthropic theory suggest we should find when we examine them with these planet finder missions?

We suspect prokaryote life is not so common as to always arise on a planet within a habitable zone, but it involves at most one really difficult step (you'll recall we can't really be sure it's critically difficult) and we consider that there is a reasonable chance that it will have arisen on some of the systems we will be able to observe. Pre-oxygenic photosynthesis is not apparently a critical step, and it arose relatively quickly on Earth after prokaryotes were established. This would enable an energetic biosphere to exist, with a characteristic atmosphere full of exotic trace gases such as hydrogen sulphide and methane as well as carbon dioxide and water. We have some hope therefore, of being able to detect a few planets that have this kind of biosphere, during the next few decades.

But oxygenic photosynthesis was a difficult step—it may be genuinely critical in the sense of the model, in which case it would occur on, at most, one in ten of the planets that took the first step, and more probably many fewer. In a sample of a few hundred planets we would be lucky to observe any that took this second step. We conclude that most likely we will not find any evidence for abundant oxygen on any of these target planetary systems.

There are some other interesting predictions that also come from the anthropic model. For example, if complex life is rare, it is likely that Earth will be found to be one of the most favourable possible spots for it to have evolved, a cosmic Garden of Eden. This leads to an interesting question: Is there anything about our Solar System that marks it out as unusual compared to most others, and might make it particularly conducive to hosting complex life? Of course it has Earth, ideally situated in the habitable zone, and it may be that few other solar systems have such planets—but we don't know as yet what the distribution of planetary systems is, so we must put that aside for the moment. Is there anything else unusual about the Solar System?

As a matter of fact there is: the Sun is an unusually bright star, brighter than more than 90% of its neighbours. Fig. 6.2 shows a histogram of the luminosity of about 250 local stars within 10 parsecs (about 33 light years) of the Sun, catalogued by the Research Consortium for Nearby Stars (RECONS). The Sun is well out on the bright limb of the distribution. This is quite a strong indication that most of the stars in our neighbourhood, which are cool and dim type-M red dwarfs, with on average only about half the mass of the Sun, are really not as suitable for hosting complex life as our bright yellow type-G star. This is all the more surprising because type M stars do have one factor that ought to make them *more* hospitable for complex life, and that is their long lifetimes. Smaller stars burn their nuclear fuel more slowly, and the lifetime on the main sequence of type-Ms is typically twice as long as the Sun's. According to our thinking and the evidence of our own planet, the shortness of the habitable period is a major factor limiting the chances that complex life develops. And yet we have awoken to find ourselves orbiting a bright, showy, short-lived firework of a type-G star, rather than a long-burning but dim and dull type-M.

We predict therefore that there is another factor or factors which make it difficult for life to develop to complexity around fainter stars,

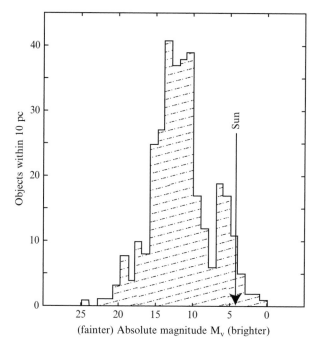

Fig. 6.2 Histogram of the luminosity of stars in a survey of objects within 10 parsecs of the Sun. Brighter stars have lower absolute magnitudes, hence the M_v axis is plotted in reverse. The position of the Sun is shown by an arrow. [Data from RECONS (the Research Consortium on Nearby Stars) http://www.recons.org/]

and explains why we don't find ourselves orbiting one. In fact there are several possible disadvantages to living round a type M star, but it's not clear as yet that any of them are so severe that they can tip the balance against them as good nurseries for life. Astro-biologists have recently been looking harder at type-M stars, as possible targets for SETI searches for example, and the discussion below owes much to some recent papers from a symposium on the subject (*13*).

The habitable zone of a red dwarf is much closer in to the parent star than is the Sun's—stars rapidly become much less bright as we go towards smaller size, and a star half the mass of the Sun emits only a few percent of its energy. A habitable planet around an M star would therefore have to huddle close to it for warmth, orbiting closer than Mercury is to the Sun. Here it will become tidally locked; its rotation slowed by internal energy dissipation until it equals the orbital period and the same face is always to the star. (The Moon of course is tidally locked to the Earth, the reason why we always see its same face.) Tidal locking will mean that there will be permanent huge temperature extremes on the planet, between the boiling, daytime, and the freezing, night time faces. If the atmosphere and ocean are not very efficient at transporting heat, this would mean that only a narrow strip around the terminator would actually be inhabitable, and we would expect that all the water on the planet would end up frozen out on the cold side by a 'cold trap' effect. However, with a sufficiently thick and mobile atmosphere, this fate could be avoided.

Another problem may simply be that small stars tend to have small planets. The mass of the central star will certainly be related to the mass of the nebula that accretes around it, hence to the size of any planets that eventually form. As we have seen, it is critically important that a planet is big enough to hold on to a sufficiently thick atmosphere, and also to generate the internal geothermal heat to power plate tectonics (tidal dissipation would help there by adding a source of heat to the interior). We don't know enough about the planetary formation process to make very firm predictions however. Current ideas tend to favour a picture of planetary formation as quite stochastic in nature, so while we certainly would expect a tendency for smaller stars to have smaller planets, perhaps there is nothing against a star half the size of the sun having a rocky planet as big as the Earth.

There is a third problem for life around a faint star, as first pointed out by Ray Wolstencroft and John Raven (*14*), and this one seems to be a really serious barrier to the development of complex life. Red dwarfs are the colour they are because they are cooler than the Sun, typically only reaching about half its surface temperature. This means the photons they emit are lower in energy. Chlorophyll makes full use of the high-energy photons emitted by the Sun, absorbing strongly in the high-energy, blue region of the spectrum, (as well as in the red, leaving the central, green wavelengths to be reflected—hence its characteristic colour). But as we've discussed, for photosynthesis to split water, a high voltage must be generated by the photosystems in plants, and even on Earth this has required that two photosystems be coupled together. Under a much cooler star, very few high energy photons would be available, and it is likely that three or even four such systems would have to be coupled together to accomplish the task of splitting water. But the evidence is that it was no simple task for evolution to arrive at water-splitting photosynthesis even on Earth, so it may be much more difficult still to accomplish this under the light of a cooler star. What this might mean is that though type-M biospheres may evolve photosynthesis, they would find it nearly impossible to evolve the water-splitting variety, and, as we'll discuss in future chapters, without oxygen, there are very unlikely to be animals, let alone intelligent animals.

6.8 Summary

Our speculations in this chapter lead us to predict that simple (prokaryote) life might be moderately abundant in the universe at large—or possibly not, depending on just how difficult evolution to this first critical step is. Whether prokaryotes are rare or common however, complex life will be rare. This idea has been named the Rare Earth Hypothesis, since it was put forward in a book of the same name by Peter Ward and Donald Brownlee (*15*). Our analysis agrees very much with their thesis, though we find their arguments frustratingly qualitative. Ward and Brownlee argue their case based on the fortunate position of the Earth, our possession of a good-sized moon, a friendly big brother planet in the shape of Jupiter etc. The difficulty with their arguments is precisely the 'self selection bias' problem that we've tried to tackle in this and the previous chapter. We can see that the Earth has these attributes and (perhaps) that they have contributed to the evolution of complex life on Earth, but with only a single example of an inhabited planet, we don't see how to decide which of these properties are really necessary for us to be here, and which are not. Maybe other solar systems have even more favourable circumstances? Anyway, we are subscribers to the Rare Earth Hypothesis, and we expect it to be borne out when we eventually start to get data from other solar systems on the atmospheres of Earth-like extra-solar planets. However, while waiting the decade or two that it is going to take before this data begin to come in, we hope that the more formal approach that we have taken here can provide some theoretical support for the 'Rare Earth' view.

We've spent two chapters developing ideas that use the simplistic critical step model as a catalyst, or a mental crutch if you like. We've been at pains to point out whenever we are in danger of believing in the model too literally, that it is just a toy. It is useful because it captures, indeed exaggerates, a particular aspect that we are interested in—the influence that a small number of rare but important events could have on the evolution of the Earth system. It's not the real system however, but a caricature of it and much is left out that is important. In particular, we have barely discussed the reorganization of the whole Earth system—the 'revolutions' of this book—because those transitions are more complex events, that can't be readily fitted into the constraints of the model.

The revolutions are however related to critical events that we have identified in this chapter. The origin of life led ultimately to the establishment of global bio-geochemical cycling and the inception of the Earth system as we know it. The invention of oxygenic photosynthesis led we think, after a time delay measured in hundreds of millions of years, to the Great Oxidation. The invention of the eukaryotic cell led after an even longer delay, to the late Proterozoic Snowball Earths and the Cambrian explosion. We'll tell the story of these latter two Earth system revolutions in Parts 3 and 4. But first we want to explore another view of the Earth system, and see if it comes to similar conclusions.

References

1. R. J. Parkes *et al. Deep bacterial biosphere in Pacific Ocean sediments, Nature* **371**, 410 (1994).

2. W. B. Whitman, D. C. Coleman, W. J. Wiebe, *Prokaryotes: The unseen majority, Proceedings of the National Academy of Sciences USA* **95**, 6578 (1998).

3. R. J. Cano, M. K. Borucki, *Revival and identification of bacterial-spores in 25-million-year-old to 40-million-year-old dominican amber, Science* **268**, 1060 (1995).

4. G. C. Dismukes, V. V. Klimov, S. V. Baranov, Y. N. Kozlov, J. DasGupta, A. Tyryshkin, *The origin of atmospheric oxygen on Earth: The innovation of oxygenic photosynthesis, Proceedings of the National Academy USA* **98**, 2170 (2001).

5. R. E. Kopp, J. L. Kirschvink, I. A. Hilburn, C. Z. Nash, *The paleoproterozoic snowball Earth: A climate disaster triggered by the evolution of oxygenic photosynthesis, Proceedings of the National Academy of Sciences USA* **102**, 11131 (2005).

6. B. Marin, E. C. M. Nowack, M. Melkonian, *A plastid in the making: Evidence for a second primary endosymbiosis, Protist* **156**, 425 (2005).

7. J. Maynard Smith, *The evolution of sex.* (Camridge University Press, Cambridge, 1978).

8. S. J. Gould, *Wonderful Life: The Burgess Shale and the nature of history.* W. W. Norton & Co., London (1989).

9. S. Conway Morris, *The crucible of creation: the Burgess Shale and the rise of animals.* Oxford University Press, Oxford, (1998).

10. S. Conway Morris, *Life's Solution: Inevitable Humans in a Lonely Universe.* (Cambridge University Press, Cambridge, 2003), pp. 464.

11. D. Hitchcock, J. E. Lovelock, *Life Detection by Atmospheric Analysis, Icarus* **7**, 149 (1967).

12. C. Sagan, W. R. Thompson, R. Carlson, D. Gurnett, C. Hord, *A Search for Life on Earth from the Galileo Spacecraft, Nature* **365**, 715 (1993).

13. J. C. Tarter *et al. A reappraisal of the habitability of planets around M dwarf stars, Astrobiology* **7**, 30 (2007).

14. R. D. Wolstencroft, J. A. Raven, *Photosynthesis: Likelihood of occurrence and possibility of detection on earth-like planets, Icarus* **157**, 535 (2002).

15. P. D. Ward, D. Brownlee, *Rare Earth: why complex life is uncommon in the Universe.* (Springer-Verlag, New York, 2000), pp. 333.

7

Playing Gaia

Imagine that you have an almost god-like position and power. The galactic council (or perhaps the supreme deity) has charged you with the task of making an uninhabited terrestrial planet very like the young Earth into a place thriving with life. It needs to be able to maintain this state in the face of various forcing agents (like a steadily brightening parent star) and perturbations (such as large meteorite impacts). You are not allowed to just magic things out of nothing, you have to work within the bounds of scientific laws, but you are given a spaceship, the 'Ark', laden with the life forms present on today's Earth and you are allowed to seed the planet with any of them. You have a scientific inclination to experiment, and a lot of time on your hands.

Your planet has settled down from its turbulent formation and has a vast ocean of liquid water at the surface—you decide to call it 'Ocean' (which as we have already noted would be logical for the Earth). Planet Ocean has volcanic activity driven by a slowly decaying internal heat source and some substantial land masses. Tectonic activity is underway. The atmosphere is dominated by carbon dioxide with some nitrogen, no oxygen and traces of methane and hydrogen. It's similar to the composition of Mars' and Venus' atmospheres today and quite unlike that of the contemporary Earth (Fig. 7.1). It's rather warm at the surface, what with all that carbon dioxide and water vapour in the atmosphere; about 50°C on average (in contrast to 15°C on the present Earth).

Clearly it's not a good idea to release humans on the surface, or indeed any other complex animal life form, because they all have a high requirement for oxygen and will suffocate. They also have a preference for cool conditions and will soon die of heat stress. Even if you give them a space suit or capsule to protect them from the heat and provide an oxygen supply, they will have nothing to eat and will soon starve to death. Ultimately, the galactic council would like planet Ocean to be made suitable for colonization by complex, intelligent life forms (human or otherwise). But the game we are about to play will not take you that far. Instead your tasks are to cool the temperature down, construct an effective planetary thermostat, and establish healthy

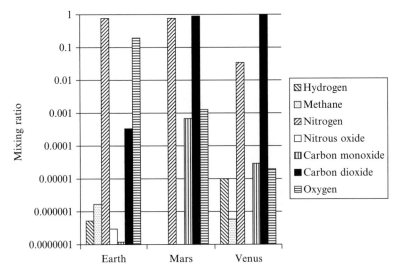

Fig. 7.1 Atmospheric compositions of Earth, Mars and Venus

recycling ecosystems with a high level of photosynthetic primary production.

In essence you are not playing God, but rather playing Gaia—in the sense in which Jim Lovelock introduced her (*1*). The central premise of Lovelock's Gaia theory is that the Earth self-regulates (most of the time, at least) in a habitable state, and your task is to create such a planetary scale self-regulating system. We have chosen to present this as a game (or thought experiment) because we want to explore with you how likely, or unlikely, it is that once life is introduced to a planet, regulation will emerge (see Box 7.1). To make a convincing case either way, we need to find a way around the problem that we have a sample size of only one Earth. Hence we have introduced planet Ocean as an additional, 'virtual world'. Recent research has increased the sample size further by simulating many virtual worlds, seeding them with 'artificial life' forms, and seeing how things turn out. In effect these represent many goes at the Gaia game. So, as your go at playing Gaia progresses, we will draw on these studies to help determine the outcome.

Box 7.1 The Gaia debate

How has life survived for so long on the Earth? There are several possible answers to this question. At one extreme is the argument that it is 'pure luck' that life has survived for as long as it has, because the Earth possesses no regulatory feedback mechanisms. Whilst on the face of it this looks an unlikely explanation, we have to be careful to acknowledge observer self-selection; our very existence requires that the Earth had a long history of life—however improbable that might be a priori. It turns out that we can discard pure luck as the explanation, but only because, as we will show, we have convincing evidence for the regulation of some global environmental variables and some well established mechanisms. This leaves us with two possibilities, which we term 'Lucky Gaia' and 'Probable Gaia'. According to Lucky Gaia it is just good luck that the Earth happens to possess predominantly regulatory feedbacks that have helped maintain a habitable state. The coupling between life and its planetary environment could equally have sent the planet into an uninhabitable state, or had a neutral effect on habitability. In contrast, Probable Gaia postulates that there are some basic principles at work that lead one to expect regulatory feedbacks to predominate on planets with abundant life, at least statistically, if not in every case. In our view, the present debate about Gaia boils down to whether Lucky Gaia or Probable Gaia is closer to the truth. This hinges on whether anyone can identify basic principles that would make Gaia a more probable outcome. In this chapter we summarize the hunt so far for those basic principles.

You might begin by wondering if there is any sort of instruction manual or rules that go with the game. Unfortunately, the instruction manual is unfinished—reflecting as it does, the current state of scientific knowledge—but we will try to present some useful extracts from it during this chapter. First it's useful to recall that prokaryotes are the most robust forms of life. They can adapt to basically any temperature consistent with the presence of liquid water. There are archaea that thrive in water over 100 °C coming out of hydrothermal vents, where the pressure prevents it from boiling. There are also bacteria that can live in super-cooled cloud droplets or in brine solutions well below 0 °C. There are prokaryotes that have adapted to extremes in salt, pH and other variables. Wherever there is free energy to be had from a chemical transformation in the environment you can bet there is a microbe that can make use of it. (This goes to the extremes of living off tiny amounts of free energy available deep in the Earth's crust.) Thus, some 'extremophiles' from the prokaryote kingdoms could certainly be successfully introduced to your planet as it is. Keeping the planet habitable for them should be the least challenging part. However, to get a thriving microbial biosphere going with a regulatory influence on the climate and recycling of essential elements is a different matter.

7.1 Feedbacks in the absence of life

Your first task is to make sure that the basic requirements for habitability by any life form are met and maintained. The planet has liquid water at its surface, which is good news, as it means the planet resides in the habitable zone around its parent star (2). It means some forms of prokary-ote life can be introduced and should be able to survive (if not thrive). However, planet Ocean's parent star is like the faint young Sun. It is about 30% less luminous than today's Sun, and will burn steadily brighter with time, tending to heat the planet and move the habitable zone out-wards. The planet's internal heat source will also slowly decay away, and there are likely to be occasional impacts from large asteroids and great volcanic outpourings that could seriously perturb the climate. These forcing factors could at some point take the planet outside the habitable zone. Are there any response mechanisms within the system that can counteract such forcing factors, or will they instead amplify the forcing? Such responses would be examples of feedback loops (see Box 7.2) and they determine the boundaries and width of the habitable zone.

Box 7.2 What are feedback loops?

To understand what a feedback loop is, think of your planet as a system comprising many components or variables with many interactions between them. You might draw a diagram of some of the important variables and interactions (Fig. 7.2). A feedback loop is a chain of cause-and-effect which can be traced in a closed circuit around the diagram, so that a perturbation to one part eventually causes a further change in that same process. Feedback loops come in two flavours, 'positive', or destabilizing, and 'negative', or stabilizing. Suppose something changes in the system, and this causes another variable to change, which causes another to change, and so on, eventually returning to affect the first variable. If the change that is fed back tends to go in the opposite direction to the original perturbation, it will stabilize the system, making the response less than it would otherwise be. This is negative feedback. On the other hand if the feedback goes in the same direction as the original perturbation, the change in the system will be reinforced. This is positive feedback. If positive feedback is strong enough the system can be so destabilized that it 'runs away', unable to find a stable state at all. In that case it will either oscillate or transit to a completely new regime.

In general, if you are trying to ensure that your planet's climate will be well behaved and stable, negative climate feedbacks will be helpful and positive ones harmful—even though 'positive' sounds like it should be more desirable than 'negative'. There are three key feedback loops that determine the boundaries and width of the habitable zone. Two positive, potentially 'runaway', feedbacks set the inner and outer bound-aries of the habitable zone. One negative feedback acts to widen the habitable zone. Let us examine each in turn.

The inner boundary of the habitable zone is set by a positive feed-back called the 'water vapour feedback' (3) (Fig. 7.2). Imagine you can

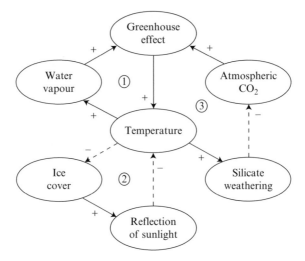

Fig. 7.2 Key interactions and feedback loops on planet Ocean in the absence of life. Positive signs on solid arrows indicate direct relationships, for example, increasing temperature increases the water vapour content of the atmosphere, which adds to the greenhouse effect, which increases temperature, thus closing a positive feedback loop (1. the 'water vapour feedback'). Negative signs on dashed arrows indicate inverse relationships, for example, increasing temperature decreases ice cover. This decreases reflection of sunlight (a direct relationship), but that increases temperature (another inverse relationship), thus closing another positive feedback loop (2. the 'ice-albedo feedback'). Whenever there are an even number of negative signs in a loop (or none), the feedback is positive. Whenever there are an odd number of negative signs in a loop, the feedback is negative—as in the third loop (3. the 'silicate weathering feedback'), which is described in the text

move planet Ocean towards its star, which is equivalent to the star getting brighter over time. This warms the planet, which increases evaporation from the oceans, increasing the water vapour content of the atmosphere. The extra water vapour in the atmosphere absorbs long wave radiation coming up from the surface and then re-emits it in all directions, some of it back down—thus acting as an extra blanket and increasing the surface temperature further. This is a positive feedback but not one that will immediately run away. Each increment of warming produces a smaller warming from the feedback and the system converges. However, as you move the planet nearer the star, a point is reached where each increment of warming will produce a larger warming from the feedback and it will run away, evaporating the oceans and creating an atmosphere of steam (4). The runaway does not happen very easily because as the water vapour in the atmosphere builds up, it acts like a pressure cooker, raising the boiling point for the remaining ocean. But a runaway greenhouse (probably) happened on Venus sometime early in its history (*4, 5*). Before it occurs another factor can set the inner boundary of the habitable zone (*2*): As you move your planet towards its star, a transition to a 'moist greenhouse' occurs in which the water vapour content of the upper atmosphere increases greatly. Up there, water molecules are split apart by high energy ultraviolet radiation and their hydrogen atoms are lost to space. This process causes the planet to lose its water (and steadily oxidizes the surface—more on this in Part 3). Importantly, there appears to be no way back from this state.

The outer boundary of the habitable zone is set by a different positive feedback called the 'ice-albedo feedback' (*6, 7*) (Fig. 7.2). Imagine you can move planet Ocean away from its star, which acts to cool the planet. At some point snow and ice caps form at the poles, which in turn tend to reflect more incoming radiation, adding to the cooling and increasing ice and snow cover. Again it looks like a dangerous recipe. At some point this positive feedback could run away and cloak the planet in ice (*8*). Actually there are potentially two phases of rapid change. When polar ice caps first start to form, the positive feedback can cause them to grow abruptly, to a size comparable with that on today's Earth (*9*). But after such small ice caps have formed things settle down—the positive feedback amplifies any cooling but it does not run away—each incremental change in ice cover produces a progressively smaller further change in ice cover, and the system converges on a stable state (this is the state we see on the Earth today). However, as you force your planet further from its star, and ice and snow cover extends away from the pole, it covers a progressively greater area for each step in latitude. If it reaches about 30 degrees from the equator there is a critical threshold where the positive feedback runs away— each incremental change in ice cover generates a progressively larger change in ice cover, and the world becomes covered in ice (*8*). This is the 'snowball' state of a planet, and it marks the outer boundary of the habitable zone.

Having established what sets the boundaries of the habitable zone is there anything that can push those boundaries further apart? To help maintain the initial state of your planet with liquid water, you should be on the hunt for a negative feedback. Luckily, one exists. It is the same one that is thought to have helped stabilize the Earth's climate and counteracted the faint young Sun. It involves the concentration of carbon dioxide in the atmosphere and the surface temperature of the planet, and it operates over millions of years. It is called the 'silicate weathering feedback' (*10*) (Fig. 7.2).

The feedback works like this: on long time scales, the source of CO_2 to the atmosphere and oceans is volcanoes and the sink is a reaction with (continental) rocks. The rocks in question are made of silicate minerals (e.g. granite), which make up most of any rocky planet's crust and mantle (including your planet and the Earth). The silicates react with carbon dioxide in the presence of water, to yield positively charged calcium (and/or magnesium) ions and negatively charged bicarbonate ions. These are washed to the ocean where they combine to form calcium (or magnesium) carbonates (see Box 7.3 for the chemical reactions). Globally, if we average over long enough times, this reaction, which is an example of chemical weathering, removes the same amount of CO_2 as comes up through volcanoes. Crucially, the rate of this silicate weathering reaction is *temperature dependent*, it goes faster as the surface temperature goes up and slower if it cools down, and it stops completely if all the water on the surface is frozen to ice. So if some factor tends to heat your planet up, such as the brightening of its star, the reaction should speed up and more CO_2 will be removed from the atmosphere. Over hundreds of thousands of years, this should decrease the atmospheric greenhouse and therefore the surface temperature. Equally if a massive outpouring of lava and CO_2 raises the atmospheric concentration of CO_2 and the surface temperature this too should be counteracted. The negative feedback works just as well to help stabilize against cooling of the planet, tending to push up the CO_2 level. This is one reason why your planet, orbiting as it does a fainter star than the present Sun, has so much CO_2 in the atmosphere. You can expect that as the star gets steadily brighter, the amount of CO_2 in the atmosphere will steadily decrease due to the operation of the negative feedback.

Box 7.3 Silicate weathering

The chemical equation for the silicate weathering reaction can be written as:

$$CaSiO_3 + 2CO_2 + H_2O \rightarrow Ca^{2+} + 2HCO_3^- + SiO_2$$

That is, a silicate mineral (here Wollastonite) reacts with carbon dioxide in the presence of water (they form a weak carbonic acid solution) to yield calcium and bicarbonate ions (in solution) and silica (a solid). In the ocean, the calcium and bicarbonate ions combine to form calcium carbonate (a solid mineral, calcite or aragonite) which is deposited at the bottom of the ocean:

$$Ca^{2+} + 2HCO_3^- \rightarrow CaCO_3 + CO_2 + H_2O$$

Carbon dioxide and water are regenerated in the formation of calcium carbonate, but one of every two carbon dioxide molecules taken up in the weathering reaction is trapped in the calcium carbonate. The overall reaction is:

$$CaSiO_3 + CO_2 \rightarrow CaCO_3 + SiO_2$$

Silicate rock and carbon dioxide is converted into carbonate rock and silica. The equivalent equations can be written for a magnesium silicate (e.g. perovskite $MgSiO_3$) being converted to a magnesium carbonate (e.g. magnesite $MgCO_3$).

So far, so good, but can this relatively slow negative feedback save your planet from the boiling fate of Venus or the freezing fate of a snowball? Such positive feedbacks operate relatively quickly, and if the planet is pushed beyond the critical threshold where either positive feedback runs away, the negative feedback cannot hold it back. Rather than removing the critical thresholds, the silicate weathering feedback makes it harder to reach them, broadening the habitable zone. The negative feedback has relatively little effect on the inner boundary of the habitable zone, because it stops working before the onset of a moist greenhouse, when all CO_2 is removed from the atmosphere. But it considerably extends the outer limit of the habitable zone, delaying the onset of a snowball state. Despite this, within the habitable zone, a sudden cooling can still potentially trigger a snowball state because the spread of ice and snow cover happens rapidly (in a matter of years) whereas the silicate weathering feedback operates over hundreds of thousands of years (*11*).

If it occurs, is there any way back from a snowball state? Well, it turns out there is, because silicate weathering is expected to shut down in a world covered in ice (*12*). This stops CO_2 being removed from the atmosphere, so as long as volcanoes do not stop adding CO_2, the CO_2 concentration in the atmosphere will build up, until the greenhouse effect gets strong enough to cause the global ice-cover to melt. Unfortunately, this rescue takes millions of years and in the meantime, the snowball state is likely to kill off any complex life that is present. However, a snowball lasting millions of years would not be fatal to all prokaryote (or simple eukaryote) life. There would almost certainly be cracks (called 'leads') in the sea ice exposing ocean water, and on today's Earth there are microbial ecosystems operating in sea-ice and in the perennial lake ice of the McMurdo Dry Valleys of Antarctica. Microbes can also survive for millions of years at the bottom of ice cores and in sediments (*13*). So, the snowball state may be neither sterile nor permanent, but it would certainly destroy a thriving biosphere leaving only a few hardy life forms hanging on.

You have not even begun to add life to planet Ocean and already there are a number of feedbacks at work, some tending to stabilize it, some to destabilize it. Keeping the planet habitable in the long term is starting to look a little tricky—you need to steer a course between the pressure cooker and the deep freeze. You have to avoid boiling your planet to a permanent (and certain) death and you want to avoid freezing it to a kind of cryogenic storage—an unpleasant torpor.

There are actually many more feedbacks at work than we have described but the three we have focused on are the most important ones. You might wonder how unconscious feedback mechanisms that don't involve life can 'know' anything about life's preferences. The answer is that they can't. They may lead to stability, but there is no a priori reason why they should stabilize *habitable* conditions, rather than uninhabitable conditions. As luck would have it, because the silicate weathering feedback depends on the presence of liquid water to operate, and liquid water is a necessary condition for life, it does tend to stabilize broadly habitable conditions that microbial life could colonize. In contrast, the water vapour and ice-albedo feedbacks can potentially create uninhabitable boiling or barely habitable frozen conditions. It is not obvious that in the absence of life the balance of these feedbacks will continue to maintain a habitable state.

7.2 Feedbacks involving life

So, can adding some robust and simple life forms to the planet somehow improve the chances of keeping things habitable? Living things have an innate tendency to grow and multiply, they inevitably alter their environment, and the state of the environment affects their growth. This should lead to feedback between life and its environment, whenever an organism alters an environmental variable that also affects its growth (*14*). However, the effect on the environment and resulting feedback could initially be positive or negative for the growth of the organisms. So, is there any reason why this should help keep things habitable?

Jim Lovelock believes so, and the idea was at the heart of the original Gaia hypothesis. In his first book on Gaia, published in 1979 (*15*), Jim asserted that life stabilized the global environment, but he wasn't very specific about how it might do this. By the time of his second book in 1988 (*16*), he had invented a simple model that went some way towards answering the question. He called it 'Daisyworld' (*17*, *18*). Andrew worked with Jim on the first paper to describe the mathematics of the model (*19*), and it has been a focus of many studies since (*20*). Daisyworld is an imaginary planet on which the only kind of life is daisies, which can be white or black in colour. If they are black they warm the planet by absorbing more sunlight than the ground on which they grow, while if they're white they cool it by reflecting light. Daisyworld was described as a 'parable'—intended to clearly illustrate what can happen as a result of strong feedbacks between life and the environment, but not to be confused with the real Earth. The daisies are a surrogate for any kind of life that can affect the global temperature—and temperature could equally well be any other environmental variable that life cares about, for instance the oxygen concentration, or pH.

On Daisyworld there is a mechanism that helps keep things stable. It is quite general, and works no matter what colour of daisies are first introduced onto the planet, provided that they have a strong effect on the temperature. It happens because the daisies also have a strong preference for 'comfortable' temperatures: they grow at their fastest at

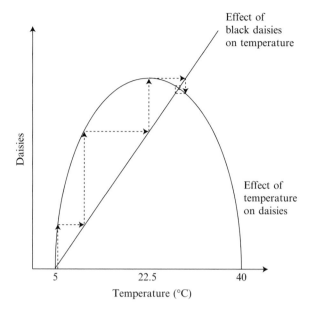

Fig. 7.3 A schematic of what happens when only black daisies are introduced to a cold Daisyworld (a little above 5°C). The dotted arrows show the trajectory of the system. Initially, a small population of daisies establishes which in turn increases the temperature, encouraging daisy growth (a positive feedback). This continues until the temperature exceeds the optimum for daisy growth, then the daisy population declines, causing cooling and the system converges on a stable state where the two functions intersect

22.5°C and either above or below this their growth slows down. At temperatures below 5°C or above 40°C, they stop growing entirely. So now suppose that black daisies are added to the planet. If it is initially cold, below their peak growth temperature, they warm things up, growing faster and covering more ground in the process, so warming even more, in a positive feedback. But once they move the temperature above their optimum, any further increase in their numbers slows their growth, so the system then settles into a stable state in which the temperature is somewhat above the daisies optimum temperature, but they provide negative feedback (Fig. 7.3). It works just as well if the daisies are white and cool the planet, but in this case the stable temperature is below the optimum temperature. So although the biological feedback can be initially either positive or negative, positive feedbacks are unstable and the system is driven out of them. Usually it is driven into a stable, negative feedback arrangement.

Daisyworld illustrates the key thing about feedbacks involving life—by definition they include life's habitability requirements. By introducing life to your planet, you can introduce what an engineer would call a 'set point' in habitable conditions. Furthermore, the behaviour of Daisyworld suggests that any positive feedback that occurs will tend to remove itself, whilst negative feedback will tend to maintain itself. This should give you some faith that negative feedbacks might win the day and predominate. If positive feedback gets the advantage initially it will drive change but the system should settle down in a habitable state.

7.3 Cooling the planet down

At this point you decide to stop theorizing and have a go at seeding the planet with some simple life forms. Your planet is too hot for complex life, so you decide to have a go at cooling things down. A natural approach is to find some organisms that can accelerate the silicate weathering reaction that removes CO_2 from the atmosphere. You reason that any land based primary producer has an incentive to promote the weathering of rocks if by doing so it can accesses limiting nutrients that are bound up in them. Phosphorus is a good candidate in this regard because on your planet (as on the Earth) it has no gaseous source from which it can be fixed. But in the process of extracting phosphorus, other rock bound ions such as calcium and magnesium are likely to be weathered out. When they are washed to the ocean they will go to form new carbonate rocks, causing a net removal of atmospheric CO_2 (following the chemistry given in Box 7.3). The resulting 'biotic amplification of weathering' is a widespread property of land life and has a large impact on the climate and atmospheric composition of the Earth (see Box 7.4).

Box 7.4 Biotic amplification of rock weathering

Soon after the importance of silicate weathering for long-term climate regulation was recognized (*10*), it was realized that life is intimately involved in the process on today's Earth (*21*). The partial pressure of carbon dioxide in soil is many orders of magnitude above the atmospheric value thanks to respiration by roots and organisms in the soil. This increases the concentration of carbonic acid in soil water, promoting the weathering reaction. Also, plant roots and their symbiotic (mycorrhizal) fungi secrete a variety of organic acids that speed up the dissolution of rocks. Lichens also accelerate weathering relative to bare rock surfaces (*22*). The land biosphere on today's Earth is estimated to increase weathering rates by at least an order of magnitude, relative to bare rock (*23*). It has also been argued that progressively stronger biological effects on weathering have evolved over Earth history, cooling the planet in the process (*24*).

You are on the hunt for some organisms that can colonize the available land surface of planet Ocean, which is a mixture of bare rock, sand and dust. They must be tolerant of hot temperatures, which in turn have produced rather arid conditions. You search around on the Ark and find some samples of sand seemingly with a thriving microbial community on them, called a 'desert crust' (25). These organisms come from hot, arid regions of today's Earth and are a hardy bunch. They can spread rapidly (relative to geologic time) and there is evidence that similar communities have been present on the Earth's surface for over a billion years (26). The mats are fuelled by photosynthesizing cyanobacteria that can switch between oxygenic and anoxygenic photosynthesis depending on available substrates. They look ideal.

Let's assume (for the sake of argument) that you can get a desert crust of bacteria to establish on the land surface and they start to accelerate weathering, lowering CO_2 and cooling the planet. Their optimum growth temperature is below the initial temperature of the planet, so as they proliferate and cool things down, they enhance their own growth—an example of positive feedback. If their effect on weathering

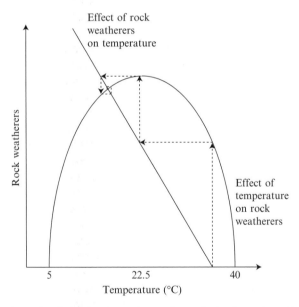

Fig. 7.4 A schematic of what happens when rock weathering bacteria are introduced to the warm planet Ocean. The dotted arrows show the trajectory of the system. Initially, a small population of rock weathering bacteria establishes which in turn decreases the temperature, encouraging their spread (a positive feedback). This continues until the temperature falls below the optimum for bacterial growth, then the population declines, causing warming, and the system converges on a stable state

were modest then the system could settle down in this positive feedback regime, still above their optimum growth temperature. However, let's assume their effect on weathering is strong and the temperature drops below the optimum for their growth. Then any further spread and cooling is self-limited—a switch to negative feedback. Sure enough, the planet cools and the system settles down in this negative feedback regime. Fig. 7.4 shows a schematic of this scenario.

The resulting conditions are not ideal for the bacteria as it is rather on the cold side of their optimum and there is less CO_2 around for photosynthesizing with. Luckily, the temperature preference of your organisms is not too cold, because if they had a really strong effect on weathering this would present the danger that they could cool the planet to the point that they triggered a runaway to the snowball state. Having noted this potential problem you decide it would be a good idea to try and warm things back up somewhat.

7.4 Rein control

On Daisyworld, there are two types of daisy, the white ones which cool the planet but which thrive when it is warm, and the black ones which warm it but grow better when it is cold. When the two types are introduced together the temperature is held at an intermediate level for a wide range of values of solar luminosity (see Box 7.5 and Fig. 7.5). You could use the same idea on your planet, introducing a second type of life that has a preference for cooler temperatures but tends to warm things

---

Box 7.5 Daisyworld with both daisy types

Daisyworld starts cold whereas your planet starts from warm conditions, but this doesn't matter—the system works either way around. Daisyworld's star is also on the main sequence, getting steadily brighter with time. The lifeless surface (or regolith) is grey, but when the temperature reaches 5°C, the first seeds germinate. The paleness of the white daisies makes them cooler than their surroundings, hindering their growth, whereas the black daisies warm their surroundings, enhancing their growth and reproduction. As they spread, the black daisies warm the planet. This further amplifies their growth and they soon fill the world. At this point, the average temperature has risen close to the optimum for daisy growth. As the sun warms, the temperature rises to the point where white daisies begin to appear in the daisy community. As it warms further, the white daisies gain the selective advantage over the black daisies and gradually take over. Eventually, only white daisies are left. When the solar forcing gets too high, regulation collapses.

(a)

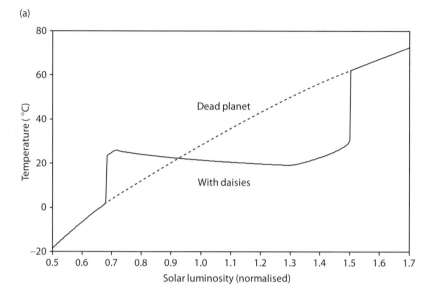

(b)

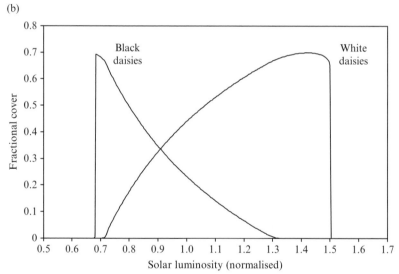

Fig. 7.5 Daisyworld with both types of daisy present showing 'rein control': (a) planetary temperature with and without daisies, (b) fractional coverage of black and white daisies

back up. Instead of removing a greenhouse gas you look for something that can create one. A reasonable candidate is methane. The organisms that make methane—'methanogens'—are a magnanimous bunch, whilst some are well known for their enthusiasm for hot conditions (*methanococcus* likes 88°C) others are most effective at low temperatures. It is the latter that you need, and you can find them on the Ark, originally from sediments of a deep, cold lake. The methanogens actually need some hydrogen or organic matter to feed on, but let's put that aside for now. The key thing is that they are 'heaters' as opposed to the 'coolers'—the rock weathering organisms you introduced first.

When you introduce the heaters to your planet they immediately proliferate in the cool temperature that suits their growth. As they warm things up, this enhances the growth of the coolers—the rock weatherers. Their spread works against the warming and a steady state is reached, between the optimum temperatures for the two types of life. As the star brightens, you expect the warming to encourage growth of the coolers, which will gradually take over from the heaters and counteract the warming. If something tends to cool the planet down, such as a large eruption injecting volcanic aerosols into the upper atmosphere, this encourages the growth of the heaters, which counteract the cooling. If there is a massive asteroid impact that kills off many of the bacteria, the populations will bounce back as long as the temperature is not shifted out of the range they can colonize.

In effect you have created a 'Bacteriaworld'—a bacterial version of Daisyworld. A planet regulated by bacteria may be less aesthetically pleasing than one regulated by daisies, but it has the advantage that it might actually have existed on the early Earth. An important requirement for it to be stable is that the heaters like a cooler planet but tend to warm it up, whereas the coolers like a warmer one but tend to cool it down. This is equally true on Daisyworld, where the local temperature of the black daisies is a few degrees warmer than their surroundings because of their dark colour, whilst the white daisies are a few degrees cooler. Unlike your bacteria, the black and white daisies live on a planet dominated by land surface and they compete for the same resource—space—but this competition is not necessary to get the model to work.

Introducing a second type of life illustrates that interacting feedbacks can lead to a different type of regulation to one feedback alone. With only e.g. coolers—rock weathering organisms or white daisies—the temperature must always track the forcing, increasing as the star gets brighter—even if it does so more slowly than in the absence of negative feedback. In your Bacteriaworld with both heaters and coolers, the temperature can remain close to constant as the star gets brighter. In Daisyworld, the temperature actually decreases for a while as the sun gets brighter (Fig. 7.5). These behaviours are a consequence of what has been called 'rein control', after the reins of a horse. The two types of life pull in opposite directions to stabilize the system keeping it heading in the right direction.

For your Bacteriaworld or Daisyworld to regulate temperature it is important that the type of life that cools things down likes it warm, and the type of life that warms things up likes it cool. In Daisyworld this is achieved by the black and white daisies experiencing different micro-environments. If this distinction is removed by making all daisies experience the same temperature, there is no regulation. The daisies respond identically to their shared environment, producing equal populations of black and white daisies that cancel one another's effects on temperature. The resulting populations respond passively to increasing solar luminosity—growing only when the temperature of a dead world permits it. A similar failure of regulation occurs in your Bacteriaworld if the heaters and coolers have the same response to temperature—they always grow together and their effects tend to cancel out.

Bacteriaworld and Daisyworld are 'thought experiments', they illustrate that self-regulation of climate involving life is feasible and it doesn't require any conscious foresight or planning on the part of the responsible organisms. There is a problem with them however, as models of how the Earth might have evolved, because of course it *did* require some conscious foresight or planning on your part to set up Bacteriaworld, and on Jim Lovelock's part to invent Daisyworld, and it is not obvious that such a well ordered arrangement can spontaneously arise. A number of critics have noted this and have invented alternative worlds that do not function so well, but equally these are teleological products. The real question is; if one had an unbiased situation (or model), could something like the rein control regulation of your Bacteriaworld or Daisyworld emerge automatically?

7.5 Adaptation world and regulatory epochs

Up to now, in order to simplify the game, we have removed the element of biological evolution by natural selection. You have acted a bit like evolution by introducing new types of life to your planet at different times. If different types of life then start competing for the same resources then selection will inevitably occur. This is what happens when the black and white daisies of Daisyworld start competing for space. However, we have not allowed new types of life to evolve with different effects on, or responses to, their environment. In reality, evolution by natural selection is inevitable once there are different organisms present with an innate tendency to replicate and faithful inheritance of variation between them. To make a less biased experiment, we should take account of the fact that life can adapt to prevailing conditions. In other words, natural selection tends to produce a world populated by those organisms that replicate best under prevailing environmental conditions.

Let us return to your Bacteriaworld of coolers (rock weatherers) and heaters (methanogens) in which the coolers like it hot and the heaters like it cold. The system has settled down and is regulating at a temperature between the optima for growth of each type. Take the cooling bacteria population, if there is variation in the optimum growth temperature of individuals within it and they all have the same maximum potential growth rate, then those individuals that prefer cooler conditions closer to the actual temperature will be selected. Similarly in the population of warming bacteria, those individuals that like it warmer and closer to the actual temperature will be selected. Over time therefore, the two populations should tend to converge on the same optimum temperature for growth—the actual temperature. The unfortunate consequence of this, if it happens, is that regulation will break down.

In fact allowing adaptation on your Bacteriaworld is rather like a model recently developed by a group (Jamie McDonald-Gibson, James Dyke, Eziquel Di Paolo and Inman Harvey) at the University of Sussex (27). We call the model 'Adaptationworld'. It is based on individual organisms all of which share a well mixed environment, but each of which can have different responses to a single environmental variable (on your planet, temperature) and can either increase or decrease that environmental variable. The organisms' growth response to the environment is the same shape but the value of the environmental variable at which growth peaks can vary due to genetic mutation. Hence despite the environment being perfectly mixed, the organisms can experience different growth rates due to their different responses to it. What typically emerges is a pattern of punctuated equilibrium (Fig. 7.6)—the environmental resource is tightly regulated for relatively long intervals, but these eventually break down leading to short intervals of rapid transition to a new stable state. If the external forcing of the environment is steadily increased, then the transitions tend to (but don't always) go in the direction of the forcing. It looks somewhat like Daisyworld (Fig. 7.5) except that instead of there being one regulatory epoch there can be a variable number of epochs and the values at which the environmental variable is regulated are arbitrary.

How can such a pattern emerge? James Dyke has provided a neat explanation (28): The model biosphere stumbles on and gets locked in rein control regimes—that are basically the same as in your Bacteriaworld or Daisyworld—where one type of life likes it hot but makes it colder whilst another type of life likes it cold but makes it hotter. Imagine that like your planet the world starts hot and the first type of life that can grow acts to cool things down. As it does so it enhances its growth acting to cool things down faster until this type of life limits itself or the conditions become more favourable for a different type of life. That type of life might also cool things down, but eventually by chance the next type of life that is favoured will be one that warms things up. This pushes the temperature back up towards a level that favours the previous type of life, and the system settles in the middle in a stable state with two reins either side that tend to tug it back there.

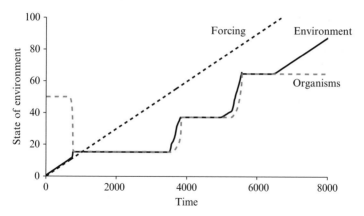

Fig. 7.6 A typical run of 'Adaptationworld' showing multiple epochs of regulation. The black dotted line is the input forcing the system, the black solid line is the state of the environmental variable, and the grey dashed line is the average state of the environment that the organisms are adapted to. 'Forcing' is analogous to insolation and 'environment' to temperature in Daisyworld. [Figure redrawn from one kindly provided by James Dyke. (*28*) of the Max Planck Institute for Biogeochemistry]

Over time the rest of the population's response to the environment can adapt and they tend to adapt towards the stable state of the environment that has emerged. If they start off liking it hotter they will evolve to like it cooler. At first the population converges on two humps—those that like it slightly hotter and those that like it slightly cooler than the emergent steady state. However, over time these humps get closer together and this weakens the regulation: the system becomes increasingly vulnerable to any type of life that drags the environment out of the stable regime and can proliferate because of this, shifting the environment further away from it. Eventually the system breaks out of the steady state and starts to transit. It follows the same random search as before until a new steady state emerges.

This result (Fig. 7.6) is beginning to look qualitatively like Earth history—long periods of stability interrupted by revolutionary changes. However, there is no true directionality to the transitions that occur—they can go up or down. The only appear to go in one direction if the forcing goes in one direction.

7.6 Can evolution screw up the planet?

Adaptationworld shows environmental regulation arising despite natural selection at lower levels. Admittedly adaptation and genetic drift ultimately lead to the breakdown of each regulatory epoch, but another is soon stumbled upon. This is somewhat surprising because Gaia has been dogged by a series of critiques from evolutionary biologists, which revolve around perceived conflicts between Gaia and evolutionary theory (*29, 30*). From an evolutionary theorists' point of view, your challenge in making a Gaia is one of maintaining planetary regulation in the face of natural selection at lower levels. They want to know: why should the organisms that leave the most descendants be ones that contribute to regulating their planetary environment?

Despite being blind, evolution by natural selection is a magnificently powerful process. If you had been given a population of habitable planets and could somehow get them to compete for a finite resource base whilst also having them reproduce—perhaps by sending out propagules onto surrounding worlds—then you could let natural selection refine the abilities of your planet to regulate and produce successful offspring (*30*). Unfortunately you don't have a population of planets to work with (leaving aside the other difficulties of making a planet into a unit of evolution).

Consequently you have had to resort to concocting a self-regulating feedback system without the help of natural selection to refine it, at least at the system level. Your original Bacteriaworld kept the two types of life separate and therefore avoided competition and selection. It would be helpful if you could turn to natural selection to help weed out the types of life that degrade their environment and promote the ones that improved theirs. However, the process of natural selection doesn't necessarily work like this. It is not interested in what is good for the environment or anything else that does not qualify as a unit of evolution, all it does is ensure that those units of evolution that produce the most descendants come to dominate. At the deepest level it seems tautological; survival of the survivors—for the 'fittest' are simply those that leave the most viable descendants. But you should not underestimate its power.

The central problem that evolution poses for your efforts to make a planet that self-regulates in a habitable state is that it will not promote a process because it improves the environment for everyone. It will only promote traits that benefit their carriers by enabling them to leave more descendants. It is possible for a trait that alters the immediate environment of an individual to aid its evolutionary success, but it is not clear how common this is in reality. It happens in Daisyworld because the daisies create their own micro-environments. But in general the problem is that the environment is shared and other non-carriers of a trait may gain its environmental benefits without incurring the costs of carrying it.

Like dark lords, evolutionary biologists love to dream up 'selfish' or 'spiteful' creatures that will wreak havoc on your nascent attempts to make a Gaia. Indeed we need look no further than our own species to see that evolution does not always concoct creatures that improve the environment for all. In the words of the late, great evolutionary biologist Bill Hamilton, we are currently a Genghis Khan species blazing a trail of destruction through the rest of the biosphere (*31*). If in degrading its environment an organism promotes its own replication over that of other more benign compatriots then it will be selected for, whether they like it or not. Clearly there is a limit to such environmental degradation—when it reaches the point that it restricts the growth or replication of the responsible organisms then self-limiting negative feedback will kick in. But a lot of damage can be done by the time this happens and a new steady state is found. Daisyworld guards against this because there is a fortuitous alignment between what is good for the individual and what is good for the global collective. If this alignment is reversed, negative feedback still predominates but there is no coexistence of the daisy types (see Box 7.6).

Box 7.6 Daisyworld when the black daisies make white clouds

In the original Daisyworld, each daisy type is assumed to alter the same environmental variable in the same direction at the individual (or patch) level and the global level. In fact this criticism was tackled by Andrew and Jim Lovelock in the paper that first presented the mathematics of Daisyworld (*19*). There, a variant of the model was produced in which the black daisies were assumed to produce white clouds that cooled the global environment—thus reversing the sign of their environmental effect from the individual to the global scale. The result is that the cloud-makers basically out-compete the white daisies, by unconsciously maintaining the shared environment in a cold state in which they have the local advantage. Their cooling effect creates an immediate negative feedback, restricting their growth, and consequently the temperature is regulated (although it does not show rein control). This variant of Daisyworld suggests that altering the environment could become part of the evolutionary game, helping an organism out-compete its compatriots. Parallels can be noted on Earth, a favourite example being peat bog plants that maintain acid and water-logged conditions in which they out-compete other vegetation types e.g. trees, that might otherwise invade. However, the critical issue becomes whether organisms would spend energy altering the environment, whether for the good or bad of others, when there is always the danger that other organisms can 'free ride' on this—benefitting from the environmental change without contributing to it.

To many evolutionary theorists, Gaia appears to demand altruism at the global scale—one organism selflessly helping another without any payback. Historically, it was a case of bad timing. The Gaia concept arrived in the popular consciousness just as evolutionary biologists were trying to rid their subject of flawed arguments that evolution might operate for the good of the species (or any other collective). Richard

Dawkins' eloquent popularization of the selfish gene (*32*) was in the ascendancy—and Gaia and the selfish gene seemed to be at loggerheads (as no doubt they still do to some readers). In the 1960s Bill Hamilton had shown that the conditions for altruism to occur are rather narrow* (*33, 34*). Subsequently the picture has become a little less bleak, with the recognition of reciprocal altruism (*35*) (where the chance of payback improves the odds for cooperation) and now at least five rules for the evolution of cooperation in different circumstances (*36*). But it is still extremely challenging to conceive of a formulation of Gaia based on organisms expending energy to improve (or even degrade) an environment that is shared by many others.

Given the way evolution operates, you would be wise to try and build your Gaia from environmental effects of organisms that emerge automatically and inadvertently as 'by-products' of other, more selfish activities (*14, 37, 38*). This 'by-product' terminology was introduced by Tyler Volk in his book *Gaia's Body—Towards a Physiology of the Earth* (and the same concept was arrived at independently by Tim and by David Wilkinson, all at a similar time). Fortunately such by-product effects on the environment abound. They are an inevitable part of life, an unavoidable consequence of metabolism.

7.7 The importance of by-products

In fact, trying to build a Gaia from by-products is exactly what you have been doing. Your experiment of adding planetary coolers—organisms that enhance rock weathering—is an example of a by-product leading to feedback on growth. In evolutionary terms, those organisms that obtain phosphorus more efficiently gain a selective advantage over their compatriots and the trait of more efficiently weathering rocks spreads through the population. The effects on the CO_2 content of the atmosphere and the temperature are by-products that are felt much later by future generations (after some hundreds of thousands of years). Because the atmosphere is well mixed, all organisms experience the same change in CO_2 concentration—the carriers (responsible for the change) and non-carriers of the rock weathering trait alike. If the carriers and non-carriers also respond in the same way to the same changes in CO_2 concentration and corresponding changes in temperature this cannot alter the selective advantage (or disadvantage) of rock weathering. However, it can ultimately affect the growth of those organisms that carry the efficient rock weathering trait, as well as their compatriots that don't.

---

* Specifically 'Hamilton's rule' states that the benefit conveyed (*b*) multiplied by the relatedness of the recipient to the donor (*r*) must exceed the cost to the donor (*c*): $r \times b > c$.

Similarly the planetary warmers you introduced—the methano-gens—are just emitting a metabolic by-product—methane (CH_4). The effects on the CH_4 content of the atmosphere and the temperature are also by-products, and again because the atmosphere is well mixed, all organisms experience the same change in CH_4 concentration and cor-responding changes in temperature. However, the planetary warmers have a different response to temperature than the planetary coolers. This means that when both types of life are present and affecting the global environment, their growth can be affected differently by changes in temperature. This gives rise to a feedback on which life form is selected for—because the carriers and non-carriers of each trait are affected dif-ferently by the presence of that trait. Should the rock weatherers cool things down too much, this favours the methanogens because they grow better in cool conditions, but they tend to warm things back up, thus favouring the rock weatherers. Both are cases of negative feedback on selection—a trait alters the world in a way that means it is selected against. In contrast, in the variant of Daisyworld where black daisies produce white clouds (Box 7.6) there is positive feedback on selec-tion—they maintain the environment in a cold state in which they are favoured over the white daisies.

An important real world example of a by-product giving rise to nega-tive feedback on selection involves the process of nitrogen fixation. Nitrogen fixation is the transformation of nitrogen from its almost inert gaseous form as the N_2 molecule to a biologically available form as ammonia (NH_3) or, in solution, ammonium (NH_4^+). This requires a large investment of free energy to split the triple bond between the two N atoms of N_2, but under nitrogen limited conditions the trait can be selected for. The responsible nitrogen fixers gain a selective advantage as they have obtained scarce nitrogen. However, ultimately this fixed nitro-gen will leak out into the environment, as a by-product of their metabo-lism, or of the cells dying. This in turn increases the level of fixed nitrogen in the environment and reduces the selective advantage of being a nitro-gen fixer (producing negative feedback on selection). This mechanism works even if the environment is a perfectly mixed lake or the global ocean (39). A steady state is reached with a modest population of nitro-gen fixers and nitrogen being slightly limiting on average.

Although feedback on selection clearly can occur, in the nitrogen fixation example, when nitrogen leaks out to the environment it is bad for the nitrogen fixers involved. An evolutionary biologist might describe the leaking out of the nitrogen as 'maladaptive' for the nitrogen fixers. For an effect of life on the environment to instead become adaptive, there must be positive feedback on selection due to the environmental alteration, and it must pay back more to the organism or genes respon-sible than it costs them. This shouldn't be a problem if the effects on the environment are truly by-products and carry no cost to the organism.

There is then the potential that the by-product might be co-opted into becoming an adaptive trait. The more localized the environmental effects are, the better the chance that they will benefit the responsible organisms (or genes) more than they cost them. However, in trying to think of adaptive environmental alterations we are being lured away from the global scale.

Your best bet for setting up a Gaia looks to be to base it on environmental alterations which are by-products. These can then give rise to feedbacks on growth and, potentially, feedbacks on selection.

7.8 The need for recycling

To construct a healthy Gaia you need to ensure that the elements essential to life are either in abundant supply in a biologically accessible form (ultimately from the planet's crust and interior). If they are not being supplied fast enough, you need to get recycling loops to establish for them. But in your experiments at seeding the planet thus far you have ignored nutrient recycling. In fact, if you introduce only primary producers to the planet (such as those in the desert crust), in the absence of recycling, their spread will most likely soon be chronically limited by a lack of nutrients. The formation of recycling loops is a key example of by-products (quite literally waste products) becoming an essential part of a thriving planet with life. Their importance for making a Gaia has been emphasized by Tyler Volk (*37*).

A recycling loop is in essence a form of positive feedback loop based entirely on the cycling of material. Let's say your world contains compounds A and B. If you put just one type of organism on your planet, which eats compound A and excretes compound B, then they will soon eat up all of compound A and leave only their waste product—compound B—unable to do anything about it. However, if you can find a second type of organism that eats compound B and excretes compound A then they recycle the food for the first type of life, and *vice versa*. If there are gradual inputs of A and B, say from volcanoes, then neither type of life would starve on their own, but their populations would be constrained by these inputs. In the presence of recycling the populations can grow in excess of these inputs.

To work out what is being supplied from your planet's interior you decide to analyze the gases coming out of volcanoes (Fig. 7.7). There are modest fluxes of carbon dioxide (CO_2), hydrogen (H_2), carbon monoxide (CO), nitrogen (N_2), and methane (CH_4) (in declining order of abundance) and traces of sulphurous gases. None of the fluxes of gases at the surface of your planet approach the values seen on today's Earth (Fig. 7.7), indicating that a great deal of recycling is happening on Earth. To work out what is being supplied from weathering of your

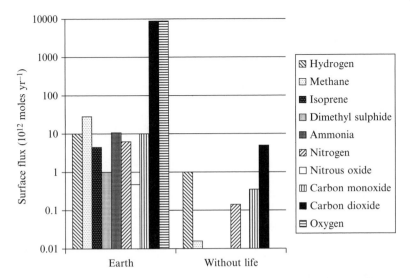

Fig. 7.7 Fluxes of gases on today's Earth and in the absence of life (from volcanic activity)

planet's crust you could analyze the composition of the rocks. They are rich in oxygen, silica, iron, and other metals, with traces of phosphorus, but very little nitrogen.

You turn to the organisms on the Ark to try and work out their requirements for different elements. It turns out that their main constituents tend to be present in rather consistent proportions (these elemental proportions are referred to as their 'stoichiometry'). The dominant elements are always carbon, hydrogen, nitrogen, oxygen, phosphorus and sulphur. Proteins are particularly rich in nitrogen, nucleic acids particularly rich in phosphorus, and structural material particularly rich in carbon. Alfred Redfield (*40*) proposed that, on average, marine plankton had carbon, nitrogen and phosphorus atoms in proportions C:N:P = 106:16:1—known to geochemists today as the Redfield ratios. Freshwater plankton show similar proportions, but land plants have a habit of accumulating a much larger proportion of carbon, which dominates their structural material. Animals tend to have more nitrogen because they are relatively protein rich. Metal elements such as iron, manganese, zinc, and copper are found in all organisms in trace amounts, and are at the reaction centres of various enzymes making them essential for catalyzing biochemical reactions.

Comparing the composition of the Ark's organisms with the composition of your planet's ocean and atmosphere there is some correspondence, which is encouraging—the necessary elements generally seem to be available. This should not be surprising, as life when it evolved under similar conditions on Earth naturally used the available elements. However, their supplies into the system are modest and the proportions of different elements don't often match what the Ark's organisms require. For example, there seems to be a gross shortage of phosphorus coming into the system. Furthermore, the elements are not always in a form that is easily available to the organisms. For example, most of the nitrogen is tied up as nitrogen gas (N_2) in the atmosphere with only a modest reservoir of easily available ammonium (NH_4^+) in the ocean.

To build a healthy biosphere you are going to need some biological processes that enhance the input of certain key elements that are in short supply. You have already introduced some weathering organisms that liberate phosphorus from rocks, and this phosphorus will leak to the ocean fuelling productivity there. But you should also introduce some nitrogen fixers to the ocean to convert nitrogen gas to available form (interestingly your desert crust organisms already include nitrogen fixers). In addition to these enhanced inputs of phosphorus and nitrogen, you also need recycling loops to establish for most of the essential elements. This is going to involve a suite of chemical transformations for each element. Happily, you have on the Ark representatives of all of what Tyler Volk has called the 'guilds' of life on Earth that perform these different chemical transformations (*37*). Furthermore, all the key transformations can be performed by prokaryotes.

So, you could just shower the planet with a variety of prokaryotes and see how readily recycling loops emerge. If you do this, it is likely that recycling loops will form. If all the transformations of a loop are already occurring then the closing of the loops will happen automatically, it just requires a container with some boundaries (a finite volume). It should happen most readily in a small container such as a pond or a lake, but ultimately the whole ocean and atmosphere is a container. There is a caveat from today's Earth that different biochemical transformations in a given recycling loop can require different conditions (e.g. the presence or absence of oxygen) and therefore the container needs to provide these different conditions (e.g. sediment as well as water above), but we will set this complication aside for now.

7.9 Flaskworld

Rather than let you shower the planet with bacteria and sit back and relax as recycling loops self-assemble, let us introduce a new rule to the game; that you are only allowed to seed the planet with a few types of life. This makes the problem of establishing recycling loops on planet Ocean more difficult. If you can't establish all the links in a closed loop, evolution will have to create the others. Picture the scene in a selection of ponds or lakes. Each starts with only one guild of microbial life.

Perhaps there is a stream flowing into each lake providing an input of nutrients and a stream flowing out removing material. Or these could be the ponds formed in craters left by the asteroid bombardment as your planet's solar system formed. Then the nutrients could be deposited in ash, dust or rain from the atmosphere, and fluid removed by seepage out of the bottom of the crater. How readily can evolution create recycling loops within containers like these?

To answer this question our colleague Hywel Williams at the University of East Anglia built a computer model called the 'Flask' model (*41*). It simulates just that; a container of liquid with a continual flow-through of liquid bringing with it some input of nutrients, call them A, B, C and D. Each flask is seeded with a clonal (genetically identical) population of microbes. Each individual carries a genome that codes for their uptake preferences for nutrients and the ratio in which they excrete waste products. The chemistry is constrained in that no organism is allowed to eat its own waste products. There is a low level of mutation creating variation from the initially identical population. In this 'Flaskworld', initially not a lot happens; the population becomes chronically nutrient limited, consuming all available inputs and leaving waste products they can't consume. That is until something evolves for which a waste product is their food. If their waste product is in turn food for one of the original organisms then a recycling loop emerges. Both types of life inadvertently promote the growth of the other and their populations increase. In multiple replicates of such systems they go through a characteristic sequence of transitions as they 'learn' to recycle each nutrient in turn until there is recycling of all of them.

So, even if you have to let evolution do the work on your planet, recycling loops should emerge, and once they have, productivity is boosted, creating the kind of thriving microbial ecosystem you are after. You may be wondering; what limits the system and stops productivity rising indefinitely? Ultimately, it is the input of free energy to the ecosystem and the fact that over time free energy is always degraded to less useful forms such as heat. In Flaskworld, the input of free energy is prescribed, as it all comes in the form of chemicals entering the flask. In other words there is no photosynthesis, or active capture of free energy from sunlight. There is also an inevitable, fixed loss rate of free energy each time there is a biochemical transformation and this prescribes a maximum efficiency that can't be beaten (in reality aerobic respiration can reach efficiencies up to about 60%). This in turn sets a 'carrying capacity' for each container, which is reliably reached once recycling is in place. The same rules will apply on your planet and consequently you can expect productivity to be capped.

Recycling, it seems, is quite robust because it is a self-catalysing process that is built on by-products. But hold on, what happened to those dark lords of evolutionary biology and their evil creations? Can evolution not create something that destroys recycling? Technically in your scenario nothing can cheat recycling—e.g. removing one of the products from the loop and saving itself energy in the process—because recycling is assumed to be built on by-products—no organism is expending energy in order to contribute to a recycling loop, they are just eating food and excreting waste products.

However, if something evolves that can make use of an available nutrient, and it shifts the environment into a state that the rest of the community are not adapted to, this can cause trouble. We call these 'rebel' species and in Flaskworld when they evolve they can cause the population to crash, recycling to be removed, and in extreme cases, complete extinction of the global ecosystem (*41*). However, in most cases after a population crash, recycling and population size progressively recover. These results suggest that recycling is robust but not invincible. Once a system stumbles on recycling it is self-reinforcing—in the jargon of the field it would be called 'autocatalytic positive feedback'. Thus on your planet (or any planet) successfully seeded with microbial life we should expect the emergence of recycling loops to be a robust feature.

7.10 Environmental regulation

With multiple recycling loops established, your world now has *bona fide* ecosystems. We pictured them emerging in crater ponds or lakes, but can expect them to spread to the shallow seas and the open ocean, or to colonize parts of the land surface. (In fact the desert crust you introduced earlier is such an ecosystem with some internal recycling of nutrients.) As well as being flushed down streams and rivers, microbial life is readily transported in the atmosphere (*42*). Cells can be injected from water into the air by a process of bubble bursting when 'white cap' waves form on lakes or the ocean. The cells are so small, with such modest sinking velocities under gravity, that once they are airborne they can stay up for long periods. Then a common exit route from the atmosphere is in rain drops or (if it gets cold enough) hail and snow. This means your nascent ecosystems on the planet's surface are exchanging material and organisms. They also retain some identity by being contained in e.g. crater ponds, lakes, circulatory eddies of the ocean, or patches of the land surface. In other words, the overall system is not well mixed, even if individual ecosystems are. With such a biosphere in place, let us return to your earlier challenge of trying to establish the self-regulation of environmental variables, such as temperature, which are not nutrients. You consciously created a planetary thermostat, but can environmental regulation emerge automatically in an evolving, spatial system?

This problem too has been tackled with the Flask model (*43*). In Flaskworld there are multiple non-nutrient environmental variables and individual organisms have preferences for each of them that can be subject to mutation, variation and selection. The organisms can also alter each of these environmental variables, and these effects can be subject to mutation. In a single, isolated flask there is generally little sign of a trend toward environmental regulation. Sometimes there are periods of environmental stability but generally the environment is being continually and apparently randomly pushed around by the microbial population which is simultaneously adapting to it. The fact that each flask is well mixed prevents any organism from creating a favourable micro-environment for itself. However, the original model lacked the dimensions of space—on your planet there are multiple eco-systems, only partially connected to one another. To recreate such a scenario in Flaskworld, Hywel Williams has connected a series of flasks together, allowing some mixing between them (*43*). Typically 10 or more flasks are connected together in a ring. There is mixing between neighbouring flasks at each time-step by dipping a cup in each and swapping the cups. The rate of this mixing can be varied by changing the size of the cups.

This scenario raises an interesting possibility that some ecosystems might be more productive than others, for example due to more efficient recycling, and thus might inadvertently send out more of their constitu-ent organisms into the surrounding world (the neighbouring flasks). If by chance all the components of a successful recycling loop arrive at a new location then the recycling system would re-establish there. Sure enough, once recycling is established somewhere it spreads like wild-fire through the whole system.

However, in the multiple flask system, something more remarkable happens. These larger systems show a tendency to regulate the non-nutrient environment and this regulation improves over time. This is measured in terms of the 'environmental error'—the mismatch between the actual state of the environment (in each flask) and the preference of the organisms—which shows a reducing trend over time (Fig. 7.8). It is hard to tell whether this is due to the organisms adapt-ing to prevailing conditions, altering the conditions to suit their pref-erences, or a bit of both. To find out, in the model world, all the organisms are given the same, fixed environmental preferences. The result remains intact—there is a robust trend towards reducing envi-ronmental error over time—and this must be due to the environment being moved toward the (now fixed) preferences of the organisms. But how can this occur?

The mechanism turns out to involve a rudimentary form of selec-tion operating at the level of whole flask ecosystems. In essence if a flask population collectively alters its environment towards its prefer-ence it will experience positive feedback on growth and become larger, whereas if it alters its environment away from its preference it will experience negative feedback on growth and be smaller. The

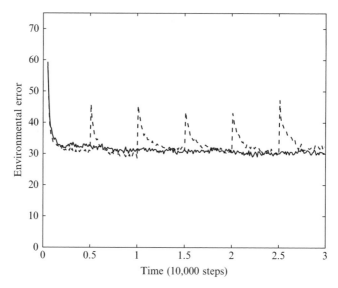

Fig. 7.8 Emergent environmental regulation in spatial Flaskworlds. The environmental error is robustly reduced over time, whether the virtual worlds are periodically perturbed (dashed line) or unperturbed (solid line). In the absence of life the environmental error is 60. [Redrawn from original (*43*) copyright (2008) National Academy of Sciences, U.S.A]

nature of the mixing between flasks is such that larger (higher density) populations tend to colonize any neighbouring smaller (lower density) populations. In this way populations (or members of them) that tend to improve the environment tend to spread at the expense of populations (or members of them) that tend to degrade the environment. There is no faithful replication of whole ecosystems so there is no ecosystem-level adaptation, but there is a form of selection occurring at the higher-level of whole flasks. The higher level selection mechanism could in principle just be selecting responsible genes in 'super-species' that are responsible for most of the environmental change, but this turns out not to be the case (all populations are highly diverse).

In this set up of Flaskworld, we have removed all chances for one type of life in a given flask to experience an advantage over another type of life in the same flask, due to the state of the environment. All organisms experience an identical environment and have an identical growth response to it. The types of life only differ in their effects on the environment and their interaction with nutrients. In fact the model systems tend to spend most of their time in a nutrient-limited regime, with lots of recycling going on, where they are not affected by the state of the non-nutrient environment. Only when the environment strays into a more extreme state does it become limiting and feedbacks on growth and higher level selection kick in to drag the system back into the nutrient-limited regime. In the Flaskworld with multiple connected ecosystems, the frequency of self-induced total extinctions is much reduced relative to single flasks, sometimes to zero. However, occasionally the multiple-flask biospheres can still be destroyed by some evolutionary innovation that emerges within them.

To explain the emergence of environmental regulation in Flaskworld we have invoked a mechanism involving a crude form of higher level of selection operating at the level of whole ecosystems. This is sure to upset some evolutionary biologists. So let us be clear that we are not invoking adaptation at this higher-level because whole ecosystems are not faithfully replicated, so the heritability of variation is weakened. Still, any appeal to higher-level selection is problematic, not because it doesn't exist, but because selection at lower levels is generally more powerful and expected to disrupt higher level units.

7.11 Increasing free-energy capture and processing

Hopefully your adventures thus far at trying to make a Gaia have convinced you that once microbial life gets started on a planet, recycling loops will robustly emerge and productivity will correspondingly be enhanced. For life to become globally dominant and take first order control over geochemical cycles on your planet, it needs more than just recycling. It also needs free energy capture from sunlight, i.e. photosynthesis. In Flaskworld, the only source of free energy is in the form of the chemical compounds added to the flasks that can be broken down. At each stage of the recycling process some free energy is lost (due to the inevitable inefficiency of metabolism) and this ultimately sets a limit on the total productivity of this biosphere. On your planet and in the real world the free energy bound up in incoming chemicals is a rather meagre source. It will restrict productivity to at most about a thousandth that of the present Earth's biosphere (as we described in Chapter 4).

To increase the free energy available to your biosphere, you need to introduce photosynthesis to your planet—the ability to capture the abundant energy available in sunlight. In fact you have already done this on the land surface where the desert crust is fuelled by cyanobacteria conducting oxygenic photosynthesis. But the vast ocean contains no photosynthesis, yet. Recall that there are two types of photosynthesis—the 'oxygenic' sort we are familiar with that uses water as a substrate (electron donor) and liberates oxygen—and various 'anoxygenic' sorts that use different substrates and correspondingly liberate different products (see Box 7.7).

Box 7.7 Different types of photosynthesis and recycling

Photosynthesis transfers the energy in sunlight to electrons, which are ultimately transferred to the carbon in carbon dioxide. Hence it requires a donor of electrons from the outside world. Suitable electron donors used in anoxygenic photosynthesis include hydrogen gas (H_2), hydrogen sulphide (H_2S), elemental sulphur (S^0), and reduced (ferrous) iron (Fe^{2+}). Just as the water and carbon dioxide used in oxygenic photosynthesis are recycled by aerobic respiration, so too the substrates for anoxygenic photosynthesis can be recycled. Fig. 7.9 gives two examples that are described further in the text.

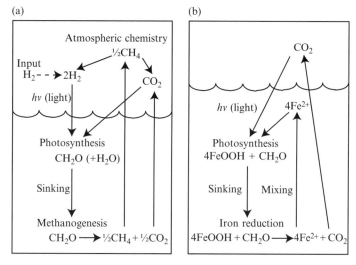

Fig. 7.9 Recycling biospheres fuelled by different types of anoxygenic photosynthesis. (a) A hydrogen-fuelled biosphere in which the hydrogen is (imperfectly) recycled by methanogenesis followed by atmospheric chemistry. (b) An iron-fuelled biosphere in which the iron is recycled by bacterial iron reduction followed by up-welling. The chemical schemes have been simplified, generally leaving out water and hydrogen ions where they appear

If you try introducing the type of anoxygenic photosynthesis that uses hydrogen gas as a substrate (see Box 7.7), your biosphere will soon run into a problem. As the photosynthesizers proliferate they will use up all the hydrogen in the atmosphere and become limited by the rather modest inflows coming from volcanoes (and metamorphic processes). The only way that life can get around this problem is to recycle the hydrogen needed for photosynthesis (Fig. 7.9a). This can be done, for example, by the methanogens you introduced earlier to the ocean, which break down organic matter and release methane as a by-product. In the atmosphere reactions triggered by sunlight convert the methane back to hydrogen. The hydrogen can then be taken up again in photosynthesis. However, such recycling is not perfect—only half of the hydrogen required can be recycled and some of this is lost, for example, when organic matter escapes breakdown by the methanogens and is buried in the crust. The productivity of your biosphere now depends on the efficiency of recycling and on the rather meagre flux of hydrogen gas from volcanoes. Although volcanic activity is greater on planet Ocean than today's Earth, your hydrogen-fuelled biosphere will still be restricted to around a thousandth of the productivity of the marine biosphere on today's Earth (*44*).

You might have rather better luck if you introduce the type of anoxygenic photosynthesis that uses reduced iron (Fe^{2+}) as a substrate (see Box 7.7), because there is a much greater supply of iron to the surface of your planet than hydrogen. In the surface ocean, these photosynthesizers produce oxidized iron and organic matter, both of which tend to sink, presenting the need for recycling (Fig. 7.9b). If you also introduce iron-reducing bacteria to the ocean they will feed on the organic matter and iron oxide as it sinks through the water column, regenerating reduced iron as a by-product. The photosynthesizers then rely on ocean mixing to return the reduced iron to the sunlit surface zone. This iron-fuelled biosphere can be much more successful but it would still be restricted to around a tenth of the productivity of the marine biosphere on today's Earth (*44*).

Despite recycling there are still considerable limits to the productivity of biospheres based on anoxygenic photosynthesis, so you decide to fertilize your planet's ocean with some cyanobacteria from the Ark. By using abundant water as a substrate they avoid this being a limiting factor, and cycling of water goes on in the absence of life. On your planet, CO_2 is also unlikely to become limiting, because it is abundant and is recycled by the breakdown of organic matter (sometimes with the help of atmospheric chemistry). So what caps the productivity of your marine biosphere? Ultimately it could simply be the sunlit volume the cyanobacteria are able to colonize and the efficiency of photosynthesis. However, it is more likely to be the availability of essential nutrients (which are still the limiting factor in the oceans of today's Earth). Nitrogen and phosphorus are the two key nutrients in this respect, and their recycling to the sunlit surface ocean also depends on physical upwelling from the deep ocean. Nitrogen is abundant as ammonium in the deep ocean, and should its supplies run short, it turns out that the cyanobacteria are among those that can fix nitrogen directly from the atmosphere. Phosphorus is supplied from the continents where weathering is being enhanced by the desert crust, but the enhancement of weathering is much less than by a biosphere with plants like on today's Earth. Consequently, phosphorus is likely to be the ultimate limiting nutrient on your planet, with the greatest need for recycling. As we will see in later chapters, there is enhanced phosphorus recycling in an oxygen-free deep ocean such as the one that prevails on planet Ocean.

In this scenario of adding anoxygenic and then oxygenic photosynthesizers, it is tempting to assume the latter would out-compete the former in most well-lit surface waters, because they are less likely to be substrate limited. However, wherever the substrates for anoxygenic photosynthesis are available, these older types of life retain the upper hand. This is because they need to capture less energy from sunlight for each molecule of CO_2 fixed. Consequently, anoxygenic photosynthesizers have a niche in lower light conditions, somewhat deeper in the sunlit zone, where iron rich water is upwelling from below.

Having established the most effective form of free energy capture for your biosphere—oxygenic photosynthesis—both on land and in the ocean, you have enabled a much more dynamic ecosystem because the oxygen liberated can be recombined with organic material in the process of aerobic respiration. This yields an order of magnitude more energy per molecule of organic matter broken down than anaerobic pathways of oxidizing organic matter such as fermentation and methanogenesis. It also recycles carbon directly to CO_2. However, a significant concentration of oxygen needs to build up for aerobic respiration to occur. The first organisms to make use of this abundant store of chemical free energy will be the cyanobacteria themselves, as they can perform aerobic respiration and they produce the oxygen required.

To make full use of the free energy store, you need to seed your planet with free living aerobic bacteria that can also consume organic matter using aerobic respiration. Their requirement for a significant concentration of oxygen means that, initially at least, they must be closely associated with the cyanobacteria. Thus you can expect to see the microbes you have added to your planet form close knit communities—microbial mats—with restricted exchange of gases and other materials with the surrounding world. In such mats, aerobic bacteria will tend to out-compete anaerobic ones because they get more energy from breaking down organic carbon. They in turn ensure efficient recycling of CO_2 to the cyanobacteria. Any organic material that escapes aerobic respiration can still go to the anaerobic bacteria, and ultimately this is also recycled to CO_2. This recycling system should become highly efficient with only a small leak of organic matter buried in new sedimentary rocks[†].

[†] On today's Earth, the flux of O_2 from plants and other oxygenic photosynthesizers is almost perfectly balanced by the consumption of O_2 by animals and other heterotrophs (correspondingly the CO_2 fluxes are also almost perfectly balanced). The imbalance is about 1 part in 1000 that gets buried as organic matter in new rocks.

That small leak of organic carbon burial equates to a long-term, net source of atmospheric oxygen. Consequently, your instigation of oxygenic photosynthesis should ultimately lead to a progressive oxygenation of the surface ocean and atmosphere of your planet, but this will not occur immediately, because initially all the oxygen reacts with the abundant reduced material, especially iron (Fe^{2+}) in the ocean and on the land surface. Then it goes to oxidizing reduced material such as hydrogen gas coming in from volcanoes and other geologic processes. Consequently, the rise of oxygen could be considerably delayed—for reasons we will examine in the following chapters. For now we note that the establishment of oxygenic photosynthesis provides a necessary (but not sufficient) condition for the establishment of an atmosphere that can support complex animal life. Thus you have set your planet on a path 'upward' (considered from the point of view of an oxygen breathing animal), but it will prove to be a long and winding one, with some nasty precipices along the way, and there is no guarantee that it will reach the 'summit' of being able to support complex life.

7.12 Can life extend the habitable period?

You have now 'won' the Gaia game—you have established a thriving microbial biosphere with healthy recycling and a high level of photosynthetic primary production. You have also established a functioning planetary thermostat and witnessed some emergent self-regulation of the environment. Finally, you have provided an essential ingredient for creating an oxygen-rich atmosphere—a biological oxygen source. Before we leave the game let us look ahead into the far future and ask one final question: could the biosphere you have created extend the period for which planet Ocean remains habitable?

As its parent star gets steadily brighter, we expect life to eventually be eliminated from planet Ocean, as from the Earth. But is there any scope for life to delay its own disappearance, by effectively shifting the inner boundary of the habitable zone towards the star? This is exactly what happens on Daisyworld (Fig. 7.5). As its star gets brighter the white daisies come to dominate and maintain a cool state when without them the planet would be too hot for them to colonize. There are two stable states for a range of values of the brightness of the star—one cool one with life, and one hot one without. Could the same phenomenon occur on planet Ocean or on the Earth?

The answer is 'yes, in principle'. On planet Ocean, as on the Earth, the net effect of life is to cool the climate. The effect is potentially large, with estimates of up to 45°C cooling by life on the present Earth (*23*). If so, the uninhabited state of the present Earth would currently be at ~60°C. This would be uninhabitable for complex life, although still habitable for some microbial life. In other words, there could be two stable states for the climate of the Earth: A cool state that supports complex life, and an alternative hot state that would not (45). If the biological amplification of weathering is particularly strong, then both states may already exist for today's Earth (Fig. 7.10). If the cooling effect is weaker, they could both become available at some point in the future. Such 'bi-stability' implies that complex life can (inadvertently) keep the planet habitable for itself for longer than it otherwise would be, thus extending its own lifespan. In the extreme case, it is already doing so on Earth.

If the habitable period of a planet can be altered by it being inhabited this alters one of the assumptions made in the critical steps model of Chapter 5—that the habitable period is independent of biology. In simulations with strong biological amplification of weathering, in which the Earth is bi-stable at the present day (Fig. 7.10) (45), without life the planet could already have become uninhabitable to complex life and observers would not have evolved. This does not violate the critical steps model, but it offers an interesting way in which the evolution of observers may be contingent on previous transitions.

There are of course other possibilities. In previous models of the lifespan of the biosphere, life tends to starve itself to death by draining the atmosphere of carbon dioxide (46–48). This happens because the silicate weathering negative feedback is assumed to be weaker in these models. In response to the brightening sun, it removes CO_2 from the atmosphere, but it goes too far. Before the biosphere is overheated, all

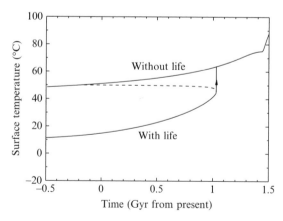

Fig. 7.10 A model in which life extends the habitable period of Earth, showing the surface temperature with and without the current biota. The dashed line indicates an unstable state and the upward arrow marks the abrupt transition when the biosphere overheats. [Adapted from Lenton & von Bloh (45).]

the CO_2 disappears, killing off all photosynthesis (not just complex plant life and the animals and fungi that depend on it). Of course, all the models used to examine the lifespan of the Earth's biosphere are based only on known types of life and known biological effects on the environment. In the future, evolution could unwittingly produce (as a by-product) a new planetary cooling mechanism that would make the models redundant. Alternatively, a self-aware species could consciously intervene to keep the planet cool.

7.13 Conclusion and predictions

We have argued that in the biosphere, recycling is an autocatalytic process—it promotes itself—and it cannot be cheated because it is built on cost-free by-products. These observations lead us to postulate an 'Autocatalytic Gaia' hypothesis; that once life is present on a planet recycling loops will robustly emerge and productivity will correspondingly be enhanced. If correct, our theoretical considerations suggest that the establishment of recycling loops is not a really difficult, or even a somewhat difficult, transition. Furthermore, once it is present on a planet, life will inadvertently (but inevitably) get entangled in environmental feedback loops, and the resulting system will tend to spend most of its time in negative feedback regimes, with occasional transitions dominated by positive feedback. It is (theoretically) possible for life to drive itself extinct, or nearly extinct, and some of the model results show this. The likelihood of such self-induced extinction is reduced (but not removed) by the presence of self-limiting negative feedbacks on the perpetrators. Finally, there may be some potential for rudimentary higher-level selection to generate a trend towards improved environmental regulation over time.

This leaves us with a 'Probable Microbial Gaia' hypothesis—once microbial life is established on a planet it is likely that there will be some emergent environmental regulation. If and when life (or some external perturbation) pushes the system to the bounds of habitability, if it is in the cold direction to a snowball state, this is not fatal to microbial life. Furthermore, the system has an in-built recovery mechanism from the snowball state—the build up of carbon dioxide that eventually melts the snowball. However, if things go in the opposite direction and the planet overheats into a runaway greenhouse effect then this is likely to be fatal. Hence there will always be some element of luck in (microbial) life surviving. Just how much of an element of luck there was in the case of the Earth we will examine in the following chapters. Still, we predict that 'Microbial Gaias' are common wherever and whenever life emerges to prokaryote form—if life can readily reach this stage than there could be as many as one per solar system.

'Complex Gaias' on the other hand are expected to be rare. Producing a planet fit for complex life to reach the point of self-reflecting observers is a much more difficult proposition. First it requires the evolution of oxygenic photosynthesis, which may be an extremely difficult evolutionary transition. Then it requires some considerable time to oxidize the planet, and this takes at least one and potentially several steps to get atmospheric oxygen to the level that can support animal-like observer species. At each step there is the danger of the planet being shifted to a barely habitable or uninhabitable state if abiotic positive feedbacks take over. After each transition, recycling needs to recover and there needs to be an overall trend towards tighter regulation of key environmental variables including the temperature and oxygen content of the atmosphere.

References

1. J. E. Lovelock, *Gaia as seen through the atmosphere. Atmospheric Environment* **6**, 579 (1972).

2. J. F. Kasting, D. P. Whitmore, R. T. Reynolds, *Habitable zones aroud main sequence stars. Icarus* **101**, 108 (1993).

3. J. F. Kasting, J. B. Pollack, T. P. Ackerman, *Response of Earth's surface temperature to increases in solar flux and implications for loss of water from Venus. Icarus* **57**, 335 (1984).

4. A. P. Ingersoll, *The Runaway Greenhouse: A History of Water on Venus. Journal of the Atmospheric Sciences* **26**, 1191 (1969).

5. A. J. Watson, T. M. Donahue, W. R. Kuhn, *Temperatures in a runaway greenhouse on the evolving Venus: Implications for water loss. Earth and Planetary Science Letters* **68**, 1 (1984).

6. M. I. Budyko, *The Effect of Solar Radiation Variations on the Climate of the Earth. Tellus* **21**, 611 (1968).

7. W. D. Sellers, *A Global Climate Model Based on the Energy Balance of the Earth-atmosphere System. Journal of Applied Meteorology* **8**, 386 (1969).

8. G. R. North, R. F. Cahalan, J. A. Coakley, *Energy Balance Climate Models. Reviews of Geophysics and Space Physics* **19**, 91 (1981).

9. G. R. North, *The Small Ice Cap Instability in Diffusive Climate Models. Journal of the Atmospheric Sciences* **41**, 3390 (1984).

10. J. C. G. Walker, P. B. Hays, J. F. Kasting, *A negative feedback mechanism for the long-term stabilisation of Earth's surface temperature. Journal of Geophysical Research* **86**, 9776 (1981).

11. K. Caldeira, J. F. Kasting, *Susceptibility of the early Earth to irreversible glaciation caused by carbon dioxide clouds. Nature* **359**, 226 (1992).

12. J. L. Kirschvink, in *The Proterozoic Biosphere*, J. W. Schopf, C. Klein, D. J. DesMarais, Eds. (Cambridge University Press, Cambridge, 1992), pp. 51–52.

13. P. B. Price, T. Sowers, *Temperature dependence of metabolic rates for microbial growth, maintenance, and survival. Proceedings of the National Academy of Sciences USA* **101**, 4631 (2004).

14. T. M. Lenton, *Gaia and natural selection. Nature* **394**, 439 (1998).

15. J. E. Lovelock, *Gaia - A New Look at Life on Earth.* (Oxford University Press, Oxford, 1979).

16. J. E. Lovelock, *The Ages of Gaia - A Biography of Our Living Earth.* L. Thomas, Ed., *The Commonwealth Fund Book Program* (W. W. Norton & Co., New York, 1988).

17. J. E. Lovelock, in *Biomineralization and Biological Metal Accumulation*, P. Westbroek, E. W. d. Jong, Eds. (D. Reidel Publishing Company, Dordrecht, 1983a), pp. 15–25.

18. J. E. Lovelock, *Daisy World—A Cybernetic Proof of the Gaia Hypothesis. The Co-evolution Quarterly* **38**, 66 (1983b).

19. A. J. Watson, J. E. Lovelock, *Biological homeostasis of the global environment: the parable of Daisyworld. Tellus* **35B**, 284 (1983).

20. A. J. Wood, G. J. Ackland, J. Dyke, H. T. P. Williams, T. M. Lenton, *Daisyworld: A review. Reviews of Geophysics* **46**, RG1001 (2008).

21. J. E. Lovelock, A. J. Watson, *The regulation of carbon dioxide and climate: Gaia or geochemistry? Planetary and Space Science* **30**, 795 (1982).

22. T. A. Jackson, W. D. Keller, *A comparitive study of the role of lichens and inorganic processes in the chemical weathering of recent Hawaiian lava flows. American Journal of Science* **269**, 446 (1970).

23. D. W. Schwartzman, T. Volk, *Biotic enhancement of weathering and the habitability of Earth. Nature* **340**, 457 (1989).

24. D. W. Schwartzman, T. Volk, *Biotic enhancement of weathering and surface temperatures on earth since the origin of life. Palaeogeography, Palaeoclimatology, Palaeoecology (Global and Planetary Change Section)* **90**, 357 (1991).

25. S. E. Campbell, *Soil stabilization by a prokaryotic desert crust: implications for PreCambrian land biota. Origins of Life* **9**, 335 (1979).

26. A. R. Prave, *Life on land in the Proterozoic: Evidence from the Torridonian rocks of northwest Scotland. Geology* **30**, 811 (2002).

27. J. McDonald-Gibson, J. G. Dyke, E. A. Di Paolo, I. R. Harvey, *Environmental regulation can arise under minimal assumptions. Journal of Theoretical Biology* **251**, 653 (2008).

28. J. G. Dyke, *The Daisyworld Control System.* DPhil thesis, University of Sussex (2009).

29. R. Dawkins, *The Extended Phenotype.* (Oxford University Press, Oxford, 1983).

30. W. F. Doolittle, *Is Nature Really Motherly? The CoEvolution Quarterly* **29**, 58 (1981).

31. W. D. Hamilton, *Ecology in the Large: Gaia and Genghis Khan. Journal of Applied Ecology* **32**, 451 (1995).

32. R. Dawkins, *The Selfish Gene*. (Oxford University Press, Oxford, 1976).

33. W. D. Hamilton, *The genetical evolution of social behaviour: I. Journal of Theoretical Biology* **7**, 1 (1964).

34. W. D. Hamilton, *The genetical evolution of social behaviour: II. Journal of Theoretical Biology* **7**, 17 (1964).

35. R. L. Trivers, *The Evolution of Reciprocal Altruism. The Quarterly Review of Biology* **46**, 35 (1971).

36. M. A. Nowak, *Five Rules for the Evolution of Cooperation. Science* **314**, 1560 (2006).

37. T. Volk, *Gaia's Body – Toward a Physiology of the Earth*. (Copernicus, New York, 1998).

38. D. M. Wilkinson, *Is Gaia really conventional ecology? Oikos* **84**, 533 (1999).

39. T. M. Lenton, A. J. Watson, *Redfield revisited: 1. Regulation of nitrate, phosphate and oxygen in the ocean. Global Biogeochemical Cycles* **14**, 225 (2000a).

40. A. C. Redfield, in *James Johnstone Memorial Volume*. (University of Liverpool, Liverpool, 1934), 176–92.

41. H. T. P. Williams, T. M. Lenton, *The Flask model: Emergence of nutrient-recycling microbial ecosystems and their disruption by environment-altering 'rebel' organisms. Oikos* **116**, 1087 (2007a).

42. W. D. Hamilton, T. M. Lenton, *Spora and Gaia: how microbes fly with their clouds. Ethology Ecology and Evolution* **10**, 1 (1998).

43. H. T. P. Williams, T. M. Lenton, *Environmental regulation in a simulated network of evolving microbial ecosystems. Proceedings of the National Academy of Sciences USA* **105**, 10432 (2008).

44. D. E. Canfield, M. T. Rosing, C. J. Bjerrum, *Early anaerobic metabolisms. Philosophical Transactions of the Royal Society Biological Sciences* **361**, 1819 (2006).

45. T. M. Lenton, W. von Bloh, *Biotic feedback extends the life span of the biosphere. Geophysical Research Letters* **28**, 1715 (2001).

46. J. E. Lovelock, M. Whitfield, *Life span of the biosphere. Nature* **296**, 561 (1982).

47. K. Caldeira, J. F. Kasting, *The life span of the biosphere revisited. Nature* **360**, 721 (1992).

48. S. Franck et al., *Reduction of biosphere life span as a consequence of geodynamics. Tellus* **52**, 94 (2000).

PART III
THE OXYGEN REVOLUTION

8
Photosynthesis

Take a deep breath...Every minute you take about fifteen, which means that if you have an average human lifespan you will take around 600 million breaths. Their prime purpose is to get oxygen into your blood stream, and in an average lifetime, your heart will beat around three billion times to pump it to the sites where it is utilized. These sites are the tiny energy factories in each of your cells—the mitochondria—where the oxygen is reacted with sugars derived from your food, to yield energy. Whether your food came in the form of animal, plant or fungal matter, the energy it contains was first captured from sunlight by a photosynthesizer—most likely a green plant—which split water, liberating oxygen gas, and fixed carbon from carbon dioxide into organic matter. Although green plants dominate the land surface today, this process of oxygenic photosynthesis first evolved in a water-borne cyanobacterium, and they are ultimately responsible for the oxygen-rich air that we breathe. It is hard to conceive of how long cyanobacteria have been on the planet, but if every beat of your heart took a year your lifetime would roughly correspond to their tenure.

Your body essentially reverses the process of oxygenic photosynthesis, recombining oxygen and organic matter to liberate carbon dioxide, water and energy in the process of aerobic respiration. You can break down organic matter other ways—if you are really exerting yourself and your breathing and heartbeat cannot supply enough oxygen to your muscles, you can switch to anaerobic metabolism—transforming one organic compound into another without the use of oxygen. But the energy yields are miserable in comparison. Aerobic metabolism yields an order of magnitude more energy per molecule of organic matter broken down compared to anaerobic pathways such as fermenting lactic acid* (*1*). This high energy yield is absolutely critical to our existence

---

* Energy yield is expressed in thousands of Joules per mole of substance 'combusted' (in this case organic matter) and is 2870 kJ/mol for aerobic respiration and 195 kJ/mol for fermenting lactic acid.

as large, mobile animals with complex nervous systems including big brains. Without abundant oxygen in the atmosphere, relying on anaerobic pathways alone, organisms simply cannot reach the size and level of metabolic activity that we enjoy (*1*).

Consequently, we have everything to thank the process of oxygenic photosynthesis for, and we should pay homage to the first cyanobacterium that evolved it. In this chapter we will see just how complex a process it is and how difficult its evolution must have been. Of course, we usually take the 21% of oxygen in the air for granted—unless you work in a submarine or other sealed vessel, there always seems to be enough oxygen available and no need to worry about its supply. There is indeed a phenomenal amount of oxygen in the atmosphere[†], but as we explained in Chapter 2, this hasn't always been the case. Oxygen has risen from essentially zero to its present levels in a series of steps, over the past few billion years. The most striking of these steps was the Great Oxidation event around 2.3 billion years ago, which neatly divides the history of the planet into two halves. Before it, the atmosphere was chemically reducing, meaning it was dominated by electron donating compounds and life relied largely on anaerobic metabolism. After it, the atmosphere was oxidizing, meaning it fuelled reactions that tend to rob electrons from substances. Paramount among these reactions is the aerobic respiration that we depend on, and which became the dominant way of breaking down organic matter after the Great Oxidation.

Thus, oxygen is essential for us and for all animals. But it is also bad for us (*2*); oxygen is toxic! This is related to oxygen's strong tendency to rob electrons from substances. The oxygen molecule (O_2) itself is actually surprisingly un-reactive. However, in the biochemical pathways of using oxygen, a bevy of unwelcome, highly reactive oxygen species are produced. These include the superoxide anion radical (O_2^-), hydrogen peroxide (H_2O_2), the hydroxyl radical ($OH\cdot$), and an excited state of the oxygen molecule ($^1O_2$). These 'reactive oxygen species' would wreak havoc in our cells were it not for elaborate defences that have evolved to deal with them. Despite these defences, some damage occurs and despite mechanisms to repair it, some of the damage is irreversible. The accumulation of this damage is the chemical cause of ageing—the gradual wearing out of organisms. Thus oxygen is for us both life giver and ultimately, destroyer: 'As little as one breath is known to produce a life-long addiction to the gas, which...invariably ends in death' (*3*).

For the organisms that had evolved in the oxygen-free Archean world, the oxygen introduced by the first oxygenic photosynthesizers

[†] Oxygen comprises 20.95% of the present atmosphere, weighing 1.2×10^{15} tonnes. Burning all the vegetation on Earth would consume barely one-thousandth of this, and even burning all the fossil fuel reserves would change it only a small fraction of a percent.

was a toxic poison. For many of them it still is—they remain 'obligate anaerobes' that are killed by exposure to oxygen. However, cyanobacteria all use aerobic respiration, and some of the key enzymes involved in aerobic respiration appear to be much more deeply rooted and widespread in the tree of life than oxygenic photosynthesis, suggesting they evolved before it did (4). The capacity to detoxify reactive oxygen species must have evolved prior to, or in tandem with, aerobic respiration, or the first aerobic organisms would have poisoned themselves. This sets up a chicken-and-egg problem; how could aerobic respiration and defences against oxygen toxicity have evolved in a world free of oxygen? The popular answer is that there was another source of reactive oxygen species on the early Earth, although this source has proved elusive (photochemical production of hydrogen peroxide has been suggested (5) and we will return to consider it later in this chapter). Whatever the trigger, it appears likely that some bacteria were able to conduct aerobic respiration, once sufficient oxygen became available from nearby cyanobacteria. (Subsequently, an ancestral eukaryote cell engulfed one of these bacteria without killing it and gave rise to the mitochondria that are still the energy factories in our cells today.)

In this chapter we focus on the mechanism of oxygenic photosynthesis. We illustrate the complexity of the process, without going into the fine detail of the biochemistry. In the next chapter we look at the evidence for *when* it evolved, the timing being important for Earth history, and also because it can potentially be informative about the likelihood that oxygenic photosynthesis would evolve on other planets—whether it might be a 'critical step' in terms of the model introduced in Chapter 5. Finally in Chapter 10 we will explore a model for the Earth system transformation that was the Great Oxidation. Did it follow immediately after the invention of photosynthesis? The evidence suggests not. Why the delay, and what exactly happened at the transition itself? Might it have been reversed, or was it a one-way ticket to an oxidized planet?

8.1 Photosynthesis, redox potential and the energy available in light

All types of photosynthesis involve the removal of electrons from some substance in the environment, the use of light energy to raise these electrons to a higher energy state, followed by sequences of reactions which enable the cell to convert some of that energy into useable chemical form. Finally the electrons are transferred to an externally sourced substance, usually carbon dioxide, converting the carbon into organic form. Chemists refer to processes in which electrons are lost as *oxidation*, gain of electrons as *reduction,* and the process as a whole as a *redox* reaction. In chemical terms therefore oxygenic photosynthesis is a redox process in which the electron donor, water, is oxidized to oxygen gas, while the carbon in carbon dioxide is reduced to sugars. Other types of (bacterial) photosynthesis use many other substrates to derive the electrons (recall Box 7.7), and don't involve the liberation of oxygen. The initial electron donor is greatly variable so that different styles of photosynthesis have very different net effects on the environment.

Redox reactions can be classified by their 'redox potential' (Box 8.1). This potential, which is measured in volts, describes their affinity for electrons. Since electrons have a negative charge they flow from more negative potentials towards more positive potentials. In photosynthesis, the energy captured from light photons is used to boost electrons to a strongly negative redox potential, after which they transfer to inter-

Box 8.1 Redox potentials

It is straightforward to work out how much energy is needed to raise an electron through the gradient required for a given type of photosynthesis. Table 8.1 below gives these energies directly if we work in units of electron volts (one electron-volt, or 1eV, is the amount of energy needed to move one electron through a potential of 1 volt, equal to 1.6×10^{-19} joules). For example in oxygenic photosynthesis, the net energy required to oxidize one water molecule and reduce one CO_2 is $+0.820-(-0.434) = 1.254$ eV. This is substantially more than required for the other electron donors used in bacterial photosynthesis and shown in Table 8.1. To oxidize two H_2O molecules, as needed to produce one molecule of O_2, needs twice this amount, or about 2.6 eV of energy. This must come from absorbing photons of light, and these have a known energy which varies with the wavelength of their light. Photons absorbed by chlorophyll *a* have a wavelength centred at 700nm, and their energy is about 1.7 eV. A single photon of this light does not therefore have sufficient energy to oxidize two H_2O molecules and release O_2, even if the process were 100% efficient. The energy from more than one photon must be harnessed together to accomplish the release of each molecule of O_2. In fact a total of eight photons are needed to oxidize two water molecules to one molecule of oxygen, four for each photosystem.

Table 8.1 Redox potentials of some substances used to donate or accept electrons in different types of photosynthesis

Electron donors	Redox potential, relative to standard E_0 (V)
$H_2O \rightleftharpoons \frac{1}{2} O_2$	+0.820
$Fe^{2+} \rightleftharpoons Fe(OH)_3$	−0.236
$HS^- \rightleftharpoons S^0$	−0.278
$1/2H_2 \rightleftharpoons H^+$	−0.414
Electron acceptor	
$CO_2 \rightleftharpoons (CH_2O)$	−0.434

mediaries down a redox gradient towards more positive values, in the process of which energy can be transferred into a chemical form for later use by the cell. Finally, the electrons are used to reduce carbon.

8.2 The key parts of oxygenic photosynthesis

We have already noted what a remarkable 'trick' oxygenic photosynthesis is. What is so neat about it is that it uses two readily available substances, water and carbon dioxide, plus light, both as a source of energy and of carbon for organic molecules. This desirable outcome does not just happen of its own accord: it is the end result of a very complex process.

Fig. 8.1 describes oxygenic photosynthesis in terms of the redox potential of its electron transfer path, and locates on that path some of the key components. The diagram is called the (Hill-Bendall) Z scheme—the resemblance to a poorly drawn Z is apparent if you hold the picture on its side. The vertical axis is the redox potential (more negative—which for electrons means higher energy—is up). The upward strokes represent the capture of light energy, used to raise the electrons up the redox potential. This happens at two separate points in the overall scheme, and the systems responsible for them are named

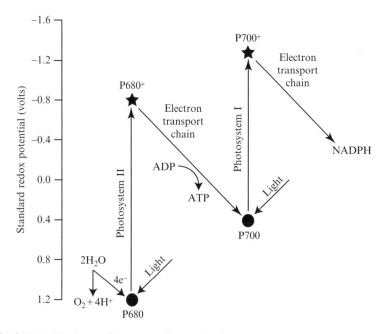

Fig. 8.1 The Z scheme of oxygenic photosynthesis

Photosystem I and **Photosystem II** (or PS I and PS II for short). Electrons and energy emerging from the first in the sequence, called photosystem II because historically it was the second to be described, are then transferred down a potential gradient, along an 'electron transport chain' to photosystem I. As they pass along the chain, they lose energy, which is used to make the energy-storage molecule ATP from ADP[‡].

Photosystem I then grabs the electrons and transfers the energy of further photons of sunlight to them. They can then pass down a second electron transport chain to be deposited in pairs on a special electron carrier called NADP⁺, making the energy-storage molecule NADPH (Nicotinamide Adenine Dinucleotide Phosphate). This molecule and ATP are then used in a separate set of reactions (the Calvin-Benson cycle, which is not light-dependent) to fix carbon dioxide into glucose, regenerating both ADP and NADP⁺ in the process. The chemistry is jaw-droppingly complex, but a crucial point to note is that *all* the intermediate compounds are regenerated, so that the *net* inputs are only CO_2 and water (and light), and the net outputs are glucose and oxygen. In total, 'light makes life and oxygen out of water and thin air' (6).

Let us now describe the key parts of this overall system in a bit more detail. **Photosystem II** can be divided into the following sub-systems:

8.2.1 The water-splitting complex (WSC)

The water-splitting complex (WSC), is sometimes also called the 'oxygen-evolving complex'. This is a molecular structure that extracts electrons from water, liberating oxygen gas as a by-product in the process. Its operation can be pictured as a sort of molecular ratchet (see the outside of Fig. 8.2). It can be cycled through five steps, of increasing energy and electric charge separation, with the last state being unstable and spontaneously decaying back to the first, using the stored energy to release a molecule of oxygen. For the intermediate steps, the progression from one state to the next requires absorption of a photon by the light harvesting apparatus (see below) which pulls an electron away from the water-splitting complex. At each step therefore, the net charge separation builds up. After the fourth electron is lost, sufficient energy is stored in the complex that, as it spontaneously decays back to the ground state, it can catalyse the (energy-expensive) oxidation of two water molecules,

---

[‡] Adenosine Tri-Phosphate, ATP is often described as a 'universal energy currency'—all organisms use it. It is made by adding phosphorus to ADP, adenosine di-phosphate. Making ATP requires energy, but this can be released by re-converting it to ADP where needed at different sites in the cell.

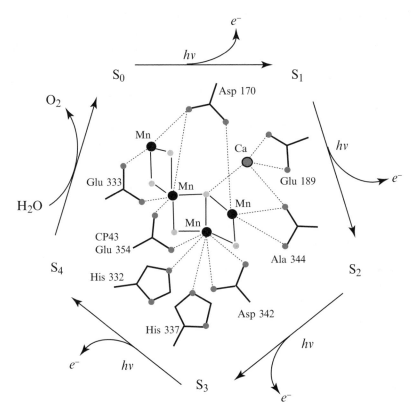

Fig. 8.2 Structure of the water splitting complex (WSC) surrounded by a schematic of the five step cycle it undergoes between states. [Redrawn from Yano *et al.* (7)]

releasing O_2, and repaying the deficit of electrons. The WSC therefore allows accumulation of the energy from four photons, without which the reaction could not be achieved. It also acts as a charge buffer, coupling the single-file electron chain needed by the rest of the photosynthetic process, to a supply reaction that produces electrons in pulses of four.

The WSC is dauntingly complex, and there is still much that is unclear about it, but its basic structure is known (Fig. 8.2, centre). At its heart, surrounded and stabilized by proteins, is a cluster of four atoms of manganese and one of calcium (7). The different possible redox states of the manganese atoms are important to the functioning of the WSC and change as it is ratcheted through its different states, but in detail the mechanism of the ratchet is not settled. What we know about the complex and the way it works has come as a result of decades of effort by some ingenious and hard-working researchers (see e.g. (8) for a review). As an example of the creativity involved and the length of time that this system has been studied, the critical experiment that showed the existence of a cyclical 'electron ratchet' was reported in 1970 (9). Preparations of chloroplasts were excited by very short light flashes, so brief that there was time for only one photon to be absorbed per reaction centre, and the oxygen produced was monitored after each flash. While we might expect that each flash would produce the same output of oxygen, the researchers found otherwise: most oxygen was evolved in a large pulse every fourth flash.

8.2.2 The light harvesting complex (LHC)

The light harvesting complex (LHC) is sometimes also called the 'photosynthetic antenna'. The LHC is a protein network containing chlorophyll pigments. They are arranged so that when a pigment molecule is raised to an excited state by the absorption of a photon, its energy can be transferred from one molecule to another, and is focussed towards particular chlorophyll molecules in a nearby reaction centre. This energy transfer process is essential for efficient photosynthesis because, at the molecular level, the energy in sunlight is quite dilute. Even in full sunlight a given pigment molecule will absorb a photon only rather rarely (a few times per second, glacially slow in molecular terms where times tend to be measured in picoseconds). The LHC enables many pigment molecules to be 'wired' to a single reaction centre, rather than each pigment molecule having a separate reaction centre and electron transport chain, which would then sit idle for most of the time.

When the excitation reaches the pigment molecule at the reaction centre it causes it to lose an electron, creating the charge separation that is stored temporarily in the WSC. As described above, for every fourth electron supplied, the WSC releases one oxygen molecule and 'repays' the electron deficit.

8.2.3 The PS II electron transport chain

The electrons and their associated chemical energy are passed via a number of intermediaries. The energy is used to pump hydrogen ions through a membrane, and the gradient so produced can subsequently be used to generate the energy-carrying molecule ATP. At the end of the chain, the electrons are transferred to photosystem I, where the absorption of a second set of photons initiates a new electron transfer chain. The intermediary compounds of photosystem II (quinones and phaeophytin) are very similar to those found in specific kinds of bacteria which perform kinds of non-oxygen producing photosynthesis (see Box 8.2). These photosystems are clearly evolutionarily closely linked, and are all classified as 'Type II'.

Box 8.2 Type I and II photosystems and some of their anaerobic carriers

Type I photosystems use ferrodoxin as a 'terminal electron acceptor' (receiving the electron ejected from the reaction centre). They can recycle electrons internally, or transfer them in a one-way process from an external donor to an internal acceptor. Carriers of a type I photosystem include green sulphur bacteria (which use, for example, hydrogen sulphide as donor and fix carbon dioxide) and heliobacteria, which though photosynthetic and capable of nitrogen fixation, cannot fix inorganic carbon and need an organic source. Green sulphur bacteria (Chlorobiaceans including the genus *Chlorobium*) tend to be dark green in colour and can be found in the oxygen-free (anaerobic) zones of eutrophic (nutrient-overloaded) lakes and at depth in the anoxic Black Sea. Heliobacteria are found in soils, particularly those of paddy fields, where they are avid fixers of nitrogen.

In type II photosystems, which use quinones as the terminal electron acceptor, electrons are transferred in a one-way chain from an external donor, and used ultimately to fix carbon from an external source. Carriers of a type II photosystem include purple bacteria (proteobacteria, including sulphur and non-sulphur types) and green filamentous (non-sulphur) bacteria (chloroflex-aceans). *Chloroflexus aurantiacus* has an unusual carbon dioxide fixation (3-hydroxyproprionate) pathway. It is a green filamentous (non-sulphur) bacte-rium, which is found in colonies around hot springs. When photosynthesizing the colonies are dark green, but they can also grow in the dark in the presence of oxygen (using aerobic respiration) and then they are dark orange.

Rhodopseudomonas is a purple (non-sulphur) bacterium, found in many marine environments and soils, which has a fantastic diversity of metabolisms; it can fix carbon dioxide or use organic carbon sources in the dark or light, and it can fix nitrogen when photosynthesizing.

8.2.4 Photosystem I

Photosystem I uses the same principle of an antenna structure to capture the light energy, a reaction centre, and a subsequent electron transfer chain, though structurally it is not closely related to photosystem II. There is no water splitting complex, since in the Z scheme the electrons are supplied by photosystem II. The 'terminal electron acceptor' that initially receives the electron liberated from the reaction centre is an iron and sulphur-containing protein called a ferrodoxin (which is much more efficient at capturing free energy than is the quinone used by PS II). From here the electron can be transferred down a path which is used to form NADPH and eventually to fix carbon. Alternatively it can be diverted onto the same transfer pathway as is used by PS II, in which case it generates ATP and thus stores energy for the cell but does not fix carbon. Since that transfer chain ends at the 'base' of PS I, this is a cyclic electron path: the electron can be re-used in PS I. Because there is no requirement for electrons from PS II in the cyclic pathway, there is no net splitting of water or production of oxygen either. When PS I is being used in this way, the overall reaction of photosynthesis is that light is used to generate energy for the cell, with no material input or output required.

Photosystem I also has clear kinship with 'Type I' photosystems in certain anoxygenic bacterial photosynthesizers, quite distinct from those carrying type II photosystems (see Box 8.2). Some anoxic bacteria therefore have type I photosystems, some have type II, but only the oxygen-producing cyanobacteria and their descendants, the chloroplasts in all plants and algae, have both type I and type II.

To summarize, the biochemical machinery of oxygenic photosynthesis has a number of special features that are essential to its function. These are:

(a) two photosystems, each of which is related to systems that can be found separately in different types of anoxic photosynthetic bacteria,
(b) the linking together of these photosystems in such a way that electrons are raised to a sufficient potential to liberate oxygen from water, and
(c) a special catalyst that enables that splitting of water.

All these features had to arise in an ancestral cyanobacterium before oxygenic photosynthesis could occur. Without a special catalyst, water could not be split to provide the electrons, and oxygen would not be liberated. Without the harvesting of sufficiently high energy photons there wouldn't be enough electrical potential to drive the removal of electrons from water. Without the coupling of photosystem II to photosystem I the electrons would not have enough energy to be used in carbon fixation, and there would be nowhere for them to go, hence the process would stop.

8.3 Putting the two photosystems together

The requirement to evolve all three features is already beginning to look like a challenge. Add to it the fact that oxygenic photosynthesis actually involves about 100 carefully ordered proteins and the challenge starts to look monumental. The total number of genes involved in programming the machinery of oxygenic photosynthesis in cyanobacteria is much greater than in any of the anoxygenic photosynthesizers (*10*). Many independent changes to the genome must have been required, making it inherently improbable. In addition, once the machinery for oxygenic photosynthesis had been created, it had to confer an advantage to its carriers to be naturally selected. In fact, each step along the way to evolving it had to provide some selective advantage—or at least, not give its carriers such a damaging disadvantage that they were selected out of the population.

In trying to come up with scenarios for the evolution of oxygenic photosynthesis, much attention has been focussed on the separate existence of photosystems that are like I and II, in other bacterial lineages conducting anoxygenic photosynthesis (see Box 8.2 for some examples). The two different types are found scattered in disparate branches of bacterial lineages (*11*), and there is no clear hierarchy that allow us to say with certainty whether PS I preceded PS II or vice versa, or that either of them independently preceded the cyanobacteria that possess both. Both types of photosystem can make use of a variety of electron donors (drawn from H_2, H_2S, H_2O_2, S^0, SO_3^{2-}, $S_2O_3^{2-}$ and Fe^{2+}), for example hydrogen sulphide (H_2S) is used in both type I and type II photosystems in different bacteria. They also make use of a variety of types of chlorophyll in their light harvesting complexes and often different types at their reaction centres. But no anoxygenic photosynthesizer possesses the water-splitting complex and hence none use water as an electron donor. Furthermore, none of the anoxygenic photosynthetic bacteria possess both photosystems together.

There are two broad possibilities to explain the distribution of photosystems in the bacterial realm: The first is 'fusion', where PSI and PSII evolved in separate organisms, with cyanobacteria eventually acquiring both sets of genes. This could happen thanks to the process called lateral or horizontal gene transfer, which is widespread in the prokaryote realm. It would then require that the photosystems were coupled together with the additional evolution of the WSC. The alternative is 'selective loss' according to which both PS I and PS II evolved in the same cell, which was therefore a 'proto-cyanobacterium'; similar to a modern cyanobacterium but anaerobic as it did not possess the WSC. Subsequently the lineages leading to the types containing a single photosystem lost one or the other, while the cyanobacteria evolved the WSC and became aerobic.

The fact that no anoxygenic photosynthesizer possesses both types of photosystem favours the fusion model, with both types of photosystem evolving separately in different anaerobic bacterial lineages. Both models could have involved lateral gene transfer, and now that the genomes of a variety of photosynthetic bacteria have been sequenced, it is clear that the genes encoding for the photosystems have indeed been exchanged between different lineages of bacteria during evolution (*6*). However, it is not clear which genes were transferred between whom. The fusion hypothesis has two variants: A cell containing photosystem I could have acquired photosystem II (*12*), or a cell containing photosystem II could have acquired photosystem I (*13*).

At face value, the development of both photosystems in a proto-cyanobacterium looks the most complicated explanation and, using Occam's razor, we might cut it out of our considerations. However, some of the most recent evidence supports it—a core set of genes involved in photosynthesis have been identified in a variety of cyanobacteria and anoxygenic photosynthesizing bacteria (*10*). These genes could have arisen in a group of (presumed extinct) anoxygenic proto-cyanobacteria possessing both types of photosystem, and might then have been passed to separate lineages. If such an organism existed, one can ask why it would bother to maintain both photosystems in working order, and a possible answer is that this would have made it more adaptable, able to switch on one or the other depending on the growth conditions. But if a mutation caused the two photosystems to be present in the same membrane this would generally be harmful, because the different types of electron transport (linear and cyclic) would interfere with one another. The situation would be transformed with the addition of a source of electrons from water (the water-splitting complex), because a continuous electric current would then flow (*6*).

So, is there any support for the existence of a proto-cyanobacterium? There is a true cyanobacterium, *Oscillatoria limnetica*, which sometimes switches off one of its photosystems: in the presence of hydrogen sulphide (H_2S) it disables PS II and reverts to using just PS I, thus switching from oxygenic to anoxygenic photosynthesis. Energetically it makes sense for it to do so, since fewer quanta of light are needed for the same amount of carbon fixed, it being easier to extract electrons from hydrogen sulphide (H_2S) than water (H_2O) (see Box 8.1 above). But there is, as yet, no known cyanobacterium that switches off PS I and just uses PS II. Still less is there evidence for an organism that has both PS I and PS II, but no water-splitting ability. However, we have probably only characterized a tiny fraction of the bacteria that actually exist, so the fact that it has not yet been found does not mean it does not exist.

For our argument, it does not particularly matter which of the above scenarios is correct. However, a serious problem is that none of them gives much clue as to how the water-splitting complex evolved: Without the WSC, the advantage of maintaining both types of photosystem seems quite marginal. We now look at what is known about the evolution of this crucial piece of biochemical apparatus.

8.4 The evolution of water splitting

Water has a strong affinity for electrons and does not give them up readily. In comparison, all the substrates used in different forms of anoxygenic photosynthesis are easier to extract electrons from. This means that less energy needs to be captured from sunlight per molecule of carbon dioxide fixed. This raises a fundamental question; how did photosynthesis ever come to use water as an electron donor?

The great advantage of using water is that it is present almost everywhere on the Earth's surface. But if a substrate that can be used in anoxygenic photosynthesis is available, then that process will be energetically favoured (this is well illustrated by *Oscillatoria limnetica* switching off oxygenic photosynthesis and reverting to using hydrogen sulphide as a substrate for photosynthesis when it is available). Thus, in many environments on the early Earth where substrates for anoxygenic photosynthesis were available, oxygenic photosynthesis would have been selected against. This would have added to the difficulty of evolving it. In terms of the energy required, a natural sequence is for photosynthesis to start with hydrogen gas (H_2) as an electron donor, then move to hydrogen sulphide (H_2S), and when these hydrogen species are exhausted to turn to iron (Fe^{2+}, as in $Fe(OH)^+$) (*14*). Only when the iron runs out is there an incentive to extract electrons from water (H_2O).

This has led a number of authors to invoke special environments on the early Earth as suitable for the evolution of oxygenic photosynthesis. Many substrates for anoxygenic photosynthesis are supplied by volcanic or hydrothermal activity, so an environment where these are absent—such as a microbial mat (*15*) in an isolated, sulphur-deficient lake (*16*)—would be suitable. But how plentiful would the alternative substrates be, for example, in the open ocean? The anoxygenic photosynthesizers would themselves have been using them up and potentially drawing them down to limiting concentrations in sunlit surface environments. In Chapter 7 we suggested that recycling loops would have readily emerged and these would have included recycling of substrates for anoxygenic photosynthesis, such as hydrogen sulphide (H_2S). However, there would inevitably be some leaks from the system, and limits to the efficiency of recycling, such that the geologic limits on the supply of substrates would ultimately cap the productivity of the anoxygenic photosynthesising biosphere. Current estimates place its maximum productivity at one tenth that of the modern marine biosphere (*17*). We think this left the environmental door somewhat wider open for oxygenic photosynthesis to evolve. For example, each spring, in the open ocean, after an early bloom phase of anoxygenic photosynthesizers that used up their substrates, a niche would open for any oxygenic photosynthesizers using water instead (assuming that other nutrients were still in sufficient abundance).

So, environmental niches for oxygenic photosynthesis already existed, but there remain two key things that are required to use water as an electron donor; having enough energy from sunlight, at the right potential, to rob the electrons from water, and having a special catalyst to actually split water in the first place. Let us consider the evolution of each feature in turn.

8.4.1 Capturing enough energy from sunlight

To get enough energy at the right potential to split water requires chlorophyll *a* (or a closely related compound) at the reaction centre of photosystem II. In fact there are two chlorophyll *a* molecules at the reaction centre of photosystem II, and another two at the reaction centre of photosystem I (but in a different arrangement). These reaction centres of PS II and PS I are also known as P680 and P700, respectively, because their absorption of red light peaks at different wavelengths, which are given by the numbers 680 and 700 in nanometres. The individual photons of red light absorbed by P680 in PS II have slightly shorter wavelength, meaning they are of higher energy, than those absorbed by P700 in PS I. This ability of photosystem II to harvest higher energy photons was essential to getting enough energy to split water.

Anoxygenic photosynthesizers, in contrast, generally capture lower energy photons (longer wavelengths of light), using a variety of related compounds called 'bacteriochlorophylls' at their reaction centres. This use of longer wavelength light may be a remnant of the evolution of photosynthesis. It has been suggested that photosynthesis evolved from an ability to detect light of particular wavelengths and move towards or away from it—an ability called phototaxis (*18*). This ability is important around hydrothermal vents, which have temperatures of up to 400°C and consequently emit infrared radiation. There, organisms use phototaxis to position themselves relative to the source of heat and light so that they don't get fatally scalded. Intriguingly, the estimated optimum wavelengths for detecting this infrared light are in the ranges of the absorption maxima of bacteriochlorophyll *a* (800–950 nanometres) and bacteriochlorophyll *b* (1000–1150 nanometres). Hence early bacteria could have used precursors of these compounds to detect and position themselves relative to the hot hydrothermal plumes. They might then have begun to use the infrared radiation, hydrogen (H_2) and hydrogen sulphide (H_2S), coming from the vents in a rudimentary form of photosynthesis. If they then found themselves washed into shallower sunlit water they could have made use of the infrared part of the spectrum of sunlight. Today there still exist photosynthetic bacteria adapted to using low-energy photons, including infrared light. The photosynthetic reaction centres of purple bacteria, for example, cover a variety of long wavelengths including P870, P850, and P800. These provide enough energy, and cover the right potential (see Box 8.1), to extract electrons from hydrogen (H_2), hydrogen sulphide (H_2S), or ferrous iron (Fe^{2+}), but not to split water.

So, how did evolution proceed, in a series of steps, to capture higher energy light and provide enough potential to rob electrons from water? Although we don't know for sure, it appears that this is one part of the machinery of oxygenic photosynthesis that is not that difficult to evolve.

A plausible scenario involves the chemical conversion of bacteriochlorophyll *a* into chlorophyll *a* (*19*). There are two main structural differences between bacteriochlorophyll *a* and chlorophyll *a*, one of which is primarily responsible for the shift to absorbing shorter wavelength (higher energy) light. The intermediary (3-Acetyl-chlorophyll *a*) is very like chlorophyll *d*, which has peak absorption at 716 nanometres and is capable of extracting electrons from water. It would be a modest further chemical step to produce chlorophyll *a*. (In fact in today's pathways bacteriochlorophyll *a* is made from chlorophyll *a* and not *vice versa*.) The precise absorption peak of the reaction centre of chlorophyll *a* would then depend on its molecular surroundings and could be fine tuned by natural selection.

8.4.2 The need for transitional electron donors

Although the ability to capture higher energy light may be relatively easy to evolve, it would be of little use to an anoxygenic photosynthesizer. In fact, having the P680 reaction centre, or something similar, would actually be dysfunctional in an organism using electron donors such as H_2, H_2S or Fe^{2+} because its reduction potential is too high and hence it would not be effective at removing electrons from them (*19*). Along with the capturing of higher energy light, evolution needed to proceed in a series of incremental steps from substrates that it is easy to rob electrons from, to water. The need for intermediary steps has led researchers to look for so-called 'transitional electron donors' on the way from H_2, H_2S or Fe^{2+} to H_2O, with intermediate reduction potential and some chemical connection with the water splitting complex. Recall that this special catalyst contains a cluster of four manganese atoms and a calcium atom (Mn_4Ca), which is quite unlike anything seen in anoxygenic photosynthesizers.

There are at least two possible candidates for transitional electron donors; hydrogen peroxide (H_2O_2) and bicarbonate (HCO_3^-):

Hydrogen peroxide (H_2O_2) (*19*) has a standard reduction potential such that it can readily have its electrons removed by the purple bacterial P870 reaction centre, and it has an intriguing chemical link to the Mn_4Ca cluster. Extracting electrons from hydrogen peroxide is similar to the mechanism of catalase enzymes, which catalyse the breakdown of hydrogen peroxide to water and oxygen. Catalases are present in nearly all organisms and one of the defences against oxygen toxicity. What is more, there is a catalase with a reaction centre containing two manganese atoms that is structurally similar to half the Mn_4Ca cluster. This led to the suggestion that an early anoxygenic photosynthetic reaction centre could have been linked up to a Mn catalase as a source of electrons and become the first photosynthesizer to liberate oxygen, using hydrogen peroxide as an electron donor (*19*).

A problem with this scenario is finding a significant source of hydrogen peroxide on the early Earth. A small amount could have been produced by the action of ultraviolet light on water in the atmosphere (5). But hydrogen peroxide is highly reactive stuff so it would be unlikely to accumulate in significant concentrations. This means other electron donors, such as abundant Fe^{2+} would be used in favour of it. Perhaps reaction of iron pyrite (FeS_2) with seawater could have formed significant concentrations of hydrogen peroxide (20). However, the reactivity of hydrogen peroxide (H_2O_2) is such that its conversion to oxygen (O_2) and water (H_2O) actually yields energy, begging the question of why any organism wanting to draw energy from this reaction would require an input of energy from light (i.e. photosynthesis)?

Bicarbonate (HCO_3^-) is formed when carbon dioxide dissolves in (and reacts with) seawater, and would have been abundant on the early Earth (as it is today). A remarkable study (21) has shown that bicarbonate is a more efficient substrate for oxygen production than water at the high carbon dioxide concentrations estimated for the Archean Earth. Furthermore, manganese and bicarbonate form clusters in seawater with two manganese atoms and four bicarbonates, $Mn_2(HCO_3)_4$. These have a reduction potential accessible to anoxygenic photosynthesizers, and are efficient precursors for the assembly of a reaction centre with four manganese atoms that could be an intermediate step towards the Mn_4Ca cluster. A drop in carbon dioxide levels is proposed as a trigger for switching to the water-splitting complex, with chlorophyll a to provide the potential. Fig. 8.3 illustrates the bicarbonate scenario.

Both the hydrogen peroxide and the bicarbonate scenarios for transitional electron donors have been envisaged as a series of three evolutionary steps (19, 21). Both scenarios start with an organism with bacteriochlorophyll *a* and an external donor of electrons, potentially Fe^{2+}. The first step involves the switch to a different donor of electrons—either a Manganese catalase providing them from hydrogen peroxide, or a Manganese bicarbonate cluster ($Mn_2(HCO_3)_4$) providing them from the

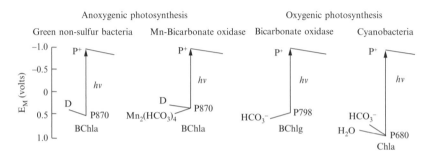

Fig. 8.3 A three step scenario for traditional electron donors involving bicarbonate. [Introduced by Dismukes *et al.* (21), redrawn from original, copyright (2001) National Academy of Sciences, USA]

environment. The second step involves a change in pigment at the reaction centre, either to chlorophyll *d* (generating more potential) or bacteriochlorophyll *g*, and some changes in the electron donation; either a closer association of the catalase, or the close association of a four manganese cluster with bicarbonate as the electron donor. The final step in both scenarios involves switching to chlorophyll *a* as a pigment, the Mn_4Ca catalyst, and water as the electron donor. Whilst both evolutionary scenarios seem broadly plausible, the likely scarcity of hydrogen peroxide in the environment and its high reactivity leads us to favour the bicarbonate scenario (Fig. 8.3).

8.4.3 Self-assembly of the water splitting complex?

Recent work has shown that a cluster of four manganese atoms could have self-assembled from natural manganese oxide (MnO_2) precipitates in the ocean (*22*). The structure of the water-splitting reaction centre (Mn_4Ca cluster) appears to be similar to that of manganese oxide minerals (*7*). Such chemical self-assembly in the ancient ocean would have eased the evolution of the reaction centre. Furthermore, manganese ions of the type present can be readily oxidized by ultraviolet light of wavelengths less than 240 nanometres, causing them to throw off an electron (in much the same way as occurs in the water-splitting complex today).

This has led to a somewhat simpler scenario: Without an ozone layer, ultraviolet light reaching the surface of the early Earth might have provided an environmental source of electrons from manganese (*6*). These could have been fed into a proto-cyanobacterium expressing a type II photosystem. Initially this would have caused a deadly log-jam of electrons but if the organism also expressed photosystem I it would have allowed them to flow, producing the basic wiring of oxygenic photosynthesis. All that was then needed was to switch from an environmental source of electrons to incorporating the manganese atoms (and a calcium atom of unknown function) into a water-splitting reaction centre. Natural selection might then have progressively refined it.

8.5 A candidate for a 'critical step'?

In summary, evolving oxygenic photosynthesis involves bringing together several major components, each of which has a number of potentially difficult steps in its evolution, and the overall scenario must somehow fit together in a way that each stage is selected for (or at least not rapidly selected *against*). Whilst some promising scenarios have been put forward, and nature may have offered a helping hand in assembling the water-splitting complex, we are still far from sure how it was done. Creating the whole looks formidably difficult, and far harder than evolving any of the forms of anoxygenic photosynthesis we see today.

When faced with trying to explain how a complex and beautiful system such as oxygenic photosynthesis arose, the challenge for biologists is to show how it could have happened by plausible evolutionary steps, and we are still a long way from reaching this goal. The complexity of the process, and the fact that there is only evidence for it ever having occurred in one single branch of the bacterial evolutionary tree (or bush, or thicket if you prefer), make us suspect that it might be a truly difficult evolutionary step, not inevitable, or even likely, to happen. However, we need to be cautious, because the complexity of a biological structure is not necessarily a good guide to how likely it was to evolve. We know that complex structures can sometimes evolve with comparative ease, if there is a navigable pathway to them on which each step confers some selective advantage over its predecessor. The most famous example is that of animal eyes, which have evolved dozens of times on slightly different patterns.

But that of course is the point: the proof that it is not difficult for animals to evolve eyes (given time, measured in tens of millions of years), is precisely that it has happened repeatedly. By contrast, oxygenic photosynthesis appears to have evolved just once. This still does not prove that its evolution was so difficult as to count as a 'critical step'—it could be for example, that once the ancestral cyanobacterium had evolved and had become established, it suppressed any tendency for other potential oxygen producers to evolve, by out-competing them before they had time to get the biochemistry right. But while a unique evolution is not a *sufficient* condition to prove that a given advance is critical, it is a *necessary* condition, and oxygenic photosynthesis satisfies it.

There is another line of evidence that can shed some light on the likelihood of oxygenic photosynthesis evolving, and this is concerned with the timing of its evolution. If it is a very difficult step, it would be unlikely to evolve as soon as all the preconditions were right for it—more likely there would be a substantial delay. Recall the critical step model (Chapter 5) in which the most likely spacing of difficult steps is of order a billion years. In the next chapter we will examine the evidence that bears on when oxygen production evolved, to see if it passes this test also.

References

1. D. C. Catling, C. R. Glein, K. J. Zahnle, C. P. McKay, *Why O_2 is Required by Complex Life on Habitable Planets and the Concept of Planetary 'Oxygenation Time'*. Astrobiology **5**, 415 (2005).

2. I. Fridovich, *Oxygen is Toxic! BioScience* **27**, 462 (1977).

3. D. L. Gilbert, in *Oxygen and Living Processes*, D. L. Gilbert, Ed. (Springer-Verlag, New York, 1981), pp. 376–92.

4. J. Castresana, M. Luebben, M. Saraste, D. G. Higgins, *Evolution of cytochrome oxidase, an enzyme older than atmospheric oxygen. European Molecular Biology Organization Journal* **13**, 2516 (1994).

5. C. P. McKay, H. Hartman, *Hydrogen peroxide and the evolution of oxygenic photosynthesis. Origins of Life and Evolution of the Biosphere* **21**, 157 (1991).

6. J. F. Allen, W. Martin, *Out of thin air. Nature* **445**, 610 (2007).

7. J. Yano et al., *Where Water is Oxidized to Dioxygen: Structure of the Photosynthetic Mn$_4$Ca Cluster. Science* **314**, 821 (2006).

8. G. Renger, T. Renger, *Photosystem II: The machinery of photosynthetic water splitting. Photosynthesis Research* **98**, 53 (2008).

9. B. Kok, B. Forbush, M. McGloin, *Cooperation of charges in photosynthetic O$_2$ evolution. Photochem Photobiol* **11**, 457 (1970).

10. A. Y. Mulkidjanian et al., *The cyanobacterial genome core and the origin of photosynthesis. Proceedings of the National Academy of Sciences USA* **103**, 13126 (2006).

11. M. Wheelis, *Principles of modern microbiology.* (Jones and Bartlett publishers, Sudbury, MA 01776, 2007).

12. F. Baymann, M. Brugna, U. Muhlenhoff, W. Nitschke, *Daddy, where did (PS)I come from? Biochimica et Biophysica Acta* **1507**, 291 (2001).

13. J. Xiong, W. M. Fischer, K. Inoue, M. Nakahara, C. E. Bauer, *Molecular Evidence for the Early Evolution of Photosynthesis. Science* **289**, 1724 (2000).

14. J. M. Olson, *Photosynthesis in the Archean Era. Photosynthesis Research* **88**, 109 (2006).

15. D. J. DesMarais, *Microbial mats, stromatolites and the rise of atmospheric oxygen in the Precambrian atmosphere. Global and Planetary Change* **97**, 93 (1991).

16. R. Buick, *The Antiquity of Oxygenic Photosynthesis: Evidence from Stromatolites in Sulphate-Deficient Archaean Lakes. Science* **255**, 74 (1992).

17. D. E. Canfield, M. T. Rosing, C. J. Bjerrum, *Early anaerobic metabolisms. Philosophical Transactions of the Royal Society Biological Sciences* **361**, 1819 (2006).

18. E. G. Nisbet, J. R. Cann, C. L. Van Dover, *Origins of photosynthesis. Nature* **373**, 479 (1995).

19. R. E. Blankenship, H. Hartman, *The origin and evolution of oxygenic photosynthesis. TIBS* **23**, 94 (1998).

20. M. J. Borda, A. R. Elsetinow, M. A. Schoonen, D. R. Strongin, *Pyrite-Induced Hydrogen Peroxide Formation as a Driving Force in the Evolution of Photosynthetic Organisms on an Early Earth. Astrobiology* **1**, 283 (2001).

21. G. C. Dismukes et al., *The origin of atmospheric oxygen on Earth: The innovation of oxygenic photosynthesis. Proceedings of the National Academy of Sciences USA* **98**, 2170 (2001).

22. K. Sauer, V. K. Yachandra, *A possible evolutionary origin for the Mn_4 cluster of the photosynthetic water oxidation complex from natural MnO_2 precipitates in the early ocean. Proceedings of the National Academy of Sciences USA* **99**, 8631 (2002).

9

The trial of the oxygen poisoners

Oxygenic photosynthesis (OP for short) is difficult. It evolved only once on Earth. But when exactly did it occur? If we could tie this down it would provide a clue as to just how difficult a transition it was, and whether it really constitutes an improbable, 'critical' step that could easily never have happened at all.

Recall why the timing is important and what it can tell us about this question: If OP was a likely event, it would happen rather quickly once the basic conditions were right. But if it was a very unlikely event there would probably have been a significant delay, of order of magnitude a billion years according to our critical step model. We know that life was established by, at the latest, 3.5 billion years ago, so if oxygenic photosynthesis also evolved near then, this would count against its being an unlikely step. If it did not evolve until some hundreds of millions of years later, that *would* be consistent with this idea, though it wouldn't constitute proof. However, suggested dates for the origin of OP range from near-simultaneous with the evidence for first life, to the time of the Great Oxidation at 2.3 billion years ago. If the earlier date is correct then there's clearly no support from this quarter for the argument that it is a critical step, whereas dates nearer the end of this span would fit the model rather well. But the huge range is, to put it mildly, not a very promising start. Can we narrow it down?

The OP organisms stand accused of poisoning much of the rest of the biosphere with their by-product, oxygen. The pre-existing organisms had evolved in an oxygen-free world and could not tolerate its toxic side effects—they were (and many still are) 'obligate anaerobes'. Subjected to indiscriminate 'bacteriacide', they lost their place in the sun with the coming of oxygen. So imagine this is a criminal trial, and it is our task to use forensic evidence to say when this crime occurred. In this case, examining the rock record really is like investigating the scene of a long forgotten crime, a cold case where the action took place billions of years ago, and the intervening eons have wiped away most of the evidence. What little evidence is left is summarized on a timeline in Fig. 9.1.

There are several lines of enquiry that can be used to try to establish the presence of oxygen producers. The simplest is the presence of microfossils

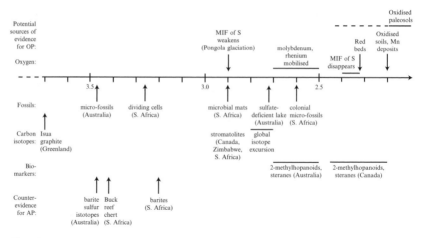

Fig. 9.1 Timeline summarizing the potential sources of evidence for oxygenic photosynthesis (OP) (together with some counter-evidence for anoxygenic photosynthesis (AP)

that look like cyanobacteria, but this is very circumstantial, evidence that any defence counsel should be able to easily discredit: bacteria mostly tend to look alike, even when they are alive, and especially when they have been dead for billions of years. Morphology—their physical structure, as revealed by fossils—is not therefore a firm guide. Two other types of evidence are carbon isotopes—which may show a characteristic fractionation due to OP, and biomarkers—fossil organic compounds that may be diagnostic for the presence of cyanobacteria or oxygen. Finally there is the geochemical evidence for the presence of oxygen itself in the atmosphere, which is the most convincing type of evidence. This only becomes unmistakable at the Great Oxidation itself, at 2.3 billion years ago, which therefore provides a 'latest possible' date. None of the evidence relating to the earlier times is fully convincing, but neither can it be dismissed.

Below we discuss the evidence, and the counter-arguments, for the presence of oxygenesis through this interval. We start with the three sources of evidence below the timeline (Fig. 9.1)—fossils, carbon isotopes and bio-markers—from which we can make a preliminary assessment. Then we look for the smoking gun; evidence of oxygen building up in the atmosphere.

9.1 The Isua formation and carbon isotopes at 3.7 billion years ago

As we saw in Chapter 1, the earliest claims for the presence of life that have withstood critique are those by Minik Rosing (*1*), based on graphite inclusions in greater than 3.7 billion year old rocks from Isua in Western Greenland. These inclusions are depleted in carbon-13 in a way that is consistent with carbon fixation by oxygenic photosynthesis using the enzyme Form-1 Rubisco ($\delta^{13}C$ of -19 ‰ down to -25.6 ‰ in the least altered samples). Form-1 Rubisco occurs in most organisms that perform OP: cyanobacteria and the great majority of eukaryote oxygen-producing photosynthesizers, whereas most bacteria that do anoxygenic photosynthesis (AP) use Form-2 which fractionates somewhat less, or some other pathway. Fractionation by different pathways leading to organic carbon in the rock record is discussed further in Box 9.1. The important point in this context is that, though the Isua graphite does have an isotopic depletion squarely in the range given by OP, it does not constitute unassailable proof that oxygen production was under way by that early date. AP, perhaps with some help from a subsequent process such as fermentation and methanogenesis, might give a similar signature.

Box 9.1 Carbon fixation pathways and $\delta^{13}C$ isotopic fractionation

Fig. 9.2 summarizes some of the metabolic pathways of importance in setting the carbon isotope depletion found in fossil organic carbon. Carbon fixation by photosynthesis, be it OP or AP, results in a depletion of the heavier ^{13}C isotope of carbon relative to the CO_2 that the cell is growing on. Rubisco, first introduced in Chapter 1, is the enzyme usually responsible for fixing carbon from carbon dioxide in photosynthesizers. There are in fact many variations to the Rubisco structure, and several major forms. The Form-1 kind is found in most oxygen evolving organisms, both cyanobacteria and eukaryotes, and gives a maximum $\delta^{13}C$ depletion of -29 ‰ (5). A second type, Form-2 is found in some anoxygenic photosynthesizers such as purple sulphur bacteria, and probably has a lower maximum discrimination of about -18 ‰ (6). Dissolved CO_2 has a thermodynamically-induced depletion of ^{13}C compared to carbonate ions, of about -7 ‰, which may add to the overall depletion of buried organic carbon compared to carbonate deposits.

There are three other carbon fixation pathways found in prokaryotes, (both photosynthetic carbon fixers and chemotrophs that use chemical energy). Two of these produce only small ^{13}C fractionations, but the third, the 'acetyl-CoA' pathway, when used by archaea that produce methane or acetate as a by-product, generates large ^{13}C depletions both in the cells themselves and in these by-products (up to -45 ‰ has been reported; -35 ‰ may be more common (6)). The acetate can be converted to methane, and the methane can be consumed by anaerobic oxidation, fractionating the carbon further before it is buried in sediments. The acetogens and methanogens are obligate anaerobes, killed by oxygen. However, one can imagine early, productive ecosystems, fuelled by AP,

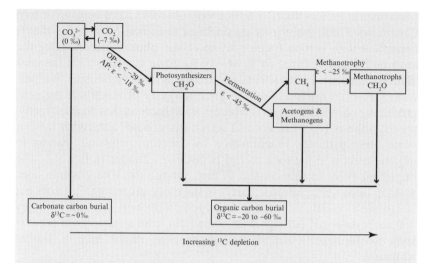

Fig. 9.2 Schematic summarizing some of the metabolic pathways of importance in setting the carbon isotope depletion found in fossil organic carbon, and their different isotopic ractionations (ε)
(OP = oxygenic photosynthesis, AP = anoxygenic photosynthesis.)

where carbon was processed through these fermentors before being buried, at a time when there was as yet no or very little oxygen available. Mixing in a variable percentage of this light carbon could give a wide range of $\delta^{13}C$ to the organic carbon being buried in sediments. (Also, some early ecosystems could have been fuelled by the direct fixation of methane from carbon dioxide and hydrogen, producing very large ^{13}C depletions of up to −95 ‰ (**7**).) Hence, though in recent geological settings, organic ^{13}C depletions of −25 ‰ to −30 ‰ are routinely interpreted as evidence for oxygenic photosynthesis, for early biospheres they are not definitive evidence for OP.

Once OP had evolved, and oxygen was available in surface waters, methane bubbling up from below could be consumed by methanotrophic bacteria that react it with oxygen and incorporate the carbon into their biomass. This can add a further isotopic fractionation to the carbon in methanotroph biomass (of up to about −25 ‰ (**8**)). If the methanotrophs then contribute significantly to the sinking flux of organic carbon, the material buried in sediments can potentially have extremely negative $\delta^{13}C$. Such methane recycling is the explanation given for very large ^{13}C depletions (−60 ‰) observed in some organic carbon from about 2.7 billion years ago (**9**) (Fig. 9.3).

There is a lot of carbon in the Isua graphite, raising a challenge for anoxygenic photosynthesizers to produce so much. However, it has been calculated that AP using molecular hydrogen as a source of electrons could have produced all the reduced carbon deposited, although there would have to have been little or no recycling (*2*).

Undaunted, Minik Rosing and Robert Frei, also at the University of Copenhagen (*3*), note the high abundance of uranium relative to thorium in the Isua sediments and suggest this was due to chemical separation of uranium from thorium in oxygenated water (*3*). However, the reduction potential of uranium does not necessarily require OP rather than AP pathways (*4*), so this too is not definitive. Isua provides good evidence for life on Earth, and persuasive evidence for a productive ecosystem, but it does not prove the presence of oxygen producing organisms.

9.2 Early Archean, 3.5–3.2 billion years ago

A long-standing claim for some of the earliest evidence for OP comes from the Warrawoona Megasequence in the Pilbara region of Western Australia. Bill Schopf of the University of California, Los Angeles, originally made this suggestion on the basis of their physical similarity to modern filament-forming cyanobacteria (*10*). As we have noted, structural similarity is no longer regarded as sufficient proof. Even Schopf concedes that his microfossils overlap in size and appearance with filamentous bacteria that conduct AP (*11*). Martin Brasier of the University of Oxford and colleagues (*12*) questioned whether these structures are even biological, showing that similar patterns can be created by the process of plating the samples for imaging. The debate continues: Schopf and colleagues have used techniques of microscopy and Laser-Raman spectroscopy to show that the structures are made of carbon and are very probably therefore biological (*11*). Current opinion is that they are biological, but not necessarily cyanobacteria.

The South African microfossils presented by Andy Knoll and Elso Barghoorn in 1977 (see Fig. 1.3) are less controversially evidence for life. The original paper reporting these points out their similarity to modern cyanobacteria, but once again they also resemble a host of other, non-photosyntheic organisms. The weakness of arguments based on structural similarity can be extended to the non-living structures, stromatolites, which are present from 3.5 billion years ago and younger. These look like the type of structure made by modern cyanobacteria but could also be formed by other processes. Some have argued that stromatolites older than 3.2 Ga are probably not biologically produced, because none seem to contain microfossils. A recent careful examination of 3.45 billion year old structures from Australia argues strongly that they are biological (*13*), but they could have been built by bacteria conducting AP.

Those sceptical of early OP present a rather different picture of the world around 3.5 billion years ago, where ecosystems are fuelled by anoxygenic photosynthesis. The same Warrawoona Megasequence that hosts Bill Schopf's structures of uncertain affinity contains 3.47 billion year old deposits called barites (barium sulphate) with microscopic inclusions of sulphide. The sulphur isotopic signature of the barites has been interpreted as evidence for an ecosystem based on sulphur reduction and AP pathways, not OP (*14, 15*).

A similar story unfolds from a study of the Buck Reef Chert, a 3.42 billion year old formation in South Africa (*16*) (that we met in Chapter 2). Photosynthetic microbial mats seem to have been present here, in shallow water environments. However, there is no geochemical evidence for oxygen being present, with iron carbonate (siderite) deposits offshore rather than iron oxides. The lack of iron oxidation indicates that iron was not donating electrons for photosynthesis, but there is no evidence of sulphur metabolism either, leaving something of a mystery as to what anoxygenic photosynthetic metabolisms were present (*2*).

Moving forward to 3.2 billion years ago and the Swaziland Sequence of South Africa, there are barite deposits with a sulphur isotope signature consistent with hydrogen sulphide (H_2S) being the electron donor for anoxygenic photosynthesis, producing sulphate. But there is no carbon deposition information to reveal more about potential carbon fixation (*2*).

9.3 Late Archean, 2.9–2.6 billion years ago

The evidence prior to three billion years ago is sparse. But from 2.9 billion years ago onwards, there is a much better sedimentary record including rocks from Canada (Steep Rock, Ontario), Zimbabwe (Mushandike and the Belingwe belt), South Africa (the Kaapvaal craton), and Western Australia (the Pilbara region). The record includes abundant banded iron formations (BIFs) containing oxidized, ferric iron, which used to be interpreted as a sign of oxygenic photosynthesis. However, they could be produced by AP just as readily (*17*). There are also many stromatolites—large carbonate reefs—including ones containing microfossils, but were they made by anoxygenic or oxygenic photosynthesizers? To see if there's convincing evidence for OP, we need to look more deeply into these rocks.

Euan Nisbet and Natalie Grassineau of Royal Holloway, University of London and their colleagues (*18*) have looked in some detail at the isotopic composition of organic carbon and carbonate in stromatolites and shales from Canada and Zimbabwe. They find a carbon isotope difference between carbonate and organic carbon (of −30 to −25 ‰) consistent with oxygenic photosynthesizers. Similar values occur in a few samples from the 2.9 billion year old Pongola Supergroup in South

Africa (*19*). As we have seen above, these are suggestive but not definitive proof of OP.

The Pongola formation is significant for other reasons: it contains the oldest evidence of glaciation on Earth, and tidal sediments, distinct from stromatolites, which show sedimentary structures that are very similar to those produced by microbial mats today. They could have been made by cyanobacteria, as they are today (*20*), but they don't have to have been. Today there are at least two mechanisms of sediment stabilization by bacteria—filamentous bacteria can form meshes that trap and bind sand grains, or bacteria can produce slime that encases sediments—and neither mechanism is unique to cyanobacteria (*21*).

Moving forward in time, Roger Buick of the University of Washington in Seattle presents evidence from 2.72 billion year old lakes containing stromatolites with abundant microfossils from the Pilbara region (*22*). These were evaporitic lakes—ephemeral, with an occasional inflow but no outflow, the water leaving instead by evaporation. They were also deficient in sulphate or iron deposits, which suggests minimal volcanic or hydrothermal sources of reduced sulphur or iron, such as needed for AP. Buick infers therefore that OP was probably the dominant carbon fixation pathway fuelling what was a productive community.

There is another line of evidence that suggests OP may have been present in these lakes. The organic $\delta^{13}C$ carbon isotope data show extremely light values (-40 to -50 ‰) associated with rather light carbonate carbon (0 to -4 ‰). Such a fractionation of organic carbon does not occur during photosynthesis, but it does occur in methanogenesis (see Box 9.1 and Fig. 9.2). The methane produced is isotopically very light, but most probably in order for this signature to end up in the organic carbon, some organisms that capture and consume the methane must have been present (*9*). Such 'methanotrophs' are well-known today; they make a living by reacting methane with oxygen, so their presence would demand the existence of OP (to make the oxygen in the first place). An anaerobic (oxygen-free) pathway of oxidizing methane has recently been recognized which could leave the same carbon isotope signature without demanding oxygen to be around (*23, 24*), but it requires the presence of sulphate, and the lakes in question were sulphate deficient. Again this is suggestive, not definitive.

The unusually negative carbon isotope values in this ancient lake from the Pilbara are part of a wider trend with a shift to particularly negative values in the global organic carbon isotope record around 2.7 billion years ago (Fig. 9.3). There is also abundant organic carbon in rocks from this time. These changes might be explained by the origin of OP providing a more widespread source of organic carbon, some of which was recycled by methanogens and methanotrophs. Jennifer

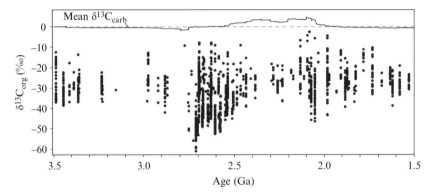

Fig. 9.3 Compilation of carbon isotope fractionation in organic matter from 3.5 to 1.5 billion years ago showing the pronounced negative values (and high variability) 2.7 billion years ago. [Taken from Eigenbrode & Freeman (25), copyright (2006) National Academy of Sciences, USA]

Eigenbrode and Katherine Freeman, then both at Penn State University (25), have taken a careful look at what happens subsequently in the ancient ocean in sediments spanning shallow-water carbonates and deepwater in a 2.72 to 2.57 billion year old section from the Hamersley Province in the Pilbara. They find progressively weaker carbon isotope fractionation over time in surface waters, which they suggest is due to a rise of aerobic respiration in the late Archean. They envisage 'oxygen oases'—patches of oxygenated surface water, perhaps in coastal seas, while the atmosphere and the deep ocean remained nearly oxygen free (25).

There is evidence from morphology too by this time: microfossils in a 2.6 billion years old stromatolite from the Nauga Formation, Campbellrand Subgroup, Prieska, South Africa show precipitation of carbonate and a strong resemblance to modern, colonial, cyanobacteria (26, 27). This is the most convincing likeness to modern cyanobacteria in the record to this point but, to repeat, it is not compelling evidence. The fossils could be the immediate ancestors of cyanobacteria for example.

Over the last decade there has been fascinating work on ancient rocks, applying the rapid advances in laboratory chemical analysis to detect organic compounds in them, as a way of studying the kinds of organisms that were present. Biomarkers—'fossil molecules'—seem to have great potential to teach us about the distant past. However, at the time of writing, their use for the question of dating oxygen production in the Archean has not provided an unambiguous answer (see Box 9.2).

Box 9.2 Biomarkers: disputed evidence for oxygen production

In the last two decades great advances have been made in the application of gas chromatography-mass spectrometry, which enables the characterization of organic compounds in tiny amounts of sample. GC-MS has found application in many fields of scientific detection, including forensic science, and drug testing in athletes for example. In Earth history, much excitement has been generated by its use in the search within ancient rocks for biomarkers—organics that effectively provide a fingerprint for particular groups of life-forms.

Australian researchers Roger Summons, Jochen Brocks and Roger Buick pioneered the use of this technique to look for evidence for the presence of oxygen in the ancient rocks from the Pilbara region, with ages ranging from 2.7 to 2.5 billion years ago (28, 29). They found compounds called 2-methylhopanoids that they argued are a biomarker for cyanobacteria performing oxygenic photosynthesis. Some of the shales studied are interspersed with BIFs containing abundant oxidized iron, consistent with them being produced by a microbial community fuelled by OP.

These organics may be mobile in the rocks, (they are effectively, ancient oils) and a problem with such studies is the possibility that they originated in younger rocks and have subsequently permeated into the more ancient formations (30). In fact, this potential contamination has long been recognized, and in the 1960s there was an initial wave of enthusiasm for chemical analysis of organics in Precambrian rocks which dissipated because of the realization of such difficulties (31). The original authors concede that such contamination could be a problem, though they considered the issue in detail to try to rule it out (32). Sceptics raise a further objection that 2-methylhopanoid biomarkers could conceivably be made by AP (4). In support of this, it has recently been found that an anoxygenic photosynthesizer can make 2-methylhopanoids and therefore they are not unique to cyanobacteria (33). Also, because the biosynthesis of these compounds does not require molecular oxygen, they do not demand the presence of OP (33).

Undaunted, Summons, Brocks and colleagues have pursued another line of biomarker evidence—the hydrocarbons from the Pilbara shales also contain steranes derived from sterols, for which the only known biosynthetic pathways require oxygen. So these biomarkers seem to imply aerobic respiration and, by implication, oxygenic photosynthesis. Sceptics speculate that an anoxygenic pathway for making sterols may exist (4, 34) but have yet to provide evidence for one. Again, the major uncertainty is the possibility that the hydrocarbons are actually much younger migrants into the ancient shale (30).

To address this issue of contamination, Dutkiewicz et al. (35) turn to much harder and somewhat younger, quartz-pebble conglomerate rocks from the Huronian Supergroup of Canada. These date from 2.45 billion years ago and contain hydrocarbon inclusions that are sealed off and include the 2-methylhopa-noid and sterol biomarkers (35). The sealing off occurred at the latest 2.2 billion years ago. The evidence for these biomarkers being contemporaneous with their host rocks is very good, but unfortunately such a late date for the presence of oxygen is not very helpful. By then the Great Oxidation had happened and all are agreed that oxygen producers were present.

9.4 An interim judgment

Having weighed up the evidence based on isotopes, morphology, and biomarkers, let us give a preliminary verdict on the timing of the origin of oxygenic photosynthesis. We are not convinced by evidence of its presence at or before 3 billion years ago. The evidence becomes progressively stronger for OP by 2.7 billion years ago. In particular, the strong depletions in organic $\delta^{13}C$ at that time suggest *something* is afoot in the carbon cycle, and a rise in methane recycling, as a result of the increasing presence of oxygen and increased organic carbon production, might be that something.

This is all very circumstantial, however. There are clearly grounds for reasonable doubt with all the evidence discussed so far, and sceptics argue that the only really convincing evidence for the presence of oxygenic photosynthesis is the appearance of significant concentrations of oxygen in the atmosphere and the surface ocean (*4*). Even low concentrations of oxygen should leave some kind of record, so let us turn now to the evidence for oxygen accumulating in the environment.

9.5 Evidence for the first appearance of oxygen in the atmosphere

The Great Oxidation is marked by the transition, in surface sediments, of redox-sensitive elements such as iron, uranium and manganese from their reduced to their oxidized forms. It is a moment when the surface rocks and soils (and, we presume, the atmosphere in contact with them), becomes stable to the presence of excess oxygen, which henceforth is a permanent presence in the atmosphere. However, it is unlikely that it marks the *first* appearance of oxygen.

Before oxygen producers evolved, the concentrations of oxygen in the atmosphere would have been measured in parts per trillion, representing the occasional generation of a short-lived molecule of O_2 by chemical reactions. However, once OP was present in the shallow waters of a lake or coastal sea we would expect some oxygen to quickly appear in the atmosphere above. We expect that the atmospheric concentration would still be relatively low—at most a couple of parts in a million (*36*), because it would have been rapidly mopped up by reactions with methane and other reduced gases.

The situation for oxygen at that time would have been analogous to that of a highly reducing compound such as methane in the modern atmosphere. Today, methane is continually produced by very active biological sources (wetlands, termites, and cows to name a few—about 500 litres a day from the flatulence of a typical dairy cow for example). In the atmosphere methane is in huge (thermodynamic) disequilibrium with oxygen, and each molecule has a lifetime of a few years only, before being destroyed by chemical reactions. Despite its intense source, this rapid removal keeps its concentration low, around a part in a million.

In the Archean atmosphere, the roles would be reversed. Oxygen would have been the trace gas and methane and other reduced compounds the abundant forms. Leaking into a reducing environment, oxygen would be rapidly removed by reaction with the chemically dominant gases. Being present at only parts-per-million, this oxygen did not produce a general oxidation, because electron-rich (reducing) gases still dominated the atmosphere.

Thus there were potentially two step changes in oxygen:

The first with the origin and spread of oxygenic photosynthesis, from parts per trillion to an intermediate level of around a part per million;

The second being the Great Oxidation with oxygen going from this intermediate level to at least 0.2% of the atmosphere (or 1% of its present level).

It is conceivable that the intermediate stage could have been by-passed (*4*) but if we can find evidence of it, in the form of a 'whiff' of oxygen in the atmosphere, then it would in turn indicate the presence of oxygenic photosynthesis. To detect the signs of such a modest 'whiff' of oxygen in rocks over 2 billion years old is a considerable challenge. However, over the last decade, important new proxies have become available that can help us solve this problem. Fig. 9.1 summarizes this evidence, above the timeline. In the next sections let's look at what we have.

9.6 Iron in paleosols

Historically, iron in ancient rocks formed from soils or loose surface sediments and deposited in contact with the atmosphere, was the first indicator to be used to establish the existence of the Great Oxidation. Paleosols exist from around 2.9 billion years ago, and all those from prior to 2.44 billion years ago show significant loss of iron during weathering. This suggests that atmospheric oxygen was not sufficient to immobilize the iron by oxidizing it from soluble ferrous (Fe^{2+}) to insoluble ferric, (Fe^{3+}) form (*37*). Paleosols from the interval 2.44 to 2.2 billion years ago give ambiguous indications regarding the level of oxygen in the atmosphere (*37, 38*), but all younger paleosols show negligible iron loss, indicating an oxidizing atmosphere containing probably at least 1% of the present level. Indeed, there are strong indications that oxygen had risen significantly by 2.32 billion years ago because rust—oxidized iron—starts to appear in the form of 'red beds' of the mineral hematite (deposited in ancient river deltas) both in rocks from Canada (Huronian Supergroup) and South Africa (upper Transvaal Supergroup). Subsequently, massive (oxidized) manganese deposits in South African rocks laid down (in the ocean) 2.2 billion years ago demand the presence of oxygen (*4*). The disappearance of uranium ore deposits after 2.0 billion years ago, confirms that oxidizing conditions were by that time persistent, causing the more soluble and mobile, oxidized form of uranium to be washed off the land into the ocean.

The lack of iron in paleosols prior to 2.44 billion years ago is not a particularly sensitive indicator of oxygen levels, because it could still occur with up to 500 parts per million of oxygen in the atmosphere (about 0.25% of its present level). To establish the first appearance of oxygen we need to look for more sensitive indicators.

9.7 Mass-independent fractionation of sulphur isotopes

We first introduced the 'MIF of S' signal in Chapter 2 (Fig. 2.2). To recap, sulphur has four stable isotopes, and all reactions of sulphur that occur in solution or in biological systems fractionate them in a predictable sequence—this is 'mass *dependent* fractionation' (*39*). However, reactions initiated by ultraviolet radiation in the atmosphere disrupt this orderly pattern. For this signal to be preserved in rocks requires a route from air to sediment that does not involve going through solution—only possible if an aerosol of solid sulphur is formed in the atmosphere and deposited at the surface. This can only occur in a reducing or neutral atmosphere, because in the presence of excess oxygen the S aerosol reacts rapidly to form soluble sulphates. The disappearance of the MIF-of-S signal has been interpreted as being co-incident with the formation of an ozone layer, at a few parts per million of oxygen. However recent modelling by Kevin Zahnle and colleagues shows it is also dependent on the concentration of methane, with low methane concentrations favouring loss of the signal and high concentrations tending to allow it to persist (*40*).

The MIF-of-S signal is present in 2.47 billion year old sulphides from Western Australian BIFs (consistent with the reduced paleosols from 2.44 billion years ago). But it is absent in 2.32 billion year old sulphides from black shales in the Transvaal, South Africa (consistent with the appearance of oxidized paleosols and red beds at this time). To pin down the disappearance of MIF of S more closely, a recent study examined the 2.45 to 2.22 billion year old Huronian Supergroup from Ontario, Canada, which contains three glaciations (*41*). MIF of S is found after the first glaciation at around 2.4 billion years ago, but it disappears after the second glaciation, by 2.32 billion years ago. This indicates that oxygen did not exceed a few parts per million of the atmosphere until after 2.4 billion years ago, but that the atmosphere was oxidizing to sulphur by 2.32 billion years ago. This has been widely interpreted as the timing of the Great Oxidation.

Such a precise timing for this, the greatest transformation that has taken place in Earth history probably oversimplifies things however. In particular, extreme climatic instabilities would cause fluctuations in the oxygen concentrations. The oldest 'snowball' event on Earth, called the Makganyene and found in South Africa, happens in this interval, with one reconstruction placing it after the three Huronian glaciations and the first red beds around 2.3 billion years ago, but before the massive manganese deposits 2.2 billion years ago (*4*). The ordering of events is debated, with some preferring to align the strata containing the third Huronian glaciation with the Makganyene snowball, making them one and the same event (*41*). It is hard to see how large amounts of oxygenic photosynthesis continued through a snowball Earth, so it is natural to infer a drop in oxygen during it. The Huronian glaciations do not show evidence of extending to low latitudes, but they might have also induced temporary oxygen decline interrupting a general rise in oxygen (*42*).

Although there remain a few who are unconvinced, the existence of a Great Oxidation that started sometime after 2.4 billion years ago and had finished by 2.2 billion years ago, is strongly supported by the available evidence. But are there earlier signs of a 'whiff' of oxygen, or a temporary rise to higher concentrations that died away?

Intriguingly, the MIF of S signal weakens around 2.9 billion years ago, in the Pongola rocks from South Africa (*19*) and in shales from the Pilbara, Australia (*43*). Could these indicate more than a few parts per million of oxygen in the atmosphere at the time, and provide the smoking gun of oxygenic photosynthesis? The original authors certainly interpreted it that way, arguing for more than a few parts per million of oxygen in the atmosphere (*19*) and perhaps even an early oxidation (*43*). However, subsequent work has looked in more detail at multiple sulphur isotopes in this interval (*44*) and shown that the data are quite consistent with a largely oxygen-free atmosphere, with possible changes in transparency of the atmosphere to ultraviolet. The MIF of S signal reasserts itself strongly by 2.7 billion years ago, indicating that whatever the change in atmospheric composition responsible, it was a transient phenomenon.

9.8 Oxygen-sensitive trace elements

As we introduced in Chapter 2, there are yet more sensitive indicators of the presence of oxygen amongst the transition elements with multiple oxidation states having different solubility. Promising candidates are the elements molybdenum and rhenium. These should be among the first to be released by the onset of oxidative weathering on the land surface. Once molybdenum is in water, isotope fractionation can occur. So both the concentration and isotopic fractionation of molybdenum in ancient sedimentary rocks should provide a measure of any oxygen accumulation in the air.

Enhanced molybdenum in sediments from 2.7 billion years, through to 2.5 billion years support a gradual (and probably regional, and variable) rise in oxygen during this period (*45–48*). Enrichment of both molybdenum and rhenium, suggests the onset of oxidative weathering of sulphide minerals on the land surface (*48*). Consistent with this, widespread activation of the oxidative sulphur cycle in the surface ocean is detected (*47*). This suggests that the surface ocean was oxygenated 2.5 billion years ago, and potentially received a supply of sulphate from the land surface. All this can happen at oxygen concentrations of only a part per million in the atmosphere and could thus provide the most sensitive indication of some oxygen in the atmosphere prior to the Great Oxidation. The 'whiff' concentration is quite consistent with our model predictions (*36*) (described in the next chapter) for the situation after the origin of oxygenic photosynthesis but prior to the Great Oxidation.

9.9 Summary

Prior to the Great Oxidation itself, there is no single observation that tells us for sure that oxygen producers had evolved, but there is abundant circumstantial evidence from which we can plausibly reconstruct when it occurred. Key to the interpretation of the evidence is the realization, from models and chemical theory, that a low and probably variable concentration of oxygen, of the order of a part-per-million, should be present once OP had evolved. Given the presence of a greater concentration of reducing gases, this was not sufficient to cause a general oxidation of the environment, but it is detectable by sensitive chemical indicators.

Prior to 3.0 billion years ago there is no hint of such traces of oxygen. Neither is there any fossil evidence that cannot readily be explained by invoking anoxygenic photosynthesis. After 2.9 billion years ago there is some slight evidence, and after 2.7 billion years ago there are increasingly convincing signs of oxygen as a trace gas in the atmosphere, demanding the presence of oxygenic photosynthesis. This is also consistent with the carbon isotope evidence which points to large and transient changes in the carbon cycle at that time, and also with the controversial evidence of OP inferred from biomarkers. We conclude that OP probably evolved after 3.0 billion years ago, but before 2.7 billion years ago. Such a relatively late origin is consistent with the evolution of oxygenic photosynthesis being a critical step.

There was then a few hundred million years delay before the Great Oxidation, sometime in, or spread over, the interval 2.4 to 2.3 billion years ago. This gap between the evolution of oxygenic photosynthesis and the oxidation of the environment indicates that this great revolution of the Earth system was a protracted interval and not a point event. However, the Great Oxidation happened (in geologic terms) rather rapidly. Furthermore, there is no evidence that it has been reversed—it appears to be a one-way or irreversible transition. These observations raise a number of questions: why didn't the planet oxidize immediately? What caused it to oxidize when it did? Why did it oxidize relatively rapidly? And what made the oxidation so difficult to reverse? In the next chapter we try to answer these questions by building up a conceptual model of the processes controlling the oxygen concentration in the atmosphere.

References

1. M. T. Rosing, *¹³C-Depleted Carbon Microparticles in >3700-Ma Sea-Floor Sedimentary Rocks from West Greenland. Science* 283, 674 (1999).

2. J. M. Olson, *Photosynthesis in the Archean Era. Photosynthesis Research* **88**, 109 (2006).

3. M. T. Rosing, R. Frei, *U-rich Archaean sea-floor sediments from Greenland-indications of >3700 Ma oxygenic photosynthesis. Earth and Planetary Science Letters* **217**, 237 (2004).

4. R. E. Kopp, J. L. Kirschvink, I. A. Hilburn, C. Z. Nash, *The Paleoproterozoic snowball Earth: A climate disaster triggered by the evolution of oxygenic photosynthesis.Proceedings of the National Academy of Sciences USA* **102**, 11131 (2005).

5. C. A. Roeske, M. H. O'Leeary, *Carbon Isotope Effects On The Enzyme-Catalyzed Carboxylation Of Ribulose Bisphosphate. Biochemistry* **23**, 6275 (1984).

6. C. H. House, J. W. Schopf, K. O. Stetter, C*arbon isotopic fractionation by Archaeans and other thermophilic prokaryotes. Organic Geochemistry* **34**, 345 (2003).

7. M. J. Whiticar, *Carbon and hydrogen isotope systematics of bacterial formation and oxidation of methane. Chemical Geology* **161**, 291 (1999).

8. L. L. Jahnke, R. E. Summons, J. M. Hope, D. J. Des Marais, *Carbon isotopic fractionation in lipids from methanotrophic bacteria II: The effects of physiology and environmental parameters on the biosynthesis and isotopic signatures of biomarkers.Geochimica et Cosmochimica Acta* **63**, 79 (1999).

9. J. M. Hayes, in *Early Life on Earth*, S. Bengtson, Ed. (Columbia University Press, New York, 1994), pp. 220–36.

10. J. W. Schopf, B. M. Packer, *Early Archean (3.3-Billion to 3.5-Billion-year -old) Microfossils from Warrawoona group, Australia. Science* **237**, 70 (1987).

11. J. W. Schopf, *Fossil evidence of Archaean life. Philosophical Transactions of the Royal Society Biological Sciences* **361**, 869 (2006).

12. M. D. Brasier et al., *Questioning the evidence for Earth's oldest fossils. Nature* **416**, 76 (2002).

13. A. C. Allwood, M. R. Walter, B. S. Kamber, C. P. Marshall, I. W. Burch, *Stromatolite reef from the Early Archaean era of Australia. Nature* **441**, 714 (2006).

14. Y. Shen, R. Buick, D. E. Canfield, *Isotopic evidence for microbial sulphate reduction in the early Archean era. Nature* **410**, 77 (2001).

15. P. Philippot et al., *Early Archaean Microorganisms Preferred Elemental Sulfur, Not Sulfate. Science* **317**, 1534 (2007).

16. M. M. Tice, D. R. Lowe, *Photosynthetic microbial mats in the 3,416-Myr-old ocean. Nature* **431**, 549 (2004).

17. F. Widdel et al., *Ferrous iron oxidation by anoxygenic phototrophic bacteria. Nature* **362**, 834 (1993).

18. E. G. Nisbet et al., *The age of Rubisco: the evolution of oxygenic photosynthesis. Geobiology* **5**, 311 (2007).

19. S. Ono, N. J. Beukes, D. Rumble, M. L. Fogel, *Early evolution of atmospheric oxygen from multiple-sulfur and carbon isotope records of the 2.9 Ga Mozaan Group of the Pongola Supergroup, Southern Africa. South African Journal of Geology* **109**, 97 (2006).

20. N. Noffke, N. J. Beukes, D. Bower, R. M. Hazen, D. J. P. Swift, *An actualistic perspective into Archean worlds—(cyano-)bacterially induced sedimentary structures in the siliciclastic Nhlazatse Section, 2.9 Ga Pongola Supergroup, South Africa. Geobiology* **6**, 5 (2008).

21. M. M. Tice, *Modern life in ancient mats. Nature* **452**, 40 (2008).

22. R. Buick, *The Antiquity of Oxygenic Photosynthesis: Evidence from Stromatolites in Sulphate-Deficient Archaean Lakes. Science* **255**, 74 (1992).

23. A. Boetius et al., *A marine microbial consortium apparently mediating anaerobic oxidation of methane. Nature* **407**, 623 (2000).

24. W. Michaelis et al., *Microbial Reefs in the Black Sea Fueled by Anaerobic Oxidation of Methane. Science* **297**, 1013 (2002).

25. J. L. Eigenbrode, K. H. Freeman, Late *Archean rise of aerobic microbial ecosystems. Proceedings of the National Academy of Sciences USA* **103**, 15759 (2006).

26. W. Altermann, J. W. Schopf, *Microfossils from the Neoarchean Campbell Group, Griqualand West sequence of the Transvaal Supergroup, and their paleoenvironmental and evolutionary implications. Precambrian Research* **75**, 65 (1995).

27. J. Kazmierczak, W. Altermann, *Neoarchean biomineralization by benthic cyanobacteria. Science* **298**, 2351 (2002).

28. J. J. Brocks, G. A. Logan, R. Buick, R. E. Summons, *Archean Molecular Fossils and the Early Rise of Eukaryotes. Science* **285**, 1033 (1999).

29. R. E. Summons, L. L. Jahnke, J. M. Hope, G. A. Logan, *2-Methylhopanoids as biomarkers for cyanobacterial oxygenic photosynthesis. Nature* **400**, 554 (1999).

30. B. Rasmussen, I. R. Fletcher, J. J. Brocks, M. R. Kilburn, *Reassessing the first appearance of eukaryotes and cyanobacteria. Nature* **455**, 1101 (2008).

31. T. C. Hoering, *Criteria for suitable rocks in Precambrian organic geochemistry,* Carnegie Institution of Washington Yearbook **65**, 365–372 (1966).

32. J. J. Brocks, R. Buick, R. E. Summons, G. A. Logan, *A reconstruction of Archean biological diversity based on molecular fossils from the 2.78 to 2.45 billion-year-old Mount Bruce Supergroup, Hammersley Basin, Western Australia. Geochimica et Cosmochimica Acta* **67**, 4321 (2003).

33. S. E. Rashby, A. L. Sessions, R. E. Summons, D. K. Newman, *Biosynthesis of 2-methylbacteriohopanepolyols by an anoxygenic phototroph. Proceedings of the National Academy of Sciences USA* **104**, 15099 (2007).

34. J. Raymond, R. E. Blankenship, *Biosynthetic pathways, gene replacement and the antiquity of life. Geobiology* **2**, 199 (2004).

35. A. Dutkiewicz, H. Volk, S. C. George, J. Ridley, R. Buick, *Biomarkers from Huronian oil-bearing fluid inclusions: An uncontaminated record of life before the Great Oxidation Event. Geology* **34**, 437 (2006).

36. C. Goldblatt, T. M. Lenton, A. J. Watson, *Bistability of atmospheric oxygen and the great oxidation. Nature* **443**, 683 (2006).

37. R. Rye, H. D. Holland, *Paleosols and the evolution of atmospheric oxygen: A critical review. American Journal of Science* **298**, 621 (1998).

38. N. J. Beukes, H. Dorland, J. Gutzmer, M. Nedachi, H. Ohmoto, *Tropical laterites, life on land, and the history of oxygen in the Paleoproterozoic. Geology* **30**, 491 (2002).

39. J. Farquhar, H. Bao, M. Thiemens, *Atmospheric Influence of Earth's Earliest Sulfur Cycle. Science* **289**, 756 (2000).

40. K. J. Zahnle, M. W. Claire, D. C. Catling, *The loss of mass-independent fractionation of sulfur due to a Paleoproterozoic collapse of atmospheric methane. Geobiology* **4**, 271 (2006).

41. D. Papineau, S. J. Mojzsis, A. K. Schmitt, *Multiple sulfur isotopes from Paleoproterozoic Huronian interglacial sediments and the rise of atmospheric oxygen. Earth and Planetary Science Letters* **255**, 188 (2007).

42. M. W. Claire, D. C. Catling, K. J. Zahnle, *Biogeochemical modelling of the rise in atmospheric oxygen. Geobiology* **4**, 239 (2006).

43. H. Ohmoto, Y. Watanabe, H. Ikemi, S. R. Poulson, B. E. Taylor, *Sulphur isotope evidence for an oxic Archean atmosphere. Nature* **442**, 908 (2006).

44. J. Farquhar et al., *Isotopic evidence for Mesoarchaean anoxia and changing atmospheric sulphur chemistry. Nature* **449**, 706 (2007).

45. C. Siebert, J. D. Kramers, T. Meisel, P. Morel, T. F. Nagler, *PGE, Re-Os, and Mo isotope systematics in Archean and early Proterozoic sedimentary systems as proxies for redox conditions of the early Earth. Geochimica et Cosmochimica Acta* **69**, 1787 (2005).

46. M. Wille et al., *Evidence for a gradual rise of oxygen between 2.6 and 2.5 Ga from Mo isotopes and Re-PGE signatures in shales. Geochimica et Cosmochimica Acta* **71**, 2417 (2007).

47. A. J. Kaufman et al., *Late Archean Biospheric Oxygenation and Atmospheric Evolution. Science* **317**, 1900 (2007).

48. A. D. Anbar et al., *A Whiff of Oxygen Before the Great Oxidation Event? Science* **317**, 1903 (2007).

10
The Great Oxidation

We think of the high concentration of oxygen in the atmosphere today as being the natural consequence of the primary driving force of the biosphere, oxygenic photosynthesis (OP). But a few moments reflection shows that this is not quite as obvious as it seems.

Today for example, oxygenic photosynthesis (OP) releases around 300 billion tonnes of oxygen in the process of fixing 100 billion tonnes of carbon annually, and removes a corresponding amount of carbon dioxide from the atmosphere. However, this source must be closely balanced by consumption of oxygen otherwise, over a geologically short time (tens of thousands of years), the O_2 concentration could drastically increase or decrease. The evidence however is that concentrations are stable, so oxygen in the atmosphere is nearly in steady state. Under different circumstances, steady state could equally well be achieved at a much lower concentration of oxygen—attaining balance does not self-evidently require that the O_2 concentration should be as high as it is today.

It is a fair bet that the processes that set the steady-state concentration of oxygen have changed radically over time, and that when oxygenic photosynthesis first arose back in the Archean, the steady state was achieved at a very low concentration—a part per million or so. This would be the 'whiff' of oxygen that appears to be present from around 2.7 billion years ago onwards. Sometime later, at the Great Oxidation, conditions changed so that oxygen molecules lasted longer in the atmosphere, resulting in much higher O_2 concentration, of about 1–10% of the present day. Finally, much later still, conditions changed again and the new steady state rose to values similar to the present day. In this chapter we describe the processes giving rise to the first two, lower steady states, while the rise to more modern values will be discussed in Chapters 13 and 14. The work described in this chapter was inspired by Jim Lovelock's simple model for the Great Oxidation (1). Much of the insight into the problems described comes from working with Colin Goldblatt (2), a former PhD student of ours at UEA who is now at the University of Washington. At the same time, Mark Claire, David Catling (both at the University of Washington) and Kevin Zahnle (at NASA Ames) came to similar conclusions (*3*).

10.1 The fast oxygen cycle in the Archean

After the origin of OP, the net source of oxygen from the biosphere to the atmosphere was potentially large—smaller perhaps than the productivity of the modern biosphere, but probably not orders of magnitude smaller. In today's world, the huge OP source is very largely counter-balanced by aerobic respiration (or aerobic heterotrophy), carried out by oxygen-tolerant organisms, from bacteria to elephants. However, in the late Archean world, after the origin of oxygenic photosynthesis, aerobic respiration will have been much less important because aerobic respiration, even in the most efficient heterotrophic bacteria, only gets under way when oxygen concentration exceeds about a hundredth of the present atmospheric level (this is called the 'Pasteur point'). At lower oxygen concentrations, fermentation becomes more efficient.

The likely pathway by which much of the carbon was recycled was by fermentation, followed by methane (CH_4) formation (Fig. 10.1). This process is dominant today in carbon-rich sediments and ruminant guts, anywhere where there is insufficient oxygen for respiration but abundant organic carbon. Some of the methane may be consumed by methanotrophs—organisms that react methane with oxygen to yield energy—but they only do so when oxygen exceeds around 0.5% of the present level. This is higher than O_2 concentrations were in the Archean atmosphere, but it is possible that such concentrations were reached locally, in the surface ocean or in the upper portions of microbial mats. Still, a large oxygen flux would have evaded the methanotrophs and escaped to the atmosphere, with a corresponding methane flux of about half as many molecules, since that is the ratio in which they are produced from CO_2 and water (see Fig. 10.1).

We don't know the size of the oxygen flux after the origin of oxygenic photosynthesis, but we can put some bounds on it. It must have been greater than organic carbon burial—which the carbon isotope record tells us was of the same order of magnitude as today, around 10 teramoles of oxygen (10^{13} mol) per year. It is likely to have been less than the net productivity of the biosphere today, which is around 10 petamoles per year (10^{16} mol). Potentially it may have been one or perhaps two orders of magnitude less, because of limitation due to lower concentrations of nutrients such as phosphate (*4*). We have lower and upper bounds therefore, which define the fluxes within a few orders of magnitude.

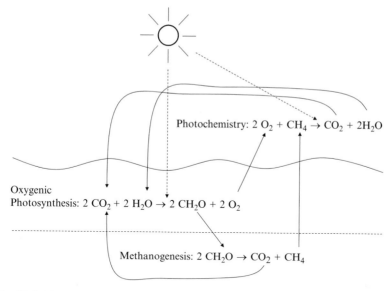

Fig. 10.1 The fast oxygen cycle after the origin of oxygenic photosynthesis, simplified to exclude methanotrophy, which has the same net effect as photochemistry. This is a closed recycling system, driven by the energy from sunlight, which fuels oxygenic photosynthesis, and also catalyses the reaction of oxygen and methane in the atmosphere

10.2 Comparing methane and oxygen in the Archean and modern atmospheres

Though the atmospheric composition was very different in the Archean from today, the modern flux of methane to the atmosphere, and of oxygen associated with it, may be not so different now from then. Today the methane flux to the atmosphere is around 35 teramoles per year (3.5×10^{13} mol yr^{-1})—representing 3–4% of the net production of the biosphere. The molecules are destroyed in the space of a few years in the atmosphere, and methane remains a trace gas, consuming some oxygen but barely denting its concentration. Back then the fluxes might well have been similar, but oxygen was the trace gas and methane the high-concentration constituent of the atmosphere.

Just as today, the methane and oxygen molecules would not have co-existed for long in the atmosphere. However, though a mixture of methane and oxygen represents an extreme thermodynamic disequilibrium, the gases do not actually react spontaneously with one another—methane gas needs a spark to ignite in air, and a catalyst is needed to start the reaction between methane and oxygen in the atmosphere. Both in the present atmosphere and in the Archean, this catalyst comes courtesy of ultraviolet light, which generates free radicals in the air that serve to initiate the attack on the stable molecules.

To work out what concentration of oxygen a given biological source would have supported in the Archean, we need to know how rapidly oxygen and methane would react together. Like many chemical reactions, this one goes faster as the abundance of either reactant is raised. It also increases with the amount of UV radiation penetrating the atmosphere, and that UV in turn depends on atmospheric composition. Most importantly, it depends on whether there is a stratospheric ozone layer or not: if there is not enough oxygen to establish an ozone layer, energetic UV penetrates to the lower atmosphere and speeds the reaction, but once an ozone layer is formed, most of the atmosphere is shielded, and the destruction reaction slows right down.

In detail, calculating the reaction rates is a complex problem requiring a model of the chemistry of the Archean atmosphere. Alexander Pavlov, James Kasting and colleagues at Penn State University made detailed studies of the chemistry of such an atmosphere in the early 2000s (5–7). Colin Goldblatt was able to use their results to derive a fairly simple function for the reaction rate, as a function of the methane and oxygen concentrations (Fig. 10.2 (a) and (b)). He was able to work out how much oxygen would be in the atmosphere for a given production rate and concentration of methane. For a wide range of conditions that would correspond to the late Archean, the oxygen is between a part per billion and a part per million of the atmosphere (i.e. a 'whiff'). The balance point increases depending on the size of the oxygen source from the biosphere, which we know only within wide limits. There is no ozone layer at these oxygen levels, so the destruction is particularly rapid and the oxygen remains a trace gas.

However, a fundamental transition can occur as the rate of production of oxygen by the biota is increased. The concentration of oxygen climbs steadily until it reaches the point where an ozone layer begins to form, typically between 1 and 10 parts per million (corresponding to around 10^{15} mol in Fig. 10.2b). Then the rate of reaction of atmospheric oxygen and methane begins to fall dramatically, further increasing the concentration of the trace gas oxygen, in a runaway positive feedback. The oxygen concentration then shoots up by several orders of magnitude, until it again reaches steady state, with its destruction rate balancing its production. In this simple model at least, the establishment of an ozone layer is the trigger that transforms the atmosphere from the late Archean, with parts per million or less of oxygen, to the early Proterozoic, with a few tenths of a percent. This instability of oxygen marks the 'Great Oxidation'.

(a)

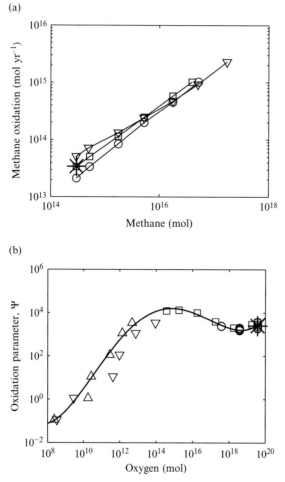

(b)

Fig. 10.2 The rate of methane oxidation as a function of (a) atmospheric methane content and (b) atmospheric oxygen content, corrected for the methane concentration (2)

10.3 The overall redox balance in the Archean

We tend to think that the inevitable result of OP must be the oxidation of the environment, but looking in more detail at the late Archean, it becomes clear that the biological cycle at that time, fuelled by OP, had the capacity to do precisely the opposite—to *reduce* the atmosphere. The net effect of the cycle is the removal of carbon dioxide from the atmosphere and its transformation, via reaction with water, into oxygen and methane. Both these gases, one of which is a strong oxidant while the other is a strong reductant, were released back into the atmosphere. For the most part once there, they reacted with one another to reform carbon dioxide, producing a closed cycle (Fig. 10.1). However, over time the net effect on the atmosphere would depend on which of them was removed more slowly by reactions with the surface and sediments. If more methane than oxygen were removed, the oxygen left behind would begin to oxidize the atmosphere. On the other hand, if more oxygen than methane were removed, the remaining mixture would become more reducing.

Which of these actually occurred? We can be reasonably sure that in the late Archean these slow-cycle reactions would remove oxygen more quickly than methane. There was a substantial amount of reduced material in sediments and dissolved in the ocean that would react readily with oxygen. We see some of this surviving in Archean rocks—the abundant banded iron formations (BIFs) of the late Archean for example (8), whose oxidation from the ferrous to ferric states represents the removal of large quantities of oxygen from the system (Fig. 10.3) (9). This would leave an excess of methane to build up in the atmosphere. Paradoxically, for several hundred million years the net effect of the invention of oxygenic photosynthesis would have been to make the atmosphere more reducing, by increasing CH_4 concentrations, while oxygen remained a trace gas with a concentration measured in parts per million at the most.

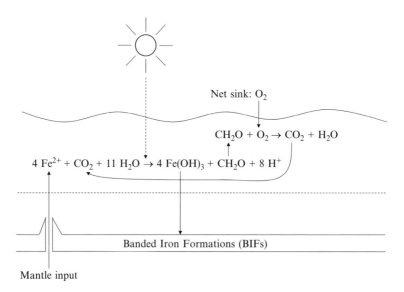

Net sink: O_2

$$CH_2O + O_2 \rightarrow CO_2 + H_2O$$

$$4\ Fe^{2+} + CO_2 + 11\ H_2O \rightarrow 4\ Fe(OH)_3 + CH_2O + 8\ H^+$$

Banded Iron Formations (BIFs)

Mantle input

Fig. 10.3 An indirect pathway for oxygen removal in the late Archean. Here it is assumed that anoxygenic photosynthesizers living at the bottom of the sunlit surface ocean were the likely oxidizers of reduced iron coming in from hydrothermal vents (9). They transferred electrons from iron to organic matter, which here is subject to aerobic respiration, consuming oxygen. Many other pathways of banded iron formation have been proposed, but in all cases the input of reduced iron has the potential to consume oxygen

The ultimate source of the reduced material that initially mopped up free oxygen was the Earth's interior. The upper mantle from which material is delivered to the surface at mid-ocean ridges, was then (and still is) more reducing than today's surface environment, with abundant ferrous iron for example (*10*), so it is no surprise that it would act as a sink for oxygen. The difficulty is rather to explain how the surface has become more oxidized since that time. To a first approximation (see Box 10.1) we might think of the 'exogenous Earth system' of continental-crust-plus-atmosphere-plus-ocean-and-sediments as a closed box, in which case it would not be possible to change the composition of the system as a whole, though there is much shuffling of electrons between carbon, sulphur and iron within it. More realistically, it is not closed, but exchanges material with the mantle. It could become more oxidized if the material being exported back into the mantle in subduction zones is consistently more reducing than that coming up at mid-ocean ridges. This is certainly a possibility but the opposite seems more likely—that the net exchange with the mantle has consistently tended to reduce the surface: it is difficult to establish either way at present (*11*). Finally, it could be that there has been loss of reduced material to space—hydrogen escape.

Box 10.1 The crustal redox reservoirs: carbon, sulphur and iron

Let us look at the processes internal to the Earth system which can control the redox state of the surface. There are three elements that have large near-surface reservoirs and which can exist in either a more oxidized, or more reduced state (Table 10.1) (*12–15*). They are carbon, iron and sulphur. Shuffling any of these from its more oxidized to its more reduced state liberates 'oxidizing power' which results in the rest of the environment becoming more oxidized. ('Oxidizing power' is just an informal way of describing a greater ability to act as an electron acceptor. Chemists measure it as a positive redox potential or higher E_h—all these are equivalent ways of saying the same thing.) For instance, burying carbon in its reduced form as organic material, where the carbon came from oxidized carbonates, releases some oxygen, and the oxygen so released might end up in the atmosphere. The same is true for shuttling sulphur atoms between sulphates (oxidized) and sulphides (reduced), and iron from ferric (oxidized) to ferrous (reduced) forms. The oxidation potential is transferred to the environment, even if no free oxygen is released by the initial reaction. Conversely, even if oxygen is released, if the environment has an excess of reduced material in sediments or dissolved in the oceans, these will quickly take it up again. For free oxygen to accumulate, these sinks must be satiated so that the uptake reactions slow or cease.

How much oxygen could these chemical shuttles potentially release? The answer is: a lot. The amount of reduced organic carbon in sediments and the crust is sufficient to account for 30 times the amount of oxygen in the atmosphere. All that oxidizing power has been released over time by making organic carbon from carbon dioxide, which in turn is derived from carbonate. Evidently however, this release of oxidized material did not all go into the accumulation of oxygen in the atmosphere. Most of the oxidative power liberated by burying this carbon must instead have gone to oxidizing other components—iron and sulphur—of the sedimentary system. To a first approximation then, if we imagine the surface system to be closed, the carbon, sulphur and iron shuttles must be in balance, because the reservoirs are so big, and shuttle so much redox power, that the atmosphere and oceans are too small to take up much of it.

We can write down chemical equations that express this balance. Two of the fathers of modern geochemistry, Bob Garrels and Abe Lerman, wrote down such an equation in a landmark paper of 1981 (*13*). Though they were unaware of it, they were actually rediscovering an approach that had first been suggested almost 150 years earlier by Jean-Jacques Ebelman, a Professor at the Ecole des Mines in France whose pioneering ideas on geochemistry have only recently been rehabilitated (*16*). The Ebelman-Garrels-Lerman reaction shows that you can make 15 molecules of organic carbon (CH_2O in geochemists parlance) from carbonate, without changing the redox state of the ocean-atmosphere, by oxidizing eight atoms of sulphur to sulphate, plus four iron atoms:

$$4\ FeS_2 + 8\ CaCO_3 + 7\ MgCO_3 + 7\ SiO_2 + 15\ H_2O$$
$$\rightarrow 15\ CH_2O + 8\ CaSO_4 + 2Fe_2O_3 + 7\ MgSiO_3$$

Garrels and Lerman made the assumption that the system of crust plus sediments, ocean and atmosphere was a closed reacting vessel, with no inputs or outputs of mass to it. We now know that it is not really closed. There is hydrogen escape from the top of the atmosphere which we discuss in the text, but there is also exchange between the mantle and the crust of the Earth, with material rising at the mid-ocean ridges, and some entrainment into the mantle of the rocks which sink under the subduction zones. We can be more quantitative today about these exchanges (*17*), which alter the picture, but not to the point of rewriting the whole story.

We know a good deal about how carbon and sulphur have shuttled between their different reservoirs, because we are able to use the stable isotopes ^{13}C and [34]S to track changes in the reservoir sizes over time (see Box 2.1). These show that through much of Earth history, changes in the carbon and sulphur reservoirs have tended to compensate for one another. Iron is missing from this story, but potentially has very large reservoirs: it also has stable isotopes which are fractionated by redox processes, and carry useful information (*18*). The evidence suggests that today, only comparatively small redox fluxes are attributable to the iron reservoirs compared to carbon and sulphur, but that iron was much more important in the Archean, as we would expect from the crucial role the reduced iron reservoir played at that time.

Table 10.1 Reservoirs of oxidized and reduced material in Earth's atmosphere, ocean, sediments and continental crust, expressed in terms of the number of moles of oxygen needed to produce or consume them.

Species	Reservoir	Size (10^{21} mol O_2 equivalent)	Reference
Oxidized species			
O_2	Atmosphere and ocean	0.037	(12)
SO_4^{2-}	Ocean	0.074	(13)
$CaSO_4$ (gypsum)	Sediments	0.19	(13)
Fe^{3+}	Sediments (oceanic)	0.025	(14)
Fe^{3+}	Continental crust	1.7 to 2.6	(14)
Carbonate	Continental crust	2.2 to 7.1	(12)
Reduced species			
Reduced carbon	Atmosphere, ocean, sediments	0.56	(15)
FeS_2 (pyrite)	Sediments	0.67	(13)
Fe^{2+}	Continental crust	2.9 to 4.6	(14)
Reduced Carbon	Continental crust	<0.78	(15)

It has long been known that hydrogen loss to space causes a steady and permanent oxidation of the planet beneath (19). Even at the present rate of hydrogen escape, twice the current inventory of oxygen in the atmosphere would have been produced over the past 4.5 billion years. However, we need to account for much more oxygen than this to explain the current oxidation state of the Earth's crust (Table 10.1). The answer may be that prior to the Great Oxidation, in an atmosphere potentially rich in methane (and other reduced gases), there was a much greater rate of hydrogen loss to space (12, 19) (Fig. 10.4). The reason is that methane, CH_4, carries hydrogen high into the atmosphere where the molecule is split by UV radiation, and the hydrogen atoms are available to escape. Today, methane is only present at a part per million or so, and hydrogen escape is very slow. Water vapour, which is today the only abundant atmospheric gas containing hydrogen, does not increase hydrogen escape, since it cannot penetrate into the upper atmosphere, but is frozen out by the atmospheric 'cold trap' effect.

In detail, it requires a little chemical thought to convert hydrogen loss to oxygen gain, but the result is clear (Fig. 10.4). Hydrogen escape leads to an irreversible oxidation of the planet beneath. By the estimates of Catling *et al.*(12), the oxygen source from H escape would lie in the range 7 to 140% of today's organic carbon burial flux (0.7-14 × 10^{12} mol yr^{-1}). Crucially, it is a permanent source, whereas organic carbon

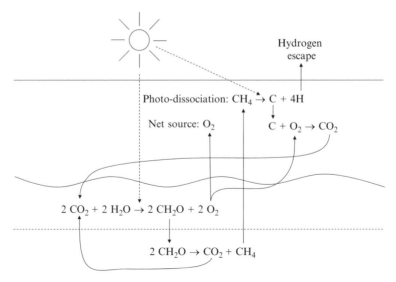

Fig. 10.4 Hydrogen escape to space as a net source of oxygen for the Earth system. Oxygenic photosynthesis followed by methanogenesis liberates oxygen and methane molecules to the atmosphere in a 2:1 ratio (Fig. 10.1). If a methane molecule reaches the upper atmosphere it is split apart by high energy sunlight (photo-dissociation), and the hydrogen atoms it contained are lost to space. The carbon left behind reacts with a molecule of oxygen, but crucially, a corresponding molecule of oxygen is left behind, representing a net source to the Earth system. The overall reaction is: $2H_2O \rightarrow O_2 + 4H(\uparrow space)$, everything else being recycled. [As outlined by Catling *et al.* (*12*)]

burial is ultimately reversed when the reduced carbon is returned to the surface system and oxidized. Taking the upper end estimate, it could produce the total amount of oxidized material in the crust and surface system in around 200 million years, which would fit reasonably well with the estimated time delay between the origin of oxygenic photosynthesis and the Great Oxidation. If methane concentrations were this high, in the range of a tenth of a percent of the atmosphere, hydrogen escape would explain the relatively oxidized state of the continental crust since the late Archean. If concentrations were lower, then we would have to invoke some net oxidation by exchange with the mantle, to explain the present state.

10.4 A model for the Great Oxidation

We have now considered the key processes for the long-term oxygen balance in the late Archean. Led by Colin Goldblatt we put these together in a simple model (*2*), and its predictions of atmospheric oxygen concentrations are shown in Fig. 10.5.

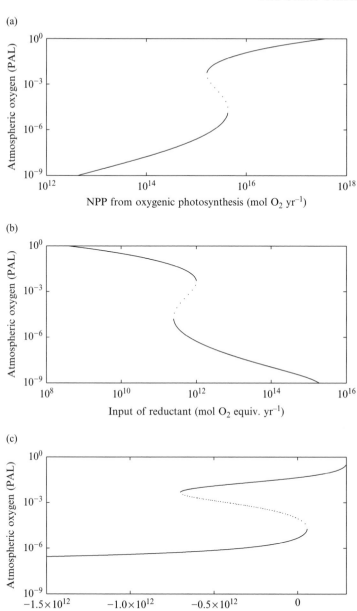

Fig. 10.5 Bi-stability of atmospheric oxygen (in fractions of Present Atmospheric Level, PAL). Predicted by our simple model, old Goldblatt et al. (2), as a function of: (a) net primary productivity (NPP) from oxygenic photosynthesis, (b) reductant input to the surface Earth system, and (c) net change in the size of the crustal organic carbon reservoir. The solid lines are stable steady states and the dotted lines are unstable steady states. [Kindly provided by Colin Goldblatt of the University of Washington (2).]

The results depict the 'Goldblatt instability'—the sudden rise of oxygen from parts-per-million before an ozone layer is formed, to tenths of a percent afterwards. In the model there are a number of parameters—'dials' that we can turn up or down—which will bring on this change, of which we've graphed oxygen against three: it can occur if the net rate of photosynthesis is increased, if the net input of reducing material from the mantle is decreased, or if there is net accumulation of organic carbon in the crust, beyond critical values. A sustained increase in the rate of hydrogen escape to space can also bring about the shift. The model has several features that seem realistic:

1) Low, stable concentrations of oxygen in the late Archean. For reasonable assumptions (somewhat less OP than today, and more net input of reduced material from the mantle) stable concentrations of O_2 between a part per billion and a part per million result. For a wide range of parameters, the oxygen and methane concentrations are 'well behaved'. Oxygen rises smoothly if OP is increased for instance, but remains a trace gas. This low oxygen steady state can neatly account for the data from 2.7 to 2.4 billion years ago. It can be maintained indefinitely, if the overall redox state of the surface is not changing, with reductive input from the mantle being balanced by hydrogen escape to space. The concentration of oxygen predicted is always less than the few parts per million of oxygen in the atmosphere that would remove the MIF of S signal, yet it is sufficient to explain the mobilization of extremely redox-sensitive elements such as molybdenum and rhenium. It also allows for some trend of increasing oxygen concentration over time, which is predicted to occur if oxygenic photosynthesis is spreading to new habitats, or if there is a progressive decline in the input of reduced material.

2) Rapid transition to a high oxygen state if the oxygen passes a critical threshold at a few parts per million in the atmosphere. As described above (Section 10.2), the transition is triggered by the formation of a stratospheric ozone layer, which shields the lower atmosphere from UV and causes the O_2-methane destruction rate to sharply decrease. The O_2 concentration then rises rapidly, and transits to a new stable steady state at around 1% of the atmosphere, the change taking about one hundred thousand years (near-instantaneous in geological terms). This high oxygen state is consistent with the data from 2.2 billion years ago onwards—it is sufficient to explain the disappearance of MIF of S, the appearance of red beds, oxidized paleosols, and manganese deposits. As a caveat, we should note however, that the level of oxygen in this new steady state depends not just on atmospheric chemistry but on how biology is depicted in the model. Specifically, it depends on the sensitivity of aerobic respiration and methanotrophy to oxygen. Both processes are assumed to speed up as oxygen rises, acting as additional negative feedbacks and contributing to stabilizing

oxygen well below the present concentration. Without them we get an unrealistically high steady state for oxygen (well above the present level). If this model is correct, the response of the biosphere to increasing oxygen was therefore important in creating the intermediate oxygen levels of the Proterozoic.

3) 'Bi-stability' of the oxygen states. The high oxygen stable state has a remarkable property. If we turn down a control dial, for example by suppressing oxygenic photosynthesis, the Great Oxidation is not reversed at the same point it occurred. Instead, the high oxygen state and the ozone layer can maintain themselves until a second, lower threshold is reached. This is an example of bi-stability—two stable states for atmospheric oxygen co-exist for a range of values of the control dials. In simulations where we do force the high oxygen state to collapse, the decline of oxygen is much slower than the rise was, taking several million years, in contrast to the hundred thousand years or so for the rise to occur. For these reasons, once the system switches to the high oxygen stable state, it is relatively robust to perturbations and not likely to quickly reverse. This could help explain why, since the Great Oxidation, the Earth system seems never to have gone back to a low oxygen state: it was not impossible to go back, but it would have required a sustained reduction in OP, well below the level that initially brought on the oxidation, to bring it about.

10.5 What caused the transition?

If we are correct, the Great Oxidation occurred when an ozone layer was formed as a result of the trace gas oxygen reaching a concentration of a few parts per million in the atmosphere. But why did it rise to these levels, some hundreds of millions of years after OP was first invented? Which of the possible control dials was being turned, and why? Four possibilities will work to flip the system: an increase in global net productivity from OP, a decline in the input of reduced material coming in from the mantle, a sustained increase of organic carbon burial, or an increase in hydrogen escape.

10.5.1 The spread of oxygenic photosynthesis

Whilst it is natural to assume that global primary productivity might have progressively increased as early forms of oxygenic photosynthesis were refined by natural selection and early cyanobacteria spread into new habitats, there is no direct evidence either way. Recall that although oxygenic photosynthesizers are dominant in today's world, when they first arose, they could have been out-competed by anoxygenic photosynthesizers in many habitats where there were sufficient supplies of alternative electron donors.

10.5.2 Reductant input from the mantle

A decline in reductant input was suggested as the trigger for the oxidation by Jim Lovelock (*1*). The Earth's internal heat source, being radioactive, decays over time, leaving less energy to drive volcanic, metamorphic and plate tectonic activity at the surface. So we should expect a gradual downward trend in the input of material from the mantle. On today's Earth, the input of reduced material at hydrothermal vents and in volcanic gases is not enough to allow a low oxygen stable state in our model—only high oxygen is stable. For the early Earth, a relevant source of data is the amount of iron leaving the surface system as banded iron formations (BIFs), which provides a lower limit on how much reduced iron was coming in at hydrothermal vents. A single BIF from the Hamersley Basin (of Australia) deposited 2.69 to 2.44 billion years ago represents a huge input of iron (*20*), sufficient to put our model in the bi-stable regime for atmospheric oxygen. Together with other deposits we estimate an input of reduced matter 2.5 billion years ago that puts the system in the low oxygen stable state but close to the bi-stable regime. The record of BIFs is rather sporadic, but after 2.5 billion years ago their abundance starts to decline (*8, 21*). If this reflects a decline in the input of reduced iron from the mantle then it would have tended to shift the Earth toward the bi-stable regime for atmospheric oxygen.

A variant on this theme is a proposed shift in the location of volcanoes in the late Archean (*22*). This relates to an important geologic change that was possibly occurring at the time—the growth of the first large continental masses. Though the time history of the growth of the continents is by no means certain, a case can be made that there was comparatively little continental crust before 3.0 billion years ago, and that major continent building first occurred in a burst of continental growth in the late Archean. Recent analyses of zircons suggests for example that major continental building first peaked at about 2.8 billion years ago (*23*) (although the opposite case has been argued on the basis of Niobium/Uranium ratios in ancient and modern basalts (*24*)). Assuming the late Archean was a time of accumulation of continental crust, there would have been a shift in volcanism from submarine to 'subaerial'—from underwater to the land. In today's world, volcanoes on land spew out a mix of gases that are less reduced than those underwater (because of the corresponding higher temperature of the melt). Consequently a shift in volcanoes to the land surface could lead to a decline in the input of reduced material.

Certainly the co-incidence of the timing of the continent-building tectonism with the oxidation is suggestive, though a potential problem is that the period of pronounced continental growth seems to end by about 2.4 billion years ago (*23*), and so somewhat predates the Great Oxidation.

10.5.3 Net accumulation of organic carbon

It has been argued that there may have been a net accumulation of organic carbon in the crust, on the grounds that, with the formation of the first continental land masses in the late Archean, shallow shelf seas would provide a location for organic carbon burial (*25*). However, any sustained increase in organic carbon burial ought to show up in the carbon isotope record of carbonates as a positive excursion (the carbonates are left isotopically heavier because of the preferential incorporation of light carbon into organic matter). We have already noted that there is no secular trend in carbonate carbon isotopes over Earth history (recall Fig. 1.5) and there is no convincing sign of such a change prior to the Great Oxidation (instead there is one afterwards).

10.5.4 Hydrogen escape

As discussed above (Section 10.3) the loss of hydrogen from the planet is a net oxidation that could bring about the transition to a high-oxygen atmosphere, if it were sufficiently rapid and sustained for sufficiently long. Catling *et al.* (*12*) suggest that most of the hydrogen loss to space went to oxidizing the crust and the ocean-atmosphere-sediment system, which as we have noted, is substantially more oxidized than the mantle. If it occurred, H escape did not apparently involve oxidation of the upper mantle, because evidence suggests that this has not changed much over the history of the Earth (*26*). This in turn requires that oxidized material is efficiently recycled back to the crust in subduction zones, and not transported into the mantle.

The H escape hypothesis is attractive, but is there any geochemical record that can corroborate or falsify the mechanism of increased hydrogen escape to space? Perhaps: Catling *et al.* predict that hydrogen would be lost preferentially relative to its heavier isotope, deuterium, hence the ocean should have become enriched in deuterium over time, and this could explain observations of Archean granites with lighter hydrogen isotope composition. Deuterium enrichment is frequently a characteristic of hydrogen escape, and such an enrichment (arising from a different cause it's true) was crucial in proving that Venus once lost substantial hydrogen by escape (*27*), so it would be worth searching for evidence for a changing D-H ratio on Earth.

10.6 Summary of the biological and geochemical changes

To summarize, we have described a simple model which can explain why, after the origin of OP, the planet did not oxidize immediately and there was a delay of hundreds of millions of years. We can understand why, once the Great Oxidation happened it was relatively rapid, and why it was apparently difficult to reverse. The model suggests that the Great Oxidation could have been triggered by a relatively small perturbation once the system had been shifted into a vulnerable (bi-stable) regime. It allows for a number of forcing mechanisms that could have shifted the system into this regime, but it cannot adjudicate between them. Indeed there is insufficient data to pin down precisely which mechanism or mechanisms were responsible. We can probably rule out a systematic, sustained increase in organic carbon burial as a cause, but we can find some support for a decline in input of reduced material from the mantle. We agree with Catling *et al.* (*12*) that hydrogen loss to space was very likely an important factor. A progressive increase in the net primary productivity of organisms performing OP could also have contributed.

10.7 Corresponding climate changes

The large changes in atmospheric composition might be expected to have effects on the climate, and we now consider these in more detail. We have already noted that the Great Oxidation happens around the time of the second of three glaciations recorded in the Huronian rocks of Canada. A snowball Earth glaciation occurs soon after (in the interval 2.3–2.2 billion years ago), and is either the same event as the third Huronian glaciation or follows all of them. The earlier Pongola glaciation (around 2.9 billion years ago) also corresponds with disruption of the MIF of S signal. So there appears to be some correlation between changes in atmospheric composition and changes in climate. Before discussing what caused the planet to cool however, we first have to work out why it was not already frozen, bringing us back to the 'faint young sun' problem.

10.7.1 The late Archean faint young sun

In the period from 3 to 2.4 billion years ago, the solar luminosity was probably in the range 80–85% of the modern value, sufficiently faint to lead to permanent global glaciation under the modern atmosphere. The traditional solution to this problem (which held sway from the 1970s to the 1990s) called on an elevated level of atmospheric carbon dioxide to provide a stronger atmospheric greenhouse. Model calculations for

2.5 billion years ago suggest around 400 times the present level of carbon dioxide would have been required to maintain liquid water at the Earth's surface (*28, 29*). The silicate-weathering negative feedback (introduced in Chapter 7) is the basic mechanism that has been assumed to be responsible for the higher CO_2: a cooler planet has a slower rate of chemical weathering of silicate minerals, reducing the long-term sink for atmospheric CO_2, which therefore builds up in the atmosphere. This weathering occurs on the continental surface, so if the area of the Archean continents was less than today, this would also act to reduce weathering and increase CO_2. Furthermore, it is very likely that there were few organisms on the land surface to amplify weathering as happens in the modern world. That too would contribute to carbon dioxide build up.

However, a problem with the enhanced CO_2 solution emerged in 1995. Constraints can be put on the carbon dioxide concentration by considering the minerals present in 'paleosols', which are the remains of soils or weathered rocks, reworked in contact with the atmosphere. Rye *et al.* (*30*) used this approach on 2.2 billion year old paleosols from the Hekpoort basalt formation in South Africa. They noted that iron, lost from the surface, was re-precipitated further down as iron silicate, but not as iron carbonate as would have occurred if the CO_2 concentration was sufficiently high. They deduced that this could only occur if the ambient CO_2 concentration was less than 100 times the present level. Subsequent work has thrown some doubt on this calculation, but an alternative approach suggests a similar (or possibly even lower) upper bound on CO_2 concentrations (*31*). Paleosols of ages 2.5 and 2.0 billion years give similar results (*31*).

The paleosol evidence has led to a resurgence of interest in other potential warming agents for the early Earth. Jim Lovelock (*1*) popularized the idea that methane was important (*32*) and Alex Pavlov, Jim Kasting and colleagues have championed it recently. In an influential paper they calculated a strong greenhouse warming effect of up to 30 °C due to elevated concentrations of atmospheric methane estimated for the late Archean (*33*). This made it seem likely that methane was a key greenhouse gas on the early Earth, and a reduction of the methane greenhouse was responsible for the Huronian cool periods, including the first snowball Earth.

The Pavlov *et al.* result (*33*) led to a bevy of papers invoking weakening of a methane greenhouse as a cause for various cold spells in Earth history. However, it has recently transpired that the original calculations were seriously in error, and methane is not nearly as effective a warming agent as suggested (*34*). Although molecule per molecule, at their present concentrations, methane is 20 or 30 times more effective than carbon dioxide as a greenhouse gas, this is because methane is scarce in today's atmosphere relative to carbon dioxide (roughly 1.7 versus 380 parts per million). If we shift to an absolute scale of radiative

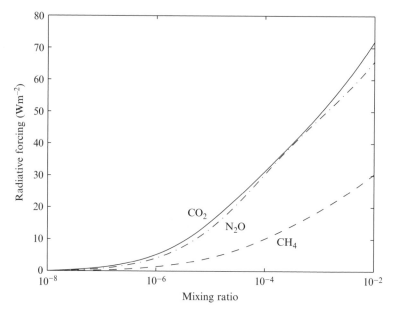

Fig. 10.6 The radiative forcing due to carbon dioxide (solid line), methane (dashed line) and nitrous oxide (dash-dot line) as a function of mixing ratio (the fraction of the atmosphere they comprise) Radiative forcing is the change in net heat flux entering the well-mixed lower atmosphere relative to 10^{-8} (1 part in 100 million) of each gas, calculated by a line-by-line model. [Results and figure kindly provided by Colin Goldblatt of the University of Washington.]

forcing versus gas concentration, calculated line by absorption line, then at the same concentration, a given change in carbon dioxide concentration is more effective than the same change in methane concentration (Fig. 10.6).

Consequently, it now appears that elevated methane concentrations are not alone sufficient to solve the faint young sun problem, and a drop in methane at the time when oxygen first appears in the atmosphere would not have had such a dramatic cooling effect. There are further problems with this mechanism as a cause for the glaciations at this time. Firstly, it is not immediately obvious that methane concentration should permanently decrease simultaneously with the oxidation: our model for example shows a transient dip, but no permanent decrease in CH_4 with the switch from low to high oxygen. Only with certain systematic changes, such as a decline in reductant input, does the methane concentration drop and stay permanently lower. Secondly, the timing seems wrong for the glaciations to be entirely caused by the oxidation, because there is an MIF of S signal *after* the first Huronian glaciation, so apparently at that point no oxidation had occurred.

It is fair to say therefore that, after a period of nearly 30 years in which there seemed to be workable solutions for the faint young sun, invoking first increased CO_2, and later increased methane, there currently is no generally accepted scenario to explain why the Earth was not frozen at this time. All the potential solutions seem to have problems. Recently, researchers have been reaching out for more exotic solutions to the faint young sun, as follows:

Hydrocarbons and haze: Jim Kasting and his student Jacob Haqq-Misra have recently focused attention on ethane and other hydrocarbons that would have been photochemically produced from methane and which could contribute strongly to warming (*34*). However, once the ratio of methane to carbon dioxide exceeds one, a thick organic haze will form in the atmosphere, which acts to cool the planet by reflecting the sun's rays back to space (*35*). The formation of such a haze has been invoked to explain the early Pongola glaciation and the weakening of the MIF of S signal at that time (*36*). However, this explanation does not work well for at least the first Huronian glaciation because the MIF of S signal is present before and after it.

More nitrogen? Colin Goldblatt and colleagues (including ourselves) have recently pointed out that the total pressure in the atmosphere might have been larger in the Archean atmosphere if some of the nitrogen now locked up in crust and mantle were in the atmosphere (*37*). Though nitrogen is not itself a greenhouse gas, higher atmospheric pressure enhances the greenhouse by broadening the range of wavelengths of individual absorption lines, for all greenhouse gases, an effect known as pressure broadening. An inventory of the likely amounts of nitrogen now stored in the Earth, does suggest that there is sufficient N_2 on the planet that there could have been two or three times as much in the atmosphere in the Archean. If the pressure of nitrogen were this much greater, the more efficient greenhouse would have been sufficient to keep the planet from freezing with relatively small amounts of CO_2 and methane in the atmosphere, which don't violate the geochemical constraints just discussed. Needless to say this requires a particular degassing history for nitrogen and by implication other volatiles, in which the proportions at the surface have been decreasing steadily over much of the history of the planet.

To summarize: to explain why the Earth remained unfrozen in the late Archean, it is still necessary to invoke an atmospheric greenhouse dependent on CO_2 or methane, or both. A less cloudy early Earth would help warm things up, but we have no data constraints on ancient cloudiness. It seems plausible that the greenhouse was enhanced by higher total atmospheric pressure, or by the presence of minor atmospheric constituents such as ethane, or perhaps carbonyl sulphide (*38*). Only by such mechanisms are we able to reconcile our present understanding with geochemical proxies indicating relatively low concentrations of CO_2.

10.7.2 A scenario for the late Archean-early Proterozoic

If CO_2 and methane were important constituents of the Archean green-house, a decrease in either of these gases remains the strongest contender to bring on the end-Archean glaciations. Methane has recently been a favourite culprit for this role, but in detail no simple explanation seems quite to fit. The oxidation could be brought on by a number of different factors, but a decrease in the input of reductant, occurring in a world in which oxygenic photosynthesis had long been established, is one of them.

So, let us consider a possible scenario (Fig. 10.7). Oxygenic photo-synthesis becomes established in the mid-Archean let's say 2.9 billion years ago, and with it the first whiff of oxygen appears. The increase in net primary productivity enhances organic carbon burial and carbon dioxide removal from the atmosphere, at least transiently, causing cooling. However, the increased organic carbon supply to deep waters and sediments fuels more methane production by methanogens, causing warming. The overall result would have been a lowering of the carbon content of the atmosphere, and a switch in its composition to a mixture of carbon dioxide and methane. We suspect this resulted in overall cooling, especially if the ratio of methane to carbon dioxide approached one

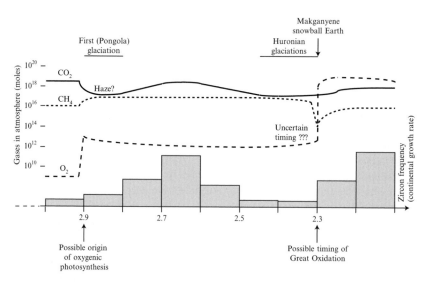

Fig. 10.7 A possible scenario for changes in climate and atmospheric composition through the late Archean and early Proterozoic. Grey boxes show the frequency of zircons—a proxy for continental growth rate—from the data of Rino et al.(23). Schematic variations in carbon dioxide concentration are based on the assumed origin of OP and the continental growth data, and are kept at or below the constraints from paleosols (31). Approximate oxygen and methane concentrations are based on results of our simple model (2)

and an organic haze started to form in the atmosphere (*39*). Thus, the Pongola glaciation might indicate the first proliferation of oxygenic photosynthesis.

The resulting system is like the 'bacteriaworld' of heaters and coolers that you created in Chapter 7. It might be expected to regulate like Daisyworld except that the warming process (methanogenesis) depends on the cooling process (oxygenic photosynthesis and carbon drawdown). Instead of two competing feedbacks, it may have boiled down to a single negative feedback on temperature (*39*): Whilst methanogenesis and therefore methane concentration increases with temperature, temperature is a peaked function of methane concentration. Initially, increasing methane increases temperature, but once the ratio of methane to carbon dioxide approaches one and an organic haze starts to form, further increasing methane lowers temperature. This will generate a negative feedback on the growth of the methanogens, stabilizing the system. It may help explain why glaciations did not recur from about 2.9 to 2.4 billion years ago. But what drove the system cooler and started the Huronian glaciations?

We think it could have been a geologically-driven decline in carbon dioxide (*40*). Critical for the carbon dioxide balance would have been the pattern of continental growth. As discussed above (Section 10.5.2), recent analyses of zircons suggest that this occurred in phases, with a major growth phase peaking at about 2.8–2.6 billion years ago (*23*) (Fig. 10.7). The continental growth phase would have increased the sink of carbon dioxide by providing a greater area for silicate weathering, but it was probably accompanied by a stronger source of carbon dioxide linked to greater tectonic activity. It appears that this early continental growth phase had settled down by 2.5 billion years ago and continental growth did not resume in earnest until after 2.3 billion years ago. This period of quiescence suggests reduced tectonic activity and a weaker source of carbon dioxide. With large continents present and being weathered, a drop in carbon dioxide concentration and planetary temperature to a lower steady state is expected in the interval 2.5 to 2.3 billion years ago. Potentially this could have contributed to causing the interval of Huronian glaciations and the Makganyene snowball Earth. The resumption of continental growth and associated carbon dioxide source after 2.3 billion years ago might then help explain why the glaciations stopped. The reduction in mantle-to-crust input might also help to explain why the Great Oxidation occurred at that time, while the bi-stable and difficult-to-reverse nature of the oxidation might be the reason why it did not reverse when continent-building resumed.

This is a very tentative scenario and not one we have tested out in great detail. We are not discounting other mechanisms contributing to cooling around the time of the Great Oxidation, although potential transient drops in methane concentrations, or intervals of haze formation don't seem to fit the available data. At present the data are simply too scarce to better constrain the problem. However, the key point is that the major transition in life and atmospheric composition also involved major climate changes, even if we do not fully understand how they are connected.

In the aftermath of the Great Oxidation, the availability of many essential elements for life would have been altered. In the ocean, the dominant form of nitrogen would have become nitrate instead of ammonia. In the process, much nitrogen would have been denitrified from the ocean to the atmosphere, providing a significant source of laughing gas—nitrous oxide (N_2O) (*41*). N_2O is a highly effective greenhouse gas (Fig. 10.6), which may help explain why glaciations stopped and did not return for over 1.5 billion years after the Great Oxidation.

10.8 Generalizing the model

Now that we have built a case that oxygenic photosynthesis was difficult to evolve, and we have a working model to explain the apparent delay between its origin and the Great Oxidation, let us take a step back and consider whether it could have taken a lot longer or a lot less time to oxidize the Earth, and what might be expected to happen on other terrestrial planets (including those around other stars).

If we are right about oxygenic photosynthesis being one of a series of three or four critical steps then, by definition, we only see it happening when it did on Earth because all the steps had to occur for us to be here as observers. By implication, even if some form of prokaryotic life were relatively common in the universe (i.e. not that difficult to evolve and not in itself involving a critical step), oxygenic photosynthesis could still be very rare. In contrast, it looks like methanogenesis is deeply rooted in the tree of life on Earth and hence relatively easy to evolve. If so, and if oxygenic photosynthesis beats the odds and evolves on a planet, it is likely that methanogenesis is already present and will return carbon as methane to the atmosphere. This in turn will promote a higher rate of hydrogen escape to space and tend to gradually oxidize the planet beneath.

However, the rate of planetary oxidation by hydrogen escape will depend on a number of competing factors that are sensitive to the size of a planet. On a smaller planet (such as Mars) there is a tendency for the whole atmosphere, not just hydrogen, to be lost, because of the lower escape velocity from a weaker gravity. Thus the surface of the planet may be oxidized, but there is little scope for oxygen to build up to the partial pressure required to support complex life. On a somewhat larger planet than Earth there will be more buoyancy to support hydrogen escape, which is limited by diffusion, and a Great Oxidation could occur somewhat sooner after the origin of oxygenic photosynthesis. In contrast, on a much larger planet, hydrogen loss to space becomes energy limited. With stronger gravity, the hydrogen escape rate would be lower, leading to a longer delay between the origin of oxygenic photosynthesis and the Great Oxidation. Thus, producing an oxygen rich atmosphere won't happen on planets that are too small as they lose their atmosphere altogether, and it won't happen on planets that are too large as they will not oxidize within their lifetime.

This adds an extra element of difficulty to the revolution we have been describing in this and the last two chapters. The upshot is that out there in the cosmos, there may be very few planets with oxygen-rich atmospheres, and thus very few that can support intelligent, observer life forms.

10.9 Life after the Great Oxidation

The Great Oxidation transformed the Earth. Suddenly there was enough oxygen for aerobic respiration to become the dominant way of breaking down organic matter in surface environments. This made much more energy available to organisms, and meant that cells could get larger and become more complex and active. The formation of an ozone layer protected those living beneath from harmful ultraviolet radiation, potentially easing the colonization of the land surface. Life no longer required the elaborate sun defences it must have developed in the Archean. But as well as these good sides, toxic oxygen was now everywhere in surface environments. A different set of defences to cope with reactive oxygen species had to be evolved in all organisms that lived in the oxygenated surface environment or tried to use oxygen in metabolism. The rise of oxygen created a stratified biosphere, with organisms that couldn't cope with its toxic effects—the obligate anaerobes—at the bottom, banished to sediments, and anoxic waters of the deep ocean. The main beneficiaries of the Great Oxidation are classically thought to have been the eukaryotes. However, the jury is still out on whether eukaryotes evolved in aerobic micro-environments before the Great Oxidation or long after it. In Part 4 we turn our attention to the rise of eukaryotes and their environmental consequences.

References

1. J. E. Lovelock, *The Ages of Gaia – A Biography of Our Living Earth.* L. Thomas, Ed., The Commonwealth Fund Book Program (W. W. Norton & Co., New York, 1988).

2. C. Goldblatt, T. M. Lenton, A. J. Watson, *Bistability of atmospheric oxygen and the great oxidation. Nature* **443**, 683 (2006).

3. M. W. Claire, D. C. Catling, K. J. Zahnle, *Biogeochemical modelling of the rise in atmospheric oxygen. Geobiology* **4**, 239 (2006).

4. C. J. Bjerrum, D. E. Canfield, *Ocean productivity before about 1.9 Gyr ago limited by phosphorus adsorption onto iron oxides. Nature* **417**, 159 (2002).

5. A. A. Pavlov, L. L. Brown, J. F. Kasting, *UV shielding of NH₃ and O₂ by organic hazes in the Archean atmosphere. Journal of Geophysical Research-Planets* **106**, 23267 (2001).

6. A. A. Pavlov, M. T. Hurtgen, J. F. Kasting, M. A. Arthur, *Methane-rich Proterozoic atmosphere? Geology* **31**, 87 (2003).

7. A. A. Pavlov, J. F. Kasting, *Mass-independent fractionation of sulfur isotopes in Archean sediments: Strong evidence for an anoxic Archean atmosphere. Astrobiology* **2**, 27 (2002).

8. C. Klein, *Some Precambrian banded iron-formations (BIFs) from around the world: Their age, geologic setting, mineralogy, metamorphism, geochemistry, and origin. American Mineralogist* **90**, 1473 (2005).

9. A. Kappler, C. Pasquero, K. O. Konhauser, D. K. Newman, *Deposition of banded iron formations by anoxygenic photrophic Fe(II)-oxidizing bacteria. Geology* **33**, 865 (2005).

10. D. J. Frost, C. A. McCammon, *The redox state of Earth's mantle. Annual Review of Earth and Planetary Sciences* **36**, 389 (2008).

11. J. M. Hayes, J. R. Waldbauer, *The carbon cycle and associated redox processes through time. Philosophical Transactions of the Royal Society B* **361**, 931 (2006).

12. D. C. Catling, C. P. McKay, K. J. Zahnle, *Biogenic Methane, Hydrogen Escape, and the Irreversible Oxidation of Early Earth. Science* **293**, 839 (2001).

13. R. M. Garrels, A. Lerman, *Phanerozoic cycles of sedimentary carbon and sulfur. Proceedings of the National Academy of Sciences USA* **78**, 4652 (1981).

14. C. Lecuyer, Y. Ricard, *Long-term fluxes and budget of ferric iron: implications for the redox states of the Earth's mantle and atmosphere. Earth and Planetary Science Letters* **165**, 197 (1999).

15. K. H. Wedepohl, *The composition of the continental crust. Geochimica et Cosmochimica Acta* **59**, 1217 (1995).

16. J. J. Ebelmen, *Sur les produits de la decomposition des especes minerales de la famile des silicates. Annu. Rev. Mines 12*: **12**, 627 (1845).

17. J. M. Hayes, J. R. Waldbauer, *The carbon cycle and associated redox processes through time. Philosophical Transactions of the Royal Society B-Biological Sciences* **361**, 931 (2006).

18. C. M. Johnson, B. L. Beard, E. E. Roden, *The iron isotope fingerprints of redox and biogeochemical cycling in the modern and ancient Earth. Annual Review of Earth and Planetary Sciences* **36**, 457 (2008).

19. D. M. Hunten, T. M. Donahue, *Hydrogen loss from the terrestrial planets. Annual Reviews of Earth and Planetary Sciences* **4**, 265 (1976).

20. H. D. Holland, *The oxygenation of the atmosphere and oceans. Philosophical Transactions of the Royal Society B-Biological Sciences* **361**, 903 (2006).

21. A. E. Isley, D. H. Abbott, *Plume-related mafic volcanism and the deposition of banded iron formation. Journal of Geophysical Research* **104**, 15461 (1999).

22. L. R. Kump, M. E. Barley, *Increased subaerial volcanism and the rise of atmospheric oxygen 2.5 billion years ago. Nature* **448**, 1033 (2007).

23. S. Rino *et al. Major episodic increases of continental crustal growth determined form zircon ages of river sands; implications for mantle overturns in the Early Precambrian. Physics of the Earth and Planetary Interiors* **146**, 369 (2004).

24. I. H. Campbell, *Constraints on continental growth models from Nb/U ratios in the 3.5 Ga barberton and other Archaean basalt-komatiite suites. American Journal of Science* **303**, 319 (2003).

25. D. J. DesMarais, H. Strauss, R. E. Summons, J. M. Hayes, *Carbon isotope evidence for the stepwise oxidation of the Proterozoic environment. Nature* **359**, 605 (1992).

26. D. Canil, *Vanadium partitioning between orthopyroxene, spinel and silicate melt and the redox states of mantle source regions for primary magmas. Geochimica et Cosmochimica Acta* **63**, 557 (1999).

27. T. M. Donahue, J. H. Hoffman, R. R. Hodges, A. J. Watson, *Venus was wet—A measurement of the ratio of Deuterium to Hydrogen. Science* **216**, 630 (1982).

28. J. F. Kasting, *Theoretical constraints on oxygen and carbon dioxide concentrations in the Precambrian atmosphere. Precambrian Research* **34**, 205 (1987).

29. J. F. Kasting, *Earth's Early Atmosphere. Science* **259**, 920 (1993).

30. R. Rye, P. H. Kuo, H. D. Holland, *Atmospheric carbon dioxide concentrations before 2.2 billion years ago. Nature* **378**, 603 (1995).

31. N. D. Sheldon, *Precambrian paleosols and atmospheric CO_2 levels. Precambrian Research* **147**, 148 (2006).

32. J. T. Kiehl, R. E. Dickinson, *A study of the radiative effects of enhanced atmospheric CO_2 and CH_4 on early earth surface temperatures. Journal of Geophysical Research (Atmospheres)* **92**, 2991 (1987).

33. A. A. Pavlov, J. F. Kasting, L. L. Brown, K. A. Rages, R. Freedman, *Greenhouse warming by CH_4 in the atmosphere of early Earth. Journal of Geophysical Research* **105**, 11981 (2000).

34. J. D. Haqq-Misra, S. D. Domagal-Goldman, P. J. Kasting, J. F. Kasting, *A revised, hazy methane greenhouse for the Archean Earth. Astrobiology* **8**, 1127 (2008).

35. K. J. Zahnle, *Photochemistry of Methane and the Formation of Hydrocyanic Acid (HCN) in the Earth's Early Atmosphere. Journal of Geophysical Research (Atmospheres)* **91**, 2819 (1986).

36. J. F. Kasting, S. Ono, *Palaeoclimates: the first two billion years. Philosophical Transactions of the Royal Society B* **361**, 917 (2006).

37. C. Goldblatt *et al. Nitrogen-enhanced greenhouse warming on early Earth. Nature Geoscience* **2**, 891 (2009).

38. Y. Ueno *et al. Geological sulfur isotopes indicate elevated OCS in the Archean atmosphere, solving faint young sun paradox. Proceedings of the National Academy of Sciences USA* **106**, 14784 (2009).

39. S. D. Domagal-Goldman, J. F. Kasting, D. T. Johnston, J. Farquhar, *Organic haze, glaciations and multiple sulfur isotopes in the Mid-Archean Era. Earth and Planetary Science Letters* **269**, 29 (2008).

40. C. Goldblatt, A. J. Watson, T. M. Lenton, paper presented at the Astronomical Society of the Pacific Conference Series (Bioastronomy 2007 meeting), 2007.

41. R. Buick, *Did the Proterozoic 'Canfield Ocean' cause a laughing gas greenhouse? Geobiology* **5**, 97 (2007).

PART IV
THE COMPLEXITY REVOLUTION

11
Life gets an upgrade

Now we have reached halfway through the history of the planet. It is 2.3 billion years ago, early in the Proterozoic Eon. There is oxygen in the atmosphere, albeit somewhat less than today. The puzzle is; why does it take another one and a half billion years or so to produce large, complex life in the form of animals? The sudden appearance of complex forms in the lowest Cambrian was a puzzle that Darwin saw as one of the principal objections to his theory of evolution. He correctly guessed that the Precambrian must actually have covered a vast amount of time and that 'during these vast periods the world swarmed with living creatures', though he could offer no explanation as to why they left no fossils (1). We now know that the majority of older fossils are microscopic and unicellular, and for this reason had not been recognized. However, despite now having a much richer (though admittedly still patchy) fossil record of those distant times, the earliest complex forms have only been pushed back a little, into the latest part of the Proterozoic, around 600 million years ago. Clearly a long and cryptic road had to be travelled to evolve complex life, and there is still a yawning chasm to bridge.

The intervening long spell of stability, especially from 2 to 1 billion years ago, has been christened the 'boring billion'. Evolution appears to have proceeded at a sedate pace, whilst the oceans and atmosphere were locked in a halfway house state, which may have impeded life. Or at least that is how it appears, given a rather sparse record of those distant times. It may be that the full biological and chemical richness of that distant world remains hidden. From what we already know, 'boring' does not really do it justice. The 'stable' or 'slow' billion would be a better description of the Proterozoic Earth system. What we can say with some confidence is that when things eventually changed, they did so radically. Complex life appeared in a time of great turbulence for the planet as a whole—known as the Neoproterozoic—that included changing oxygen levels and extreme ice ages.

In this part of the book we examine this complexity revolution of the Earth and what led up to it. We start in this chapter by considering the biological aspects. In particular, how did eukaryotes evolve from prokaryotes? And what are the key features of eukaryotes that would later allow them to produce large, complex, differentiated, life forms? In the next chapter we assess; when did eukaryotes evolve? Then in Chapter 13 we turn to the atmosphere and especially the ocean, to see what evidence there is that the Earth system got locked in an unproductive state. We look at the mechanisms that could have held oxygen levels down, and how the unusual ocean of the time may have inhibited life. Finally in Chapter 14 we take a whole system view of the Neoproterozoic, and explore some hypotheses for how different, advanced eukaryotic life forms could have been causes and consequences of the radical global environmental changes at that time.

11.1 Eukaryotes from prokaryotes

Eukaryote cells are so different from prokaryote cells that it used to be thought by some biologists that they had independent origins—in other words, life had evolved twice on Earth (or nearby). The advent of molecular biology put paid to that idea, showing unequivocally that all life on Earth has a common ancestor. However, how eukaryotes evolved from prokaryotes remains one of the great puzzles of biology. Pieces of the jigsaw are in place but the placing of large parts is argued over, and much of the picture remains empty. In a book published when he was 96, Ernst Mayr, one of the greatest figures of 20th century evolutionary biology, is unequivocal about the importance of the development of the eukaryotes. This event, he says, was the greatest transformation in the history of life (2). The differences traditionally used to distinguish eukaryotes from prokaryotes are listed by Mayr, and we summarize them in Table 11.1. They are massive, and fundamental.

The evolution of eukaryotes from prokaryotes is clearly a candidate for a critical step—an event so improbable that it might not be expected to occur during the lifetime of a planet (even one that had got as far as evolving prokaryotes). The evolution of one from the other must have proceeded by a chain of small steps, but we are very far from understanding the detail of how it occurred. Each step must have brought the resulting organism an evolutionary advantage (or at least no significant disadvantage), but we can only guess at what most of these steps are, because the intervening billions of years of evolution have separated the eukaryotes so far from the prokaryotes that few traces are left of the transitional organisms.

However, it is worth noting at the outset that, one by one, the essential differences that mark out eukaryotes from prokaryotes are turning out to be rather less fundamental than we used to think. In some cases, the classical definition turns out to be wrong—prokaryotes *do* have

Table 11.1 The 'classical' differences between prokaryotes and eukaryotes

Property	Prokaryotes	Eukaryotes
Cell size	Small, ~1–10 μm	Large, usually 10–100 μm
Nucleus	Absent	Present
Endoplasmic membranes	Absent	Endoplasmic reticulum, golgi apparatus present
DNA	Single circular chromosome, not complexed with proteins.	Multiple linear chromosomes contained in nucleus
	No introns, few transposable elements	Many introns, transposable elements
Cytoskeleton	Absent	Present
Organelles	Absent	Present (mitochondria, hydrogenosomes, chloroplasts etc)
Metabolism	Diverse	Mostly aerobic
Cell wall	Protein wall in (eu)bacteria	Cellulose or chitin, none in animals
Genetic recombination	Lateral gene transfer	Meiosis/sex
Cell division	Binary fission	By mitosis
Flagella	Rotating	Undulating cilia
Respiration	On membranes	Mitochondria
Environmental tolerance	Wide range of temperatures, pressures, chemical tolerance, resistance to desiccation.	Narrower range of tolerances

Adapted from Mayr (2)

some features that used to be thought of as unique to the eukaryotes, although they are much simpler than in eukaryotes. In other cases, particular organisms are being found that have a foot in both camps. They may be prokaryotes, yet have a eukaryote characteristic. So for example, bacteria have been found that have linear chromosomes (3), multiple chromosomes (4), giant cells, etcetera.

The two groupings may seem to be so fundamentally different from one another that the road between them is impossible to trace, but the existence of these special organisms at least indicates that many of the individual steps can be understood. The difficulty is just a rather extreme example of one that is commonplace in evolution, of gaps due to the extinction of intermediate stages, and the lack of fossils to record their passing. We would face an equal problem if we had no intermediates between say, an earthworm and a human. But these two species have evolved apart for only a quarter of the time that separates modern prokaryotes from eukaryotes. Similar problems have always challenged our understanding, usually because we tend to underestimate the subtlety and inventiveness of evolution.

From our perspective in this book, the primary innovation of the eukaryotes was their elevation of cell differentiation to a high art form, making complex creatures out of multiple different types of cells, so our question is: what enabled them to do that? We will focus just on the most fundamental changes relevant to our story. These were a revolution in the way information is carried and transmitted between generations, a revolution in the ability of organisms to acquire materials from their environment, and the acquisition of once free-living organisms as internal components of complex cells.

11.2 The information revolution

Most importantly, multi-cellularity could flower because the eukaryotes evolved a new and much larger genome. It could contain and use orders of magnitude more information than could the prokaryote genome. This was a necessary pre-condition, because multi-cellular life needs to store much more information in its genome than does single-celled life. A human being for instance, with a hundred or more cell types, must nevertheless have the blueprints for all the cell types stored in a single genome, as well as the instructions for how each cell interacts with its neighbours. Multi-cellular organisms are more complex than single cells, and they need a correspondingly bigger library in which to store their descriptions. The Neoproterozoic revolution was therefore based on an information revolution.

Living in the early twenty-first century, we know all about information revolutions. We are in the middle of one, and it parallels in some ways the move from prokaryote to eukaryote genetics. Before about the mid-1980s, computers had little ability to 'remember' programs when they were turned off. Useful programs had to be loaded by what now seem antiquated methods, after the computer had been turned on. Up until the 1970s this was done using punched cards or tape. Then along came plastic floppy disks, which greatly increased the amount of information that could conveniently be loaded into the machines. A decade or so later internal hard disks were introduced on personal computers, so that software was automatically available once the machine was turned on. Each hard disk had the capacity of hundreds of floppies. Since then the storage capacity of hard disks has continued to increase exponentially. The disk in the computer on which this is being written, in the year 2008, has about ten thousand times the capacity of those first hard disks, and about ten million times that of the first floppy disks. And of course all the other parts of the computer have similarly had to increase their capacity, in order to make use of the larger programs and data storage.

The difference between the eukaryote and prokaryote genomes is something like the difference between a computer in the 2000s and one in the early 1980s. The basic code in which information is stored remained the same in prokaryotes and eukaryotes, and many of the individual genes—fragments of program—were also conserved, but the organization of the chromosome, the operating system, so to speak, was radically upgraded. The upgrade allowed orders of magnitude more storage and faster copying. Eukaryote genome sizes vary widely, but ours is typical. Just in terms of the number of base pairs, it contains about 3.5 billion, which is about a thousand times larger than a bacterial genome. This crude size comparison is rather misleading however. If we count the number of genes that actually code for proteins, we have perhaps 25 000, which is larger by only a factor of six than *E. coli*. Apparently, the increased information needed to make a human is not hundreds or thousands of times greater than that needed to make a bacterium. The discrepancy is because the great majority of the DNA in our genome is 'non-coding'. All eukaryotes have such apparently non-functional DNA, though the amount varies hugely from organism to organism. More accurately, we should say we that it may well have a function, but we don't yet know it: it's quite humbling to have to report that we don't know why 98% of our DNA exists, but that is the current situation. One prominent view is that it has no function at all, but is 'junk' DNA—a useless but harmless parasite that is along for the ride. On this theory, we carry this material precisely because it has no detrimental effect on us. It doesn't affect our fitness in any way, and therefore natural selection is powerless to remove it. The junk DNA view is almost certainly partly correct, but it is surely not the whole story, since there are now some well documented cases of organisms which because of special life styles—usually living inside other, larger cells—have seemingly got rid of non-coding DNA.

It seems that eukaryotes evolved hugely increased capacity for storing and handling of DNA, but that the increased complexity that this allowed was almost incidental, a by-product of a process which favoured a larger amount of DNA for some different reason, not as yet fully understood. The increased information capacity was not the main function of the new machinery, and used only a small fraction of it, but, by accident so to speak, it enabled the eukaryotes to carry the extra information that allowed them to go for multi-cellularity in a big way. Such accidents are often the way with evolutionary innovations. Evolutionary biologists would describe the eukaryote genome as having been 'pre-adapted' to carry large amounts of information, as a by-product of some other function. Conceptually, this is the same kind of process by which fins in our fish ancestors turned out to be useful for walking on dry land, and forelegs subsequently became adapted for flight in birds, pterosaurs and bats. By-product it may have been, but it nevertheless would eventually be crucial in the story of life.

To understand more about the roots of this information revolution, we now delve into how bacteria and eukaryotes copy their genes, and the differences between them.

11.3 Gene copying in prokaryotes and eukaryotes

A bacterial chromosome is usually one giant circular molecule of DNA*. It consists of two strands, wound around each other in the famous double helix. The mechanics by which this is copied are illustrated in Fig. 11.1. At the commencement of replication, the two strands of DNA are unpicked at a point called the origin, identified by a particular sequence of bases. A complex of proteins called the 'replisome' assembles around the origin. The replisome remains nearly fixed in position near the mid-point of the cell, and the chromosome begins to spool through it. As it does so its strands are separated and copied, and two new chromosomes emerge. First the origin emerges as two new origins, and these rapidly migrate to opposite quarters of the cell, where they are fixed to the cell wall. The rest of the chromosome then spools through, until finally the terminus, the end point of copying, is reached. When the terminus has been copied, the two new chromosome loops are separated. The cell then divides at its centre, with one of these loops in each daughter cell.

Imagine the bacterial DNA molecule as like a zip fastener. Unlike a fastener on your clothes, it's a *circular* zip fastener, one end joined to the other to form a continuous loop—there's not much call for such a design in human clothing, but it could be used for instance to hold the circular door of a tent in place. Imagine then you have such a fastener—it's simpler to think of it as just the fastener tapes themselves, before they are sewn into a garment or a tent. Now imagine you start to unzip it, from a given point (the origin of replication). You unzip it with *two* sliders, one going in each direction—eventually they will meet again on the other side of the circle, at the terminus of replication. This is how the two strands of DNA are separated. In order to incorporate into this image how two new double helices are formed, we must imagine that, as the two sides of the fastener are separated, new teeth assemble from the surroundings. These line up with the newly unzipped teeth to form two new zip fasteners. Finally imagine that the unzipping proceeds by pulling the tapes through the sliders, rather than moving the sliders along the tapes. The two sliders remain stationary and are actually phys-

---

* How big is 'giant'? If stretched out, a bacterial chromosome would be around a millimetre long, which is several thousand times longer than a typical bacterium. The DNA helix is about 2 nanometres wide, so the molecule is, in round numbers, a million times longer than it is wide.

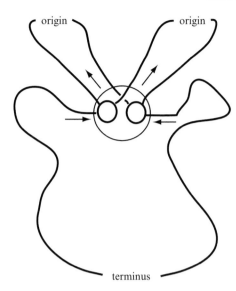

Fig. 11.1 Schematic of gene copying in prokaryotes. Replication starts at a single origin and ends at a terminus. The double strand is unwound by replisomes (circles). Either two independent replisomes 'crawl' along the parent DNA ejecting two loops of daughter DNA, or (as shown here) the replisomes form a complex through which the parent DNA spools

ically joined to one another, to form the replisome complex. If you can imagine all this, then you've a picture of the basic geometry of bacterial chromosome replication.

Central to the process is the near stationary replisome, through which the entire chromosome must pass, from its single origin to its single terminus. It used to be thought that the chromosome remained stationary and the replisome crawled along it like a train on a track, but it now seems that the chromosome itself moves. It's not entirely clear how this rapid and well organized movement is brought about. Perhaps it is provided by separate motion-inducing proteins—a cytoskeleton—which in a much altered form is the method used by eukaryotes. One of the exciting discoveries of the last few years is that bacteria do have such a cytoskeleton (5). It used to be thought that only eukaryotes could move their internal cell contents, but this is now known to be incorrect: the bacterial cytoskeleton is more difficult to see than that of eukaryotes, but well organized mechanical activity has been shown to exist, and is probably involved in the movement of the chromosome during replication.

The movement is probably also controlled by the organized coiling of the DNA itself. Older pictures in textbooks usually show the giant circular molecule haphazardly distributed throughout the cell. Recent research has shown however that the chromosome is not randomly jammed in, but neatly packed and ordered. No surprise perhaps, if we consider that this is a filing system which is being constantly accessed and regularly duplicated, and that the copying involves physically re-arranging it. Without some such organization, these operations would be defeated by tangles in the DNA, just as a tangled mass of string soon becomes unusable, but if stored on a spool it can be unwound and rewound indefinitely.

The string analogy is useful too, to understand some of the specific mechanisms that have evolved to manage the copying process. If you've ever tried unpicking a ball of tangled string, you'll appreciate some of the problems that the molecular machinery had to solve. For example, the DNA is double-stranded with the strands coiled around one another, and they have to be un-twisted in order for it to be copied. Imagine pulling apart two pieces of string twisted together: as they come apart, the as yet uncoiled part backs up in ever tighter twists. This problem is solved by having enzymes that work ahead of the replication point. Periodically, they cut one of the strands, allowing the other one to pivot around to remove the twists, after which the cut strand is re-joined.

Copying a bacterial chromosome takes time. In round figures, about 1000 base pairs per second can pass through the replisome. Copying the four million base pairs of *E. coli* takes about an hour. There's not a great deal of scope for speeding up this process, but bacteria do what they can. When a cell is growing at its fastest, the daughter chromosomes begin to duplicate, ready for the next-but-one generation, before they are themselves completely copied. But even so, a minimum of about half an hour per generation is required. This means that bacteria are limited in the complexity of their genome: it can't be much bigger if they are to be able to grow rapidly when the opportunity arises, and it seems therefore that natural selection has exerted a powerful pressure to keep the genome small.

The fundamental problem is that there is only a single origin of replication per chromosome. Is there a reason why they could not have many origins? It seems likely there is, but it is not obvious just where the problem lies. It is significant however, that in bacteria, rep-lication of DNA occurs simultaneously with cell division. In the bac-terial process the newly replicated origins, which are the first parts of the chromosome to be replicated, move to take up their positions in what will become the two daughter cells near the beginning of the process. The rest of the duplicated chromosomes follow them. The origins therefore have two roles: they are the sites at which copying begins, but they also help guide the physical separation. It could be that because they have this second task, it is impossible to evolve in a few simple steps to having a multitude of replication sites. Or per-haps the single circular chromosome design is fundamentally unsuited to having many different chromosomes all replicating simultaneously. Bacteria are known that have two chromosomes, but not dozens, as eukaryotes can have.

Eukaryote DNA copying and cell division is a very different process from its bacterial equivalent (Table 11.1). Some of the most important differences are: eukaryotes have many linear strands of DNA, rather than a single circular one, and they are confined in a nucleus within the cell. Accompanying proteins called 'histones' help organize them, winding the long molecules onto well defined spools (histones are also found in some archaea, and these are probably ancestral to eukaryote histones). Perhaps most importantly, duplication of the DNA occurs from many different origins at the same time, and copying occurs before, rather than simultaneous with, cell division. The ability of eukaryotes to copy using many origins is what allows them to have much larger genomes than bacteria, and yet be able to copy them quickly enough for rapid cell division.

Copying of the chromosomes takes place in the nucleus, simultaneously from hundreds of origins, before the cell shows any visible signs of preparing for division. Following this, a complex and beautiful intracellular ballet, called mitosis, begins. Its purpose is to ensure that one copy of each chromosome will be segregated to each daughter cell. The details vary between classes of eukaryotes, but in animals for example, mitosis begins with the 'condensation' of the DNA. While being copied the molecules were dispersed and invisible, but now they wind into clumps, lengths of 'supercoiled' material, so thick that the individual chromosomes can be seen in the nucleus under a microscope. Each chromosome having already been duplicated, they arrange themselves, in sister-pairs, bound together at a central point so that they look like miniature 'X' shapes. Simultaneously, outside the nucleus, structures form: two nodes, called 'centrosomes' appear each radiating protein microtubules like a wheel of ropes. The centrosomes migrate to opposite ends of the cell, and some of their microtubules extend, growing towards the nucleus. Then the membrane around the nucleus dissolves, and the growing microtubules attach to the chromosomes at their central points, one from each centrosome to opposite sides of the chromosome. The microtubule ropes then tighten, pulling the chromosomes into a central plane halfway between the centrosomes. Then the sister-chromosomes part from one another, and are pulled by their microtubule attachments towards opposite ends of the cell.

Why, and how, did the very different eukaryote chromosomes evolve from the prokaryotes, with their revolutionary new DNA replication and new cell division process? This is a big, and little understood question, and one that is entwined with the wider question of the origin of the eukaryote cell itself. We can't offer a definitive step-by-step account of how eukaryotes evolved from prokaryotes, because most of the steps are not understood or agreed. We can however describe some views of what constitutes the fundamental step at which the two lineages diverge. Is there a 'eureka moment' before which there are no eukaryotes, and after which there are?

11.4 The flexible cell membrane—a revolution in the acquisition of materials

The fundamental innovation, according to Tom Cavalier-Smith of Oxford University, was to get rid of the prokaryotes' 'rigid corset'—his description of the cell wall that is characteristic of bacteria. They need this wall to counter osmotic pressure, the tendency for solvents such as water to transfer into them and dilute their internal cytoplasm. This steady dilution can be prevented by allowing the pressure inside the cell to increase so that it pushes the water back in the opposite direction, but to do this, requires a sturdy pressure-resistant casing, hence the cell wall. Most archaea and all true bacteria have such a wall, but it differs in its structure between the two domains. Eukaryotes in general have a much more flexible outer wall (and as a consequence they need active pumping mechanisms to counteract osmosis).

The eukaryote cell wall can be physically moved by an advanced cytoskeleton—the system of molecular motors, tubules and filaments that gives movement to the inside of the cell. As we've described, the cytoskeleton is also employed to manipulate chromosomes, and the many other internal structures that eukaryotes have. Since we now know that bacteria have a primitive cytoskeleton too, it seems that it was not invented *de novo* by eukaryotes, but it was greatly elaborated, and redeployed to manipulate their outer membrane. Perhaps it was only the restriction of their rigid cell wall structure that prevented the more widespread development of cellular movement by bacteria. In Cavalier-Smith's scenario, once the rigid corset had been shed, a more advanced cytoskeleton evolved within the eukaryote lineage.

The great advantage of a flexible outer membrane steered by the cytoskeleton was that it enabled the process of *endocytosis*, by which food particles or prey could be surrounded by the cell, enclosed by it, engulfed, taken inside and digested. This is a much more efficient way of getting nutrients than the only alternative open to bacteria, which is to excrete digestive juices around the food particle and then take up the dissolved organics so released through the cell wall. The flexible membrane also opened the way for the cell to develop internal membrane-bound structures, of which modern eukaryotes have many, by taking parts of its cell wall into the interior.

Internal symbiotic organelles such as the mitochondria and chloroplasts could also be captured from free-living bacteria, and imported into the cell. As a general rule, prokaryotes cannot engulf one another. Very rare cases are known in which one type of bacteria lives inside another (*6*), but in this special circumstance, both live inside an insect, the mealybug (another example of the 'Russian doll' principle in biology). Hence this can't be considered an example of a free-living bacterium capturing another. It does indicate though, that endocytosis is not always necessary if one cell is to end up inside another.

According to Cavalier-Smith's theory then, the origin of the nucleus, the linear chromosomes contained in it, the complex cell cycle including mitosis, and sex (the regular pair-wise division and recombination of the genotypes of individual organisms between generations), were all subsequent innovations that could follow once the habit had developed of using the cytoskeleton to manipulate membrane-bound organelles inside the cell. In this theory the eukaryotes are a well defined lineage (branch) in a deeply rooted tree of life, and their key innovations evolved within the lineage.

11.5 Mixed relations

Until recently, most researchers have assumed that Darwin's tree of life stretches down far enough to show a clear branching between archaea, bacteria and eukaryotes. The early molecular genetic studies of Carl Woese and others placed eukaryotes closer to archaea than to bacteria, leading to the view that the ancestor to eukaryotes was an early archaean. If this was the case, you might expect that today there would be a branch of the archaea that is more obviously related to eukaryotes than are other archaea. An analogous case is that of the mitochondria whose genes reveal them to be clearly derived, not just from bacteria in general, but from a particular group (the alpha proteobacteria) within the bacteria. However, no such clear ancestral relationship is visible between eukaryotes and archaea.

Cavalier-Smith has promoted an alternative view that the ancestor to eukaryotes does not lie within the archaea, but rather that eukaryotes and archaea are sister groups (or branches), with a common ancestor further back in time, before the innovations and symbioses which created the modern eukaryotes. In some respects this is an attractive hypothesis. Sisterhood would explain why there is a general relationship between them, but not a very specific one. But the sisterhood proposal also has its disadvantages. It would suggest that we might hope to find organisms descended from the ancestral eukaryote, but which did not undergo all the changes leading to modern eukaryotes, but none is currently known. More problematically, in Cavalier-Smith's scheme, archaea-like eukaryotes evolved less than a billion years ago, hence methanogens and thermophiles, both of which are archaea, are recent. This contrasts with the consensus view that they were among the most anciently evolved organisms on the planet. We disagree completely with Cavalier-Smith on this, on biogeochemical grounds—we cannot see how to make a working model of Earth's early biogeochemistry, one that does not bury far too much carbon, without methanogens. Furthermore, there is indirect isotopic evidence that methanogens had evolved by 3.5 billion years ago (7). (Still, the mere fact that such a view can be robustly defended illustrates how much uncertainty can still be accommodated within our present knowledge of the tree of life.)

The revelation from recent molecular genetic studies, using ever larger and more diverse databases, is that eukaryotes contain a substantial amount of genetic material from both bacteria and archaea. The conventional way of explaining this, without clear-felling the tree of life, is that genes from acquired organelles (mitochondria, plastids), which were once free living bacteria have been transferred to the nucleus. In terms of the tree metaphor, some smaller bacterial branches entered a larger branch that was already leading to eukaryotes. However, the latest work brings this interpretation into question, indicating that the link to bacteria stretches far deeper. In fact, some researchers argue that no single tree of life can explain the mixed genetic make-up of eukaryotes (8). The way out of this seeming paradox, is if the origin of the first eukaryote involved a symbiotic fusion between an archaea and a bacterium.

11.6 Fusion to create a chimera?

In this evolutionary scenario, two quite different types of life became conjoined, to the extent that their genes combined and they became one organism; the first eukaryote. The composite organism, as it was first created along this evolutionary path, is labelled a 'chimera', the word coming from Greek mythology: Chimera was a monster with the body and head of a lion, a goat's head rising from the centre of its back, and a tail that ended in a snake's head. Whilst the word 'chimera' has come to be associated with impossible or foolish fantasies, the possibility that all our cells came originally from the fusion of quite different organisms could neatly explain their mixed genetic make-up. If it happened, the fusion event would have been the defining moment at which the eukaryotes came into existence.

There are several fusion hypotheses, which differ over what the specific partners in the chimera were, and which organelles or features of the eukaryote cell they gave rise to. In some versions of the fusion hypothesis, the archaean partner in the chimera becomes the nucleus, incorporated into a host bacterium. But in one of the most meticulously worked out scenarios, championed by Lynn Margulis and her colleagues (*9*, *10*), it is an archaean that is the host which gets 'invaded' by bacteria (spirochetes). The bacteria end up providing a propulsion system for the eukaryote cell, contributing to and tethering its newly-formed nucleus, and giving motility to its cytoskeleton by providing microtubules. These microtubules are in turn essential for the formation of a bipolar spindle and hence the process of mitosis. We describe this scenario in more detail in Box 11.1, not least because it has an intriguing link to the changing chemistry of the Proterozoic ocean, which we will discuss in Chapter 13.

Margulis' scenario differs fundamentally from that of Cavalier-Smith in that the advanced cytoskeleton of eukaryotes and the origin of the nucleus and mitosis rely on a founding symbiotic fusion event, rather than evolving in an already created eukaryote lineage. Although theirs are only two of over twenty hypotheses for the origin of eukaryotes, they serve to illustrate the two main types of hypothesis: either that eukaryotes are a primary lineage in a tree of life, or that they are a chimera originating from the fusion of pre-existing prokaryotes.

To our untrained eyes, the latest molecular genetic analyses seem to be leaning towards a fusion origin for eukaryotes, because no single tree of life can explain the mixed genetic make-up of eukaryotes (*8*). But what were the partners in the original chimera? Eukaryote genomes suggest three conflicting trees, which link them, in order of strength, to cyanobacteria, alpha proteobacteria, and a group of archaea; thermoplasmatales. The strong link to cyanobacteria can be explained by a later event; their symbiotic acquisition as plastids to form the ancestor of all algae and plants followed by the transfer of numerous cyanobacterial genes to the eukaryote nucleus. This leaves alpha proteobacteria and archaea as likely partners in the original chimera. Although Margulis' scenario (Box 11.1) includes the right type of archaea, the bacterial partners (spirochetes) are not alpha proteobacteria. This directs attention to alternative fusion hypotheses involving the acquisition of mitochondria, which were once free-living alpha proteobacteria, and form the energy factories in many eukaryote cells today.

11.7 Internalizing nutrient cycling and energy generation

The theories of Cavalier-Smith and Margulis both suggest that mitochondria were captured at a later stage, after the first eukaryote formed. If so, eukaryotes might still exist that don't have mitochondria, descended from the period before they were captured. While the great majority of eukaryotes possess mitochondria, some do not. They are all single-celled protists, and all are anaerobes. They include for instance, some parasites that live in the anaerobic insides of larger animals.

Box 11.1 A sulphur cycling chimera

In Margulis' scenario (Fig. 11.2) (*9, 10*), the archaea partner is likened to a modern-day organism called *Thermoplasma*, which has a unique membrane in place of a cell wall, and a rudimentary cytoskeleton. The postulated bacterial partners were free-living spirochetes—organisms which propel themselves through viscous fluids with a spiralling motion. The two are argued to have struck up a recycling relationship in which *Thermoplasma* converted globules of elemental sulphur into sulphide, and spirochetes converted the sulphide back into sulphur. If the swimming spirochetes attached to the archaea the resulting consortium may have more efficiently reached the carbon sources it also needed as food. This kind of close metabolic association is seen today in a consortium known as *Thiodendron* (involving spirochetes and *Desulfobacter*), which is found just below the oxygen-sulphide interface in anoxic waters, and sometimes coating sediments. In this evolutionary scenario, a chimera was formed when the spirochete partly entered the host cell, perhaps forced there by an environment that became enriched in oxygen and thus more hostile to it (and the sulphide it depended on). The next step was the forming of a proto-nucleus containing their combined DNA, which was tethered by two spirochetes, their tails still wriggling in the environment as a source of propulsion. This tethered proto-nucleus is likened to a feature called the 'karyomastigont', which is observed in some protists (single-celled eukaryotes) that lack mitochondria and all inhabit anoxic environments. In this scenario, the acquisition of mitochondria (and other cell organelles) came later in an environment increasingly dominated by oxygen.

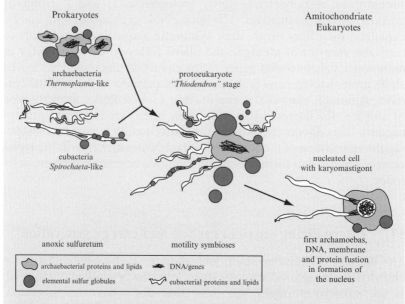

Fig. 11.2 Lynn Margulis and colleagues' scenario for the origin of eukaryotes. [Redrawn from (*9*), copyright (2000) National Academy of Sciences, USA]

Loss of the wriggling flagella that provided cell motility (the former spirochetes) released the nucleus and led to different lineages including plants and animals (raising the question; did they give up flagellation for sex?!).

The intriguing fact that eukaryotic cells today, including our own blood cells, still have the capacity to convert sulphide into sulphur, is argued to be an evolutionary leftover of the chimera (sulphur does not build up in our cells because the mitochondria present rapidly oxidize it further to sulphate). A key test of the hypothesis will be if spirochete DNA can be identified in the eukaryote genome.

Examples in humans include *Giardia,* and *Entamoeba*, both of which infect the intestinal tract, and *Trichomonas vaginalis* that causes sexually transmitted disease. However, despite the absence of mitochondria, these organisms usually do have either 'hydrogenosomes' or the more recently discovered 'mitosomes', which are internal organelles that help them to derive energy by anaerobic processes.

Cavalier-Smith grouped the non-mitochondrial eukaryotes together as 'archezoa', and proposed that they were ancient, pre-mitochondrial eukaryotes, and on this point, Margulis agrees. The archezoan hypothesis predicted that the non-mitochondrial eukaryotes should have branched from the rest of the eukaryotes a long time ago, and that they should never have possessed mitochondria, rather than having lost them at some subsequent date. They would then be one of those 'intermediate' groups marking the evolutionary pathway to eukaryotes.

The archezoan hypothesis enjoyed some initial success, but it has recently run into trouble and has now been largely discarded. The trouble is that is has been found that these non-mitochondrial eukaryotes all seem to retain a memory, in their genes, of a time when they *did* have mitochondria. In ordinary eukaryotes, most mitochondrial genes have been transferred to the nucleus, where they can be identified for instance by their similarity to bacterial genes. Some of these same mitochondrial genes have now been found in the non-mitochondrial eukaryotes. The explanation must be either that they once possessed mitochondria and lost them, or that there is a common ancestor of both hydrogenosomes and mitochondria. The mitosomes meanwhile have been found to be very highly reduced forms of mitochondria.

The current situation then, is that no eukaryote that *never* had mitochondria-like organelles has been found. This has led some researchers to suspect that mitochondria-like energy factories are essential to eukaryotes, and that even the first eukaryote had them—in other words that the origin of eukaryotes occurred when mitochondria were first acquired. But what did they join with?

One suggestion (*11*), known as the 'hydrogen hypothesis' is that an anaerobic archaean, dependent on free hydrogen for its survival, formed a close association with a hydrogen-producing bacterium—just such close associations are known to exist today. Then, as oxygen appeared in the atmosphere and made life hard for the archaean, the response was an ever closer dependence on the bacterium, culminating in the bacterium being taken into the archaean and forming the first internal organelle.

An alternative scenario, is that a sulphur-metabolizing archaean related to *Thermoplasma* joined with the ancestor of mitochondria (*12*). This involves one partner from Margulis' scenario, and one from the hydrogen hypothesis. Like those scenarios, each partner would have fed on the others' waste products, deriving energy and in this case recycling sulphur. As environmental conditions became harsh for one partner, the energy generation and recycling system became internalized, and the partners fused to form one organism. This scenario may be the closest to explaining the mixed make-up of eukaryote genomes (*8*).

Both these mitochondrial fusion hypotheses explain some aspects of the puzzle rather neatly. They can explain for example why most eukaryotes are oxygenic, but accommodate the existence and significance of a variety of anaerobic eukaryotes (*13*). However, none of the fusion hypotheses explain how the engulfment of the symbiont by the host happened. This appears to require that the host cell was already capable of endocytosis, so Cavalier-Smith would claim that it was already a eukaryote, since it was the acquisition of that talent that marked their true origin.

11.8 Summary

To summarize, there is plenty of lively debate, no shortage of ideas, but as yet no agreed theory for the first steps on the road from prokaryotes to eukaryotes. If it was a fusion event that formed the first eukaryote, then the 'tree of life' no longer seems an apt metaphor at or below the level of eukaryote origins. Some researchers have replaced it with a ring of life out of which a tree of eukaryotes emerges. Others talk of a network of life, or a tangled web or bush of life (see Box 3.1). In fact, the one type of complex life whose 'branches' (hyphae) genuinely fuse together are fungi. Networks of fungal hyphae are found in forest soils, which attach to the roots of trees. So, perhaps the new metaphor should be a fungal network of prokaryote life, out of which the tree of eukaryotes grew.

Despite the uncertainties surrounding their origin, the key features that mark out the eukaryotes from the prokaryotes are clearer.

Their ability to ingest other cells enabled them to build a much more complex cell, including vital components that were once free-living prokaryotes. Added to this, their much larger genome with its rapid copying, gave eukaryotes the capacity to create many different types of cell and assemble them together into large, complex, multi-cellular creatures. We now turn our attention to the fossil record to see what clues it holds as to when the eukaryotes evolved, and when they began to realize their potential to create complex life forms.

References

1. C. Darwin, *On the origin of species by means of natural selection, or the preservation of favoured races in the struggle for life.* (John Murray London, 5th edition, 1869).

2. E. Mayr, *What evolution is.* (Basic Books, New York, 2001), pp. 318.

3. C. M. Fraser *et al., Genomic sequence of a Lyme disease spirochaete, Borrelia burgdorferi. Nature* **390**, 580 (1997).

4. E. S. Egan, M. A. Fogel, M. K. Waldor, *Divided genomes: negotiating the cell cycle in prokaryotes with multiple chromosomes. Molecular Microbiology* **56**, 1129 (2005).

5. Y. L. Shih, L. Rothfield, *The bacterial cytoskeleton. Microbiology and Molecular Biology Reviews* **70**, 729 (2006).

6. C. D. von Dohlen, S. Kohler, S. T. Alsop, W. R. McManus, *Mealybug beta-proteobacterial endosymbionts contain gamma-proteobacterial symbionts. Nature.* **412**, 433 (2001).

7. Y. Ueno, K. Yamada, N. Yoshida, S. Maruyama, Y. Isozaki, *Evidence from fluid inclusions for microbial methanogenesis in the early Archaean era. Nature* **440**, 516 (2006).

8. D. Pisani, J. A. Cotton, J. O. McInerney, *Supertrees disentangle the chimerical orgin of eukaryotic genomes. Molecular Biology and Evolution* **24**, 1752 (2007).

9. L. Margulis, M. F. Dolan, R. Guerrero, *The chimeric eukaryote: Origin of the nucleus from the karyomastigont in amitochondriate protists. Proceedings of the National Academy of Sciences USA* **97**, 6954 (2000).

10. L. Margulis, M. Chapman, R. Guerrero, J. Hall, *The last eukaryotic common ancestor (LECA): Acquisition of cytoskeletal motility from aerotolerant spirochetes in the Proterozoic Eon. Proceedings of the National Academy of Sciences USA* **103**, 13080 (2006).

11. W. Martin, M. Muller, *The hydrogen hypothesis for the first eukaryote. Nature* **392**, 37 (1998).

12. D. G. Searcy, W. G. Hixon, *Cytoskeletal origins in sulfur-metabolizing archaebacteria. BioSystems* **25**, 1 (1991).

13. M. Mentel, W. Martin, *Energy metabolism among eukaryotic anaerobes in the light of Proterozoic ocean chemistry. Philosophical Transactions of the Royal Society B-Biological Sciences* **363**, 2717 (2008).

12
When did eukaryotes evolve?

Given the disagreement over *how* eukaryotes evolved, it is not surprising that there is great uncertainty about *when* it occurred.

We know eukaryotes only evolved once, and from this and the complexities of their origin, we can be fairly sure it was a difficult evolutionary transition. If we could tie down when it occurred, this would give a vital additional clue as to just how difficult the transition was, and whether it represents a 'critical' step that could easily have not happened at all. But as we have been at pains to point out, the critical steps model of Chapter 5 is just a mental crutch. In the case of eukaryote origins, its simplifying assumptions really struggle to bear the weight of reality. As we have highlighted in the last chapter, it proves fiendishly difficult to identify a single key feature, or corresponding step, that marks the origin of eukaryotes. So, not surprisingly, when looking back through deep time, instead of a simple 'now you see them, now you don't', the picture of eukaryote origins is blurred. When the first eukaryote evolved depends on what you believe constitutes a eukaryote.

Happily our emphasis is on the revolutions of the Earth system, and for the complexity revolution it was not so much the origin of eukaryotes that mattered, but when they began to realize their potential to make differentiated, multi-cellular life forms. Consequently, in this chapter we address the evidence both for when eukaryotes first appear, and for when different multi-cellular lineages arise. Fig. 12.1 summarizes some of this evidence on a timeline.

12.1 The emergence of eukaryotes

As we touched on in Chapter 6, there are three main techniques for recognizing when eukaryotes first appear, but they give different answers and each has its own problems.

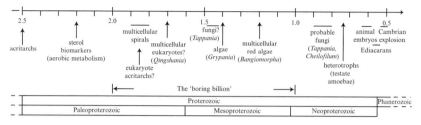

Fig. 12.1 Proterozoic timeline summarizing the candidate fossil evidence for eukary-otes and for multi-cellular life, with question marks denoting uncertainty about affinities

12.1.1 Fossil evidence

The most obvious source of evidence is the fossil record, especially of macroscopic and multi-cellular life, which now extends back into the Precambrian (Fig. 12.1). But unfortunately (though not surprisingly), paleo-biologists disagree over what represents an unequivocal eukaryote fossil. The most stringent judges, such as Tom Cavalier-Smith, demand clearly identifiable traces of all the characteristics of a eukaryote—a nucleus, an internal membrane system, and a cytoskeleton, which might show up as fossil cells with some discernable protrusions or branches. To preserve all this in ancient rocks is asking a lot. However, some criteria are required because historically some fossils have been rather too read-ily assigned as eukaryotes, by researchers eager to claim they have found the oldest evidence for them. Examples are some spirals found in rocks from a mine in Michigan and initially dated to about 2.1 billion years ago (*1*). These are fascinating and important fossils, but they are not necessarily eukaryotes. They are large, about 2 cm across, and were ini-tially identified as the same group as more recent, multi-cellular, fossils called *Grypania*, dated to about 1.4–1.6 billion years. However, the more recent fossils have thicker coils and resolve traces of internal detail that makes their identification as eukaryotes more secure. In the older spirals, no finer detail is visible, and sceptics point out that such large structures can sometimes be made by filaments made up of modern prokaryote cells (*2*). The older fossils were probably also wrongly dated—a more accurate age is 1.8–1.9 billion years (*3*).

There are a host of small, often roughly spherical fossils present from before two billion years ago, and throughout the Proterozoic. These are bundled together under the umbrella term 'acritarchs'—a taxon of assorted discontinued lines whose name means 'of uncertain origin' and which can't be convincingly assigned to any particular part of the tree of life. The oldest of these, very recently reported, are just two specimens of a spine-covered cell-like body, found in 2.5 billion year old rocks in

Australia (4). A popular characterization of many acritarchs is that they are 'cysts'—inactive resting stages of microorganisms. They are larger than most structures made by bacteria, and many are near spherical, some with symmetric ornamental patterns and are argued to be the resting stages of eukaryote algae—but bacteria sometimes make cysts and spores, and so a possible prokaryote origin can't be ruled out. Experts on fossil life argue that some acritarchs from about 1.85 billion years and later are eukaryotes. At 1.7 billion years there is a larger structure, *Qingshania*, that might be a multi-cellular eukaryote (5).

So, something is stirring on the tree of life, but just what? Things become clearer in a 1.43 billion year old northern Australian shale: strikingly shaped, irregular cells assigned to a taxon called *Tappania* with branches extending from their outer surface and bulbous protrusions that suggest vegetative reproduction by budding of offspring (6). These tiny fossils, up to about 0.1 mm across, clearly possess a cytoskeleton. They are the earliest confidently dated specimens of *Tappania*, but *Tappania* are found elsewhere too, notably in a 0.9–0.8 billion year old north western Canadian shale. The Canadian specimens are beautifully preserved, and show multi-cellularity and differentiation—the cells within an individual organism are of recognizably different types, a characteristic of multi-cellular eukaryotes. Some of the branches coming from the cells re-fuse together, forming patterns that are better described as nets than branches. Nick Butterfield of the University of Cambridge points out that this is characteristic of fungi, which often have hair-like structures called hyphae extending from their cells, and which sometimes fuse in this manner. He argues persuasively that these cells are fungi and that the earlier Australian specimens may indicate a long record of fungal development in the Proterozoic (7). In the same 0.9–0.8 billion year old Canadian rocks, Butterfield has described fossil fragments of a different organism that he names *Cheilofilum hysteriopsis* (8), which is also fungal-like in many of its characteristics.

Even more convincing evidence of eukaryotes comes from fossils from about 1.2 billion years ago that Nick Butterfield has assigned to a still existing phylum of red algae—a multi-cellular eukaryote related to seaweed (9). These fossils, preserved in exquisite detail, look like members of the modern genus *Bangia*, right down to the detail that a cross-section through the stems of the fossils reveals a 'pie' divided into six that is unique (in today's world) to *Bangia*. Correspondingly, Nick calls his fossils *Bangiomorpha pubescens*. They are remarkable in showing multi-cellularity, differentiation of cells into different types, and evidence of sexual reproduction in the form of separate spore and gamete production.

To the arch-sceptic, none of these fossils fulfil all three eukaryote criteria of clearly identifiable traces of a nucleus, an endomembrane system, and a cytoskeleton. Correspondingly, they would place the earliest unequivocal eukaryote evidence at about 0.75 billion years ago, where fossils of amoebae show all these features (*10*). Cavalier-Smith takes the radical view that the critical step in eukaryote evolution happened shortly before these fossils appear, around 0.85 billion years ago, in an episode of 'quantum evolution'. Whilst this would be consistent with the view that eukaryote evolution was especially difficult, we find the earlier fossil evidence persuasive that eukaryote evolution was already reasonably well advanced by the middle of the Proterozoic.

12.1.2 Molecular fossils

Fossilization is always a rare, fortunate event. Until life invented shells a mere 0.55 billion years ago, all the products of living things were made exclusively of soft organic material. As such it almost never fossilized; even if it was preserved in sediments, it would be quickly broken down by even moderate pressure and heating of rock, so the fossil record of the earliest eukaryotes is inevitably indistinct. There is a different kind of fossil however that can help to date the origins of eukaryotes. These are 'molecular fossils'—chemical traces of metabolism, otherwise known as biomarkers (see Box 9.2). The idea here is simple: Eukaryotes have some structures, such as their cell membranes, that have no equivalent in the bacterial world, and they need therefore to manufacture some compounds that bacteria do not. The presence of these compounds or their immediate breakdown products in organic material, such as oil in ancient rocks, would then betray their presence, even if no actual fossils could be found.

Sterols, for example, are compounds that are critical in the construction of the membranes of eukaryotes. (The best known sterol is cholesterol, which all animals make—it has a bad press because too much cholesterol in the blood can cause cardiovascular disease, but none of our cells would function without it.) So traces of sterols, or steranes which are their breakdown products, have been used as biomarkers for eukaryotes. In addition, all known pathways for the production of sterols require some oxygen, so sterols and steranes have also been interpreted as diagnosing the presence of at least traces of oxygen (see Box 9.2). In a landmark study, Jochen Brocks and his colleagues found them in traces of oil contained in 2.7 billion year old shales from the Pilbara Craton, Australia, and argued that eukaryotes, or at least their metabolic ancestors, were present at that time (*11*). If the inference is correct they can claim the current record for the oldest direct evidence of eukaryotes.

Unfortunately, things are not quite so simple. Some bacteria, it turns out, do produce sterols. Among these is *Gemmata obscuriglobus*, which is also remarkable because it is the only bacterium to possess a prototype nucleus (*12*). A few other types of bacteria make sterols too. The situation is not hopeless however, because some of the sterols found in the Pilbara shales do seem to be made *only* by eukaryotes and not by any (known, modern) bacteria.

There is another problem though. Contamination with more recent hydrocarbons is a possibility, difficult to completely discount in these rocks through which oils can migrate (see Box 9.2). The presence of at least one compound, dinosterane, thought to be characteristic of a group of plankton that evolved much later, hints at such contamination. To get around this problem Brocks *et al*. had to invoke unknown ancient eukaryotes able to manufacture it. But this kind of ad hoc appeal to unknowns is double-edged, and by a similar argument their critics constructed an alternative explanation for the entire sterol signature—as yet unknown bacterial and anaerobic pathways of sterol manufacture (*13*). When scientific debate reaches the point that both sides of an argument are invoking unknown causes there is clearly something wrong (philosophically, if not scientifically). Sure enough, a recent study including Jochen Brocks as a co-author comprehensively dismisses the biomarkers from 2.7 billion year old rocks as much younger contamination (*14*).

The contamination problems have been solved in some impressive recent work, which analysed oils from slightly more modern (but much harder) rocks, still some 2.45 billion years old (*15, 16*). In this material, the ancient oil is contained in well-sealed microscopic inclusions, so that all possible contamination could be removed before crushing the rock to release the signal. These oils were also found to contain the sterol biomarkers thought to be unique to eukaryotes. The tiny capsules holding the oil were sealed off before about 2.2 billion years ago, and this currently represents the earliest biomarker evidence for sterols.

12.1.3 Molecular clocks

A third and last approach to dating the emergence of eukaryotes uses what have come to be called molecular clocks, where the molecules in question are DNA or RNA, or the proteins they encode. We have already met the study of molecular taxonomy, where the degree of similarity between genetic sequences is used to deduce family trees that chart the relatedness between groups of living things. The molecular clock technique goes one stage further, using the amount of genetic difference between different groups to deduce *when* the groups diverged. The critical assumption is that the rate at which mutations accumulate along a particular line is constant—this provides the regular 'ticking' of the clock. The mutations in question have to be neutral—that is, they are changes that don't actually affect the organism at all, so that they are not subject to natural selection which would influence whether they persist or not. The rate of ticking of the clock first has to be calibrated, and this is done by counting the number of mutations between two branches of life whose divergence time is already known from the recent fossil record, for example the split between birds and mammals, which has an agreed date of about 0.33 billion years ago. This calibration is then used to place the older divergence times on an absolute time scale.

As you can probably guess, the molecular clock method also has plenty of problems and uncertainties. The assumption that the mutation clock always ticks at the same rate is not always good, and more recent refinements employ a 'relaxed clock' which allows for variation in speed on different branches of the tree of life. There is a more fundamental problem however, in that a family tree has to be assumed and it must have the right structure, but it has proved problematic to sort out the true structure of the eukaryote family tree (*17*). What makes this so difficult is that occasional genetic transfer can occur between distantly related groups. Sometimes this is through lateral gene transfer, which we've already seen can occur between bacterial groups, but can also occur in eukaryotes, and sometimes via acquisition of endosymbionts, which come as a package and bring some of their own genes with them. Such transfers do not have to have occurred very often to be a problem—one such occurrence in the history of the particular gene being analysed can change the result completely. The end result is that the structure of the family tree can look very different depending on which genes have been sequenced.

So it will come as no surprise that there is controversy and debate about the methods and results of molecular clocks, particularly when applied to eukaryotes. Historically, absolute divergence times from molecular clocks have tended to be overestimated. In the case of the split of prokaryotes and eukaryotes, molecular clock estimates range from 3.97 billion years ago (*18*) (one could hardly get much older and still be on Earth!) to 2.2 billion years ago (*19*). For the earliest divergence of major eukaryote lineages the estimates range from 2.3 billion years ago (*20*) to 1.1 billion years ago (*21*). Great insight should eventually come from this work, but just now it seems the uncertainties of about a factor of two in age are too great to give us much help in dating the rise of the eukaryotes.

12.1.4 Summary

Let's now summarize the evidence for when eukaryotes originated. It is obvious that there is no single date for the origin of modern eukaryotes—unlike the simplistic assumption of the critical steps model (Chapter 5), they do not suddenly appear fully formed, but rather their characteristics become increasingly widespread over a long period. We find the fossil evidence at 1.45 billion years (*Tappania*), and certainly by 1.2 billion years (*Bangiomorpha pubescens*) compelling: near-modern eukaryotes were in place by then, and fungi and algae had probably diverged. Given the complexity of the fossils a much earlier origin is possible. We take the biomarker evidence at 2.45–2.2 billion years ago as a good indication that aerobic metabolism was in place in what may be a 'proto-eukaryote'—but this does not mean that the key features of eukaryotes were in place by then. Tentatively, we place the key event, the evolution of phagocytosis, followed relatively quickly by the acquisition of symbionts, at around 2 billion years ago. We follow others in concluding that trying to date the split of prokaryotes and eukaryotes using molecular clocks is pushing the methods too far (*17*).

12.2 The divergence of multi-cellular eukaryotes

Happily for our story, it is the appearance of cell differentiation in eukaryotes that is most important for the complexity revolution, and there is a little more evidence for that. Here we give a brief description, concentrating on the origins of the major multi-cellular lineages of plants, animals and fungi. Our focus on these groups means that we pay scant attention to the great majority of eukaryotes; the rich, diverse, mostly single-celled miscellany that used to be called the protists.

Eukaryote taxonomy is a large, and far from settled research area. Molecular sequencing techniques have revolutionized the subject, but also revealed new problems. There are many technical difficulties, but there is also a fundamental conceptual problem: evolutionary trees (or 'cladograms' to give them their technical title) assume that genetic material is transmitted vertically—by inheritance—only. But lateral gene transfer, rife among prokaryotes, has also been important in eukaryotes, especially endo-symbiotic gene transfer from incoming organelles to the nucleus. Hence a completely accurate universal tree does not exist. Trees constructed from different sets of molecular analyses will have different structures, and (just as with the origin of eukaryotes) branches may sometimes fuse together. Whole genomes might occasionally merge, as in endosymbiosis followed by transfer of genes to a single genome. From here on, we will mostly ignore these difficulties, being concerned only to know the basic genealogy and timing of emergence of the major multi-cellular groups.

Fig. 12.2 shows a tree, with estimates of divergence times constructed by analysis of protein sequences. The tree is based on that published by Douzery and colleagues (*21*), in which they used a 'relaxed molecular clock' approach. Their estimates for divergence times are for the most part comparable to those derived from fossils, and are much more recent than estimates previously made using constant-rate assumptions for the molecular clock. These earlier analyses suggested, for instance, that the divergence of animals and fungi occurred about 1.9 billion years ago (*18*), whereas Fig. 12.2 suggests it was about 1 billion years ago. As we've indicated above however, all such trees and divergence times should presently be treated as provisional. In particular, the positioning of the 'root' of the tree is uncertain, and where it is placed affects dating of the divergence points. Construction of the divergence times *assumes* a particular tree structure, and if this structure turns out to be wrong, the dates are likely also to need revision.

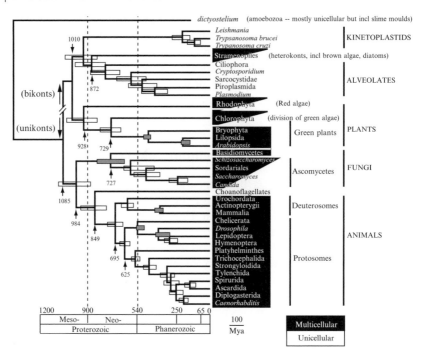

Fig. 12.2 A phylogenetic tree for eukaryotes with 'relaxed' molecular clock dates for divergence times. The time scale is given at the bottom, and best dates for divergence points (in millions of years BP) are also shown with arrows. Rectangles around divergence points give 95% confidence intervals, and grey rectangles indicate divergences constrained by calibration points taken from the fossil record. Black shading indicates groups that are multi-cellular and black/white diagonal division indicates both uni- and multi-cellular organisms. [Redrawn with additions from Douzery et al. (*21*), copyright (2004) National Academy of Sciences, USA.]

Fig. 12.2 also shows where amongst the eukaryotes multi-cellular organisms are to be found. It's obvious that, once the genetic machinery was in place to allow the development of multi-cellularity, it occurred repeatedly. It happened independently in several different lineages, including several times in algae, red, green and brown—for example the marine giant kelp, the largest of the algae, is in the huge group of Stramenopiles. However, animal differentiation, which is more fundamental in kind than the others, probably occurred once, at the point on Fig. 12.2 (arrow labelled '849') where the choanoflagellates, single-celled protistans, split from the line leading to animals. This tree dates this occurrence in the interval 0.95–0.75 billion years ago. However, a recent study (*22*) suggests that ctenophores (comb jellies) may be the most basal of living animals, that is, the closest to the last common ancestor, whereas it had previously been supposed that this honour belonged to sponges. We might expect this new information to change the date of origin of the animals as derived by molecular clocks.

In the current classification, the deepest split in the eukaryotes occurs between the 'unikonts' and the 'bikonts', the 'uni' and bi' referring to the number of flagella that individual cells have. Each of these classifications has several major groups in it. The unikonts contain the fungi and animals (metazoa) which together form a super-kingdom, called the opisthokonts. 'Plants' of most descriptions, are bikonts, including the land plants, red algae, green algae and other eukaryote plankton and seaweeds. (Note how very different all this is from the traditional classification, according to which fungi, plants and algae were studied by botanists, in one university department, while animals were studied by zoologists, in another.) According to Douzery *et al.*'s dating, the unikont-bikont split occurred sometime after 1.2 billion years ago and red and green algae separated 1–0.9 billion years ago. The origins of land plants are sometime later than about 0.72 billion years ago, when they split from the green algae, but well before about 460 million years ago, a date constrained by the fossil record for the appearance of bryophytes (mosses and liverworts) (see Chapter 16).

How do these divergence times compare with the fossil record (Fig. 12.1), discussed above? The oldest fossil with clearly differentiated cells that we know of is Butterfield's *Bangiomorpha pubescens*, a red alga forming filaments a millimetre or so long, which is dated to 1.2 billion years ago. This is inconsistent with the relaxed molecular clock (Fig. 12.2), in which the red algae evolve, at the earliest, about 1 billion years ago. If the 1.43 billion year old *Tappania* specimens are fungi this too is inconsistent with Fig. 12.2 in which the fungi evolved at the earliest about 1 billion years ago (although Butterfield's more detailed *Tappania* specimens at 0.9–0.8 billion years old are consistent). What is clear is that relaxing the molecular clock means that it is no longer consistently (and sometimes wildly) over-estimating divergence times when compared to the fossil record. There are still significant discrepancies, but some agreement between 'the clocks and the rocks' is emerging: The major groups of eukaryotes began to diverge in the mid-Proterozoic, roughly 1.5–1 billion years ago.

12.3 Early animals?

What of animals, the metazoa—do they have a cryptic history in the mid-Proterozoic? There are a very few reports of 'trace fossils'—markings in sediment that might be made by animals, from the early or mid-Proterozoic. For example, Rasmussen *et al.* (*23*) describe traces a centimetre or so long from the Stirling range in Australia, at least 1.8 billion years old (*24*). They are presented as evidence for 'vermiform, mucus-producing, motile organisms'—scientific language for slimy, wriggly, worms. Needless to say, the claim is controversial. Simon Conway Morris for example, argues that the marks were made by a microbial mat that was disrupted or folded by wind or wave action (25). He asks why, if animals existed so far back in time, did they take so long to proliferate? Rasmussen and colleagues blame the lack of oxygen in their reply (26)—in verse:

> *A slimy young worm in the making*
> *Found a gal he considered worth taking*
> *But she cried in despair*
> *The Precambrian air*
> *Is too stuffy—my neuron is aching!*

The idea that these very ancient tracks were due to animals has been dealt a recent blow with the discovery that giant amoeba (single-celled eukaryotes) living at the bottom of the ocean today make very similar trails (27). Conceivably their ancestors could have been responsible for such tracks in the Proterozoic.

There is one other report that we know of, from a 1.1 billion year old sandstone from India, in which substantial burrows are apparent (28). Again, other explanations are conceivable, for instance that they were made by rising gas bubbles or a subsequent invasion of the rock by tree roots. The balance of opinion is currently against ascribing these almost unique early traces to animals. What we can say is that, if animals did exist at this time, they must have been rare and inactive, because they have apparently left so little fossil evidence.

The situation in the mid-Proterozoic contrasts with the very end of the Eon, roughly 600 million years ago, when trace fossils, usually small tracks or faecal pellets of sediment dwelling worm-like creatures, do appear unmistakably in the fossil record (29, 30). Starting perhaps as early as 635 million years ago, recently described finds from the Doushantuo phosphorite formation in Southern China show exceptional preservation of embryos and small fossils that are definitely animals, and can be tentatively related to existing groups, for example, sponges and jellyfish (cnidarians) (31) (Fig. 12.1). After 580 million years ago, there also appear much more widespread, altogether stranger, multi-cellular creatures, known as the Ediacarans, who flourish for tens of million years. The fossil assemblage becomes richer and more varied. Then at 542 million years ago, the first 'small shelly fossils' appear, marking the end of the Precambrian and the start of the Cambrian explosion.

12.4 A quantum leap?

We will return to describe the flourishing of multi-cellular life in the ocean in Chapter 14. For now we note that something remarkable happened between about 900 and 600 million years ago that allowed complex multi-cellular life to flourish. Through most of the Proterozoic, probably for over a billion years, the development of eukaryotes proceeded incredibly slowly. A popular school of thought is that an unfavourable environment held back their flourishing. Others counter-argue that the process of evolving complex life, especially animals, was just inherently slow, and it simply took that long for all the necessary chance mutations to occur. We now turn to the development of the environment to see what support it lends to the first view.

References

1. T. M. Han, B. Runnegar, *Megascopic Eukaryotic Algae from the 2.1-Billion-Year-Old Negaunee Iron-Formation, Michigan. Science* **257**, 232 (1992).

2. J. Samuelsson, N. J. Butterfield, *Neoproterozoic fossils from the Franklin Mountains, northwestern Canada: stratigraphic and palaeobiological implications. Precambrian Research* **107**, 235 (2001).

3. D. A. Schneider, M. E. Bickford, W. F. Canner, K. J. Schulz, M. A. Hamilton, *Age of volcanic rocks and syndepositional iron formations, Marquette Range Supergroup, implications for the tectonic setting of Palaeoproterozoic iron formations of the lake Superior region. Can. J. Earth Sci.* **39**, 999 (2002).

4. W. L. Zang, *Deposition and deformation of late Archaean sediments and preservation of microfossils in the Harris Greenstone Domain, Gawler Craton, South Australia. Precambrian Research* **156**, 107 (2007).

5. Y. Yan, Z. Liu, *Significance of eukaryotic organisms in the microfossil flora of Changcheng system. Acta Micropalaeontologica Sinica* **10**, 167 (1993).

6. E. J. Javaux, A. H. Knoll, M. R. Walter, *Morphological and ecological complexity in early eukaryotic ecosystems. Nature* **412**, 66 (2001).

7. N. J. Butterfield, *Probable proterozoic fungi. Paleobiology* **31**, 165 (2005).

8. N. J. Butterfield, *Reconstructing a complex early Neoproterozoic eukaryote, Wynniatt Formation, arctic Canada. Lethaia* **38**, 155 (2005).

9. N. J. Butterfield, *Bangiomorpha pubescens n. gen., n. sp.: implications for the evolution of sex, multicellularity, and the Mesoproterozoic/Neoproterozoic radiation of eukaryotes. Paleobiology* **26**, 386 (2000).

10. S. M. Porter, A. H. Knoll, *Testate amoebae in the Neoproterozoic Era: evidence from vase-shaped microfossils in the Chuar Group, Grand Canyon. Paleobiology* **26**, 360 (2000).

11. J. J. Brocks, G. A. Logan, R. Buick, R. E. Summons, *Archean molecular fossils and the early rise of eukaryotes. Science* **285**, 1033 (1999).

12. A. Pearson, M. Budin, J. J. Brocks, *Phylogenetic and biochemical evidence for sterol synthesis in the bacterium Gemmata obscuriglobus. Proceedings of the National Academy of Sciences USA* **100**, 15352 (2003).

13. R. E. Kopp, J. L. Kirschvink, I. A. Hilburn, C. Z. Nash, *The paleoprotero-zoic snowball Earth: A climate disaster triggered by the evolution of oxygenic photosynthesis. Proceedings of the National Academy of Sciences USA* **102**, 11131 (2005).

14. B. Rasmussen, I. R. Fletcher, J. J. Brocks, M. R. Kilburn, *Reassessing the first appearance of eukaryotes and cyanobacteria. Nature* **455**, 1101 (2008).

15. A. Dutkiewicz, H. Volk, S. C. George, J. Ridley, R. Buick, *Biomarkers from Huronian oil-bearing fluid inclusions: An uncontaminated record of life before the Great Oxidation Event. Geology* **34**, 437 (2006).

16. S. C. George, H. Volk, A. Dutkiewicz, J. Ridley, R. Buick, *Preservation of hydrocarbons and biomarkers in oil trapped inside fluid inclusions for > 2 billion years. Geochimica et Cosmochimica Acta* **72**, 844 (2008).

17. A. J. Roger, L. A. Hug, *The origin and diversification of eukaryotes: problems with molecular phylogenetics and molecular clock estimation. Philosophical Transactions of the Royal Society B-Biological Sciences* **361**, 1039 (2006).

18. S. Hedges *et al.*, *A genomic timescale for the origin of eukaryotes. BMC Evolutionary Biology* **1**, 4 (2001).

19. D. F. Feng, G. Cho, R. F. Doolittle, *Determining divergence times with a protein clock: Update and reevaluation. Proceedings of the National Academy of Sciences USA* **94**, 13028 (1997).

20. S. B. Hedges, J. E. Blair, M. L. Venturi, J. L. Shoe, *A molecular timescale of eukaryote evolution and the rise of complex multicellular life. BMC Evolutionary Biology* **4**, 2 (2004).

21. E. J. P. Douzery, E. A. Snell, E. Bapteste, F. Delsuc, H. Philippe, *The timing of eukaryotic evolution: Does a relaxed molecular clock reconcile proteins and fossils? Proceedings of the National Academy of Sciences USA* **101**, 15386 (2004).

22. C. W. Dunn *et al.*, *Broad phylogenomic sampling improves resolution of the animal tree of life. Nature* **452**, 745 (2008).

23. B. Rasmussen, S. Bengtson, I. R. Fletcher, N. J. McNaughton, *Discoidal impressions and trace-like fossils more than 1200 million years old. Science* **296**, 1112 (2002).

24. B. Rasmussen, I. R. Fletcher, S. Bengtson, N. J. McNaughton, *SHRIMP U-Pb dating of diagenetic xenotime in the Stirling Range Formation, Western Australia: 1.8 billion year minimum age for the Stirling biota. Precambrian Research* **133**, 329 (2004).

25. S. C. Morris, *Ancient animals or something else entirely? Science* **298**, 57 (2002).

26. B. Rasmussen, S. Bengtson, I. R. Fletcher, N. J. McNaughton, *Ancient animals or something else entirely?—Response. Science* **298**, 58 (2002).

27. M. V. Matz, T. M. Frank, N. J. Marshall, E. A. Widder, S. Johnsen, *Giant Deep-Sea Protist Produces Bilaterian-like Traces. Current Biology* **18**, 1849 (2008).

28. A. Seilacher, P. K. Bose, F. Pfluger, *Triploblastic animals more than 1 billion years ago: Trace fossil evidence from India. Science* **282**, 80 (1998).

29. M. D. Brasier, D. McIlroy, *Neonereites uniserialis from c. 600 Ma year old rocks in western Scotland and the emergence of animals. J. Geol. Soc.* **155**, 5 (1998).

30. T. P. Crimes, The record of trace fossils across the Protoerozoic-Cambrian boundary. In *Origins and Early Evolution of Metazoa*, J. H. Lipps, P. W. Signor Eds. (Plenum Press, New York, 1992), pp. 177–202.

31. S. H. Xiao, X. L. Yuan, A. H. Knoll, *Eumetazoan fossils in terminal Proterozoic phosphorites? Proceedings of the National Academy of Sciences USA* **97**, 13684 (2000).

13
The not-so-boring billion

We have just sketched the development of eukaryote life, without much reference to the environment in which it was occurring. But it raises a major puzzle: Why did eukaryote evolution proceed so slowly through the so-called 'boring billion' (from roughly 2 to 1 billion years ago), and then take off at the end of the Eon? To fully understand eukaryote development, we have to know what the Earth's oceans and atmosphere were like during the Proterozoic and how they were changing. Central to this story, once again, is the history of atmospheric oxygen, and especially the oxygenation state of the ocean, where eukaryotes were evolving.

The conventional view is that eukaryotes evolved in an oxygenated environment, because the great majority of eukaryotes need oxygen and those that do not are thought to have subsequently adapted to an oxygen-free life. In this view, the critical step in eukaryote evolution must have occurred after the origin of oxygenic photosynthesis, and it seems most likely (though not absolutely necessary) that it occurred after the Great Oxidation, when oxygen became widespread. However, this traditional view stems, at least in part, from an over simplified picture of the rise of oxygen that has become enshrined in biology textbooks—one in which the deep ocean is oxygenated throughout the Proterozoic. Instead we now think that atmospheric oxygen hovered at intermediate concentrations through much of the Proterozoic, and the deep ocean remained largely anoxic and sulphidic (*1*). Fig. 13.1 adds recent evidence for this to the fossil evidence for eukaryotes.

An alternative view, consistent with the theories for eukaryote origins described in the last chapter, is that the first eukaryotes had anaerobic metabolism and they evolved in this anoxic and sulphidic ocean. This could explain growing evidence that organisms with anaerobic metabolisms are widely distributed among the eukaryote family tree (*2*). We know that high levels of oxygen were a necessary condition for the development of the large, energy hungry, fast moving animals that evolved in the ocean at the end of the Proterozoic. So, something must have eventually caused oxygen to rise. Here we consider the evidence

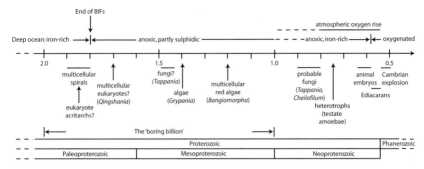

Fig. 13.1 Timeline of the Proterozoic. Adding evidence for the oxygenation state of the atmosphere and ocean to the fossil evidence for the evolution of eukaryotes (Fig. 12.1)

for an anoxic deep ocean in the Proterozoic. Then we consider the mechanisms that were influencing atmospheric oxygen at the time. In Chapter 14 we consider what could eventually have increased it.

13.1 Evidence for a low-oxygen Proterozoic

The traditional view is that the deep ocean became well oxygenated during the Proterozoic, and that atmospheric oxygen levels were uncertain (and possibly variable) but could have been as high as today for most of the last two billion years. This view was built largely on the observed disappearance of banded iron formations (BIFs)—huge deposits of iron oxides, laid down in regular and thick layers. They must have been deposited from oceans that contained high concentrations of iron, and (since the oxidized form of iron is insoluble) this means those oceans were anoxic. BIFs are common in the Archean, but they disappear early in the Proterozoic, the last being the Gunflint and Rove formations in Ontario, laid down about 1.8 billion years ago. The classical explanation of this disappearance, due to Dick Holland, also seems the most obvious: that the deep oceans became oxygenated at the time, and all the iron precipitated out—in simple terms, the oceans rusted out their iron. This broadly fitted with the picture of a continuously oxidizing surface environment since the close of the Archean. It gave little support to the idea that oxygen levels were lower than present during the Proterozoic: if the deep ocean were oxygenated throughout, oxygen concentrations might, for example, have been as high as today for most of the last two billion years. Holland's treatment of the history of oxygen included this as a possibility, and left a wide grey area of uncertainty for Proterozoic oxygen levels (*3*).

13.1.1 The Canfield ocean

This view has been challenged over the last decade by an alternative story, that the deep oceans remained low in oxygen through most of the Proterozoic. In that case, an alternative to Holland's explanation is needed for why BIFs ceased to form. Don Canfield suggested, on the basis of sulphur isotope data, that while the surface ocean would have contained oxygen, the deep oceans were 'euxinic'—profoundly anoxic and containing sulphate, rather than oxygen, as their main source of oxidizing power (4). They would have contained significant concentrations of sulphide, mostly as hydrogen sulphide (H_2S) and hydrosulphide ion (HS^-). The sulphide would precipitate out the iron, just as efficiently as would oxygen, but the insoluble product would be pyrite (FeS_2) rather than iron oxides. Detailed examination of the chemistry of the Gunflint and Rove formations seems to confirm that they were probably laid down in a sulphidic environment (5).

A modern analogue for the 'Canfield Ocean' is the Black Sea, which has a sill at the Bosphorus that prevents the exchange of deep water with the Mediterranean. Today the Black Sea seems perfectly normal at the surface, but its poorly ventilated deep waters are euxinic. Fortunately for the residents of its coasts, the hydrogen sulphide never reaches the surface. Marine anoxic basins exist elsewhere in the world, including the Framvaren and Mariager fjords in Norway and Denmark, which do occasionally vent hydrogen sulphide into the surface waters, resulting in fish kills and bad smells.

Why should the oceans have become sulphide-dominated in this way, early in the Proterozoic? Canfield suggested that the cause was an increase in the chemical weathering of sulphur on the continents as a result of the new appearance of free oxygen in the atmosphere. This would have converted sulphide in continental rocks into sulphate, which is much more soluble. From there it would have washed to the oceans, increasing their sulphur content. If there had also been abundant oxygen in the atmosphere so that the deep ocean was well oxygenated, it would have stayed as sulphate, much as it does in the modern ocean. But if atmospheric oxygen was low so that the deep ocean was anoxic, things would be different. High sulphate in the absence of oxygen are the conditions under which bacteria that reduce sulphate to sulphide can thrive, using it as an electron acceptor—a source of oxidizing power. The end result would be high sulphide concentrations, as in the Black Sea today. Thus, the 'Canfield Ocean' is a hypothesis, which explains the absence of BIFs, but also suggests a low upper limit for oxygen in the atmosphere. It makes specific predictions about the chemistry of the Proterozoic oceans, in particular that they would have contained a lot more sulphide than they do today.

13.1.2 Biomarkers

Over the past few years researchers have been working hard to find more evidence that would either confirm or refute the existence of the Canfield Ocean. Two lines of investigation have recently emerged that are consistent with it. One is the finding of biomarkers from a formation of mid-Proterozoic rocks, laid down probably in a shelf sea (6). This is from the 1.6 billion year old Barney Creek Formation in Northern Australia, which includes some of the oldest 'live' (volatile and gas rich) oil so far found on Earth. The extracted biomarkers include compounds distinctive for both purple and green sulphur bacteria. As we discussed in chapter 8, these are the major groups of anoxygenic photosynthesizers. They are both, of necessity, anaerobic, and excluded by oxygen. Their presence is very revealing therefore. Since they need sunlight, but cannot tolerate oxygen, they are found today only in rare locations where very low oxygen-water is close enough to the surface to receive light. They can be found for example in the Black Sea at the very bottom of the sunlit layer, confined to a narrow depth range that is shallow enough to receive a little light, but deep enough for oxygen to be absent. Poisoned as they are by oxygen, they are excluded from the rest of the oceans today.

These biomarkers have only so far been found at one location in the mid-Proterozoic. The rocks could perhaps have been laid down in an unusual environment (as unrepresentative of the Proterozoic as the Black Sea is of the modern ocean, for example) so we can't be sure from this evidence alone that the Proterozoic ocean was low in oxygen and high in sulphide. But corroborative evidence also comes from a quite different source, courtesy of a somewhat exotic metal, and this does seem to be global in distribution.

13.1.3 Molybdenum

This element (Moly to mechanics, who use it in lubricants, Mo to chemists), is a transition metal—another one of those elements with several oxidation states. Under reducing conditions it is insoluble, but its oxidized form dissolves readily, and today there is plenty to be found dissolved as molybdate ion, (MoO_4^{2-}) in the oceans. In these respects it is similar to several other transition metals, for example uranium. Like uranium, we would expect that, when the atmosphere became oxidizing in the late Archean, the ocean concentration of molybdenum would have increased as it was weathered and dissolved out of continental rocks and washed to the sea. But molybdate also has some additional useful chemistry: it reacts with reduced sulphur to form thiomolybdate ions, (like molybdate but with some of the oxygen atoms replaced by sulphur) and these are not very soluble. Today, even though very little (perhaps 2%) of the global ocean is anoxic at the bottom, thiomolybdate formation in this small percentage is sufficient to be the main removal mechanism for the ocean's Mo. Molybdenum therefore offers us a tool to probe specifically the Canfield ocean idea. If the oceans were oxygenated from top to bottom in the Proterozoic, Mo concentration should have been high then as it is today. But if oxygen was low and there was appreciable sulphide, even if it were only in 10 or 20 percent of the ocean, then that would be enough to strip most of the molybdenum out of the oceans, and its concentrations should have been low. It's another example of those rare but happy coincidences in science that seem almost too good to be true, where nature has provided a tool that, when we read it correctly, gives us just the information we'd like to have.

A recent paper in *Nature* (*1*) presented a record of molybdenum concentrations from black shales, dating from about 2.7 billion years ago to the present (Fig. 13.2). The most striking feature of the record is the step increase in Mo occurring near the end of the Proterozoic. Once the increase has occurred, it is continuous up to the present day. It's hard to escape the conclusion that the oceans were notably sulphidic through most of the Proterozoic, and that they changed to their present day, fully oxygenated mode, in a step late in the Neoproterozoic. A precise dating of the transition is not available at the moment—the time resolution is coarse because black shales are not continuous in time, but the data indicate it was between 663 and 551 million years ago.

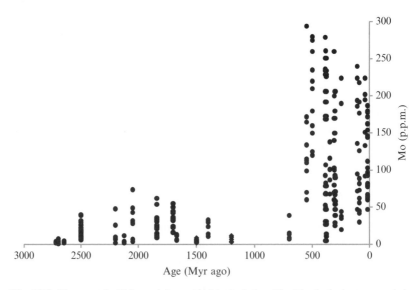

Fig. 13.2 The record of Mo enrichment in black shales. The black circles are euxinic shales; black diamonds refer to non-euxinic, organic rich shales. [Redrawn from data kindly provided by Clint Scott of the University of California, Riverside (*1*).]

As we'll see in the next chapter these dates encompass the final Neoproterozoic snowball Earth, called the Marinoan (ending 635 million years ago), and a later short and probably non-global glaciation called the Gaskiers (around 580 million years ago). It also encompasses the period when we have the first unequivocal evidence for the development of animals (7).

The molybdenum record is the clearest indication we have that there was a step rise in atmospheric oxygen at the end of the Proterozoic. It is hard to make any other cause fit such a decisive change. We can imagine other possible reasons for the sudden disappearance of sulphidic deep waters, but none that can so convincingly explain the fact that Mo increases dramatically and permanently across the divide. Another candidate explanation might be that more oxygen was supplied to the deep ocean by an increase in the rate of formation of deep ocean waters. Alternatively, perhaps tectonic movement isolated the global ocean from a large deep anoxic basin at that time. But such processes might be expected to happen also at other times in Earth history. If one of them were the cause, shouldn't we expect changes back and forth between the high and low Mo states more frequently during geologic time? What we see is just one really large change, occurring simultaneously with a suite of others that mark the end-Proterozoic transition. A major rise in atmospheric oxygen would fit the bill nicely.

In fact, the molybdenum data don't require that we believe the most extreme version of the Canfield Ocean, in which the entire deep ocean was oxygen-free, all of the time, during the Proterozoic. Removal of Mo only requires that anoxia was more common than it is today, when it is virtually absent: say 20% of the ocean is anoxic on average, rather than 2%, as today. The relative scarcity of organic rich black shales, on which the measurements were made, suggests that the nature of the anoxia then may have been different to the modern version. Black shales are associated with periods when the oceans have been anoxic in relatively recent times, because the low oxygen preserves more organic carbon in the sediments, and this is what gives the rock its characteristic colour. However, black shales are relatively unusual in the mid-Proterozoic. This may indicate that the Proterozoic ocean was rather low in productivity compared to today, so that there were rarely conditions where organics were able to build up in the sediments. This in turn could be due to a shortage of limiting nutrients for productivity in the ocean.

Shortage of nutrients has a direct bearing on the 'Canfield Ocean' model. Both atmospheric oxygen level and the concentration of limiting nutrients (that drive productivity in the surface ocean), jointly determine whether the deep ocean is oxygenated or anoxic. Anoxia is generated by an excess of oxygen demand over oxygen supply to the deep waters, and it is limiting nutrient levels that ultimately determine oxygen demand. Canfield's original model assumed that nutrient levels in the Proterozoic ocean were comparable to today and therefore a modest reduction in atmospheric oxygen readily produced an excess of oxygen demand over supply and hence anoxia. However, if limiting nutrients were much less abundant in the Proterozoic ocean (and hence productivity lower), then to produce anoxia in the deep waters would have demanded even less oxygen in the atmosphere.

13.2 Mechanisms for stabilizing oxygen at a low level

So, several lines of evidence and argument point to low (but relatively stable) levels of oxygen in the atmosphere and ocean through much of the Proterozoic. To try and disentangle what was going on, we now turn to look in more detail at the mechanisms that could have maintained such a state. Relatively little modelling work has been done on this, so what we have to say should be treated as work in progress. We start by considering how the fundamental process that had been oxidizing the planet would have changed in the aftermath of the Great Oxidation.

13.2.1 Feedback on hydrogen escape

In the Archean, the origin of oxygenic photosynthesis was essential to the oxidation of the Earth, but not *just* photosynthesis. The origin of cyanobacteria did not at first result in an increase in atmospheric oxygen, because overall, the Earth surface was reducing in the Archean. Photosynthesis did not change that: all it did was to release oxygen at the expense of producing yet *more* reduced material (organic carbon). Most of this carbon was oxidized straight back to carbon dioxide by the oxygen that had been released, doing nothing for the overall oxidation state. A small quantity escaped this fate because it was buried in crustal rocks, but was nevertheless oxidized on the slower time scale, of a couple of hundred million years, when those rocks were recycled by the tectonic mill. The long delay time between the origin of photosynthesis and the oxidation is most readily explained if the surface system started in a substantially reduced state. There was not enough capacity for carbon sequestration in the crust to oxidize the near-surface system to the point where free oxygen could persist. If this had been the whole story therefore, oxygen would never have appeared as a major constituent in the atmosphere.

To bring about the permanent oxidation, photosynthesis had to be coupled with fermentation and methane production. Methane then drove hydrogen escape from the Earth. Hydrogen escape, ultimately fuelled by photosynthesis, and sustained over hundreds of millions of years, was what provided the powerful, permanent oxidation of the planet. It 'titrated' the atmosphere, oceans and crust to the point where free oxygen could begin to build up in the atmosphere.

But once it reached that point, hydrogen escape would begin to be self-limiting (Fig. 13.3). When free oxygen appeared in the atmosphere, the rate of hydrogen escape would be reduced. We can give two specific reasons for this. First, the archaea that made the methane were anaerobes, unable to tolerate oxygen, and were driven underground. The methane they produced would have to diffuse up into the oxygen-containing upper sediment or water column. In the process, methane would be used as a rich source of energy and carbon by other bacteria, methane-oxidizers (methanotrophs) who would consume it by reaction with the oxygen, reducing the outward flux. Second, any methane that did get into the atmosphere would be much more quickly destroyed in the now oxygenated air than previously, because methane and oxygen, in the presence of sunlight and water vapour, react steadily to CO_2 and water. The hydrogen-containing water vapour would be confined by the 'cold trap' to the lower atmosphere. It could not penetrate to the upper atmosphere and could not therefore contribute to hydrogen escape.

Hydrogen escape therefore was ultimately a self-limiting process. Once the atmosphere passed the 'Goldblatt instability' and oxygen was suddenly a permanent constituent, escape would have throttled down to levels just sufficient to sustain that low oxygen concentration, leaving an atmosphere with potentially only a few parts per thousand of oxygen in it. It is our guess that this is one reason why the O_2 in the atmosphere stayed at such a comparatively low level for a long time. The very process which had been mainly responsible for the oxidation had a built-in natural feedback that tended to favour low oxygen concentrations, sufficient to limit the hydrogen-containing gases such as methane and H_2 in the upper atmosphere.

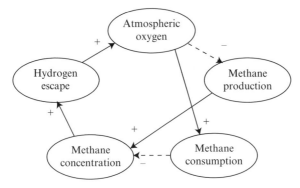

Fig. 13.3 Negative feedbacks on atmospheric oxygen from self-limitation of hydrogen escape

Hydrogen escape probably did much of the oxidizing needed to bring the surface from its original quite reduced state, to the point where an oxygen atmosphere was stable. And hydrogen escape has left an important legacy: today, the continental crust is considerably more oxidized than the mantle of the Earth (8). Much of this oxidation power is carried in an increased content of ferric iron. It is quite difficult to account for this in terms of processes that are still going on today. It would imply that today there is a net export of reducing power, presumably buried carbon, from the surface into the mantle. Since overall the crust is more oxidized than the mantle, this would require that the reducing components of the surface are somehow preferentially returned as tectonic plates are re-incorporated into the mantle after subduction beneath a continent's edge. There is not much evidence that this is occurring to any great extent today. Reduced organic carbon goes into the subduction zones, but it seems to be largely oxidized there, with the products returned to the surface as volcanic gases (9).

How then did the surface get its presently oxidized state? It's likely that this is mostly a leftover from the large amounts of hydrogen escape that occurred during the Archean, perhaps topped up during the Proterozoic, when the escape rate was diminished but still large enough to ensure the atmosphere stayed oxidizing. The self-limiting property of hydrogen escape can help explain the low atmospheric oxygen content during the first part of the Proterozoic. However, it is not the only mechanism critical to determining oxygen levels, and it cannot explain what eventually caused atmospheric oxygen to rise to higher concentrations in the late Proterozoic. We now turn to consider a second key source of atmospheric oxygen in the Proterozoic and the feedbacks controlling it.

13.2.2 Feedback on organic carbon burial

Burial of reduced organic carbon, derived from oxidized carbon dioxide and carbonates, is today the main long-term net source of atmospheric oxygen (see Box 10.1). It represents the small difference between two large fluxes; photosynthetic oxygen production minus respiratory oxygen consumption. The net source of oxygen from carbon burial is in turn counterbalanced by a net sink; the reaction of oxygen with old organic carbon released from the crust, much of it in the process of oxidative weathering. However, the dominant view of geochemists is that the process of oxygen consumption by oxidative weathering is insensitive to oxygen levels—as soon as there is some oxygen it proceeds very efficiently to completion. Hence our search for mechanisms to stabilize, and ultimately increase, oxygen in the Proterozoic now focuses on what would have controlled organic carbon burial during that Eon.

A potentially potent regulator of atmospheric oxygen in the Proterozoic would have existed if the consumption of organic carbon by respiring organisms in the ocean (both prokaryotes and evolving eukaryotes) increased with oxygen concentration, which in turn reduced the flux of organic carbon getting buried into new sedimentary rocks, and hence this source of oxygen (Fig. 13.4). Microorganisms are certainly sensitive to low oxygen levels and they generally switch from aerobic to anaerobic respiration when oxygen concentrations fall below the 'Pasteur point' of about 1% of the present atmospheric level (PAL). As oxygen rises above 1% of PAL we might expect aerobic respiration to increase and hence organic carbon burial to decrease. The resulting negative feedback is included in our model of the Great Oxidation and it helps hold down atmospheric oxygen levels in its aftermath. However, we mustn't forget that organic carbon can also be effectively recycled within an ocean devoid of oxygen, by organisms that convert it to methane, which then reacts with oxygen (either in the atmosphere or via methanotrophs). In a sulphidic 'Canfield ocean', biological methane production would be suppressed, but instead sulphate reducing organisms would recycle organic carbon. Hence it is not clear that the amount of organic carbon escaping consumption can be strongly modulated by the response of a diverse bunch of consumers to changing oxygen levels.

Instead current theories focus on what controls the net production of organic carbon by photosynthesizers, as this does affect the amount that gets buried. In today's ocean the flux of organic carbon sinking out of the surface ocean is determined by the supply of limiting nutrients into the surface waters, and the flux of organic carbon leaving the ocean as a whole in new sediments is limited by the supply of nutrients to the ocean as a whole. There is some flexibility, because the ratio of carbon to nutrient in the material that is eventually buried can vary. However, we start with the simple idea that the supply of nutrients to life in the

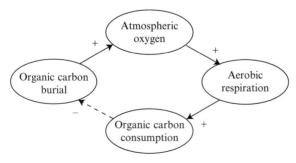

Fig. 13.4 Potential negative feedback on atmospheric oxygen involving aerobic respiration

Proterozoic ocean ultimately determined organic carbon burial and the corresponding source of oxygen.

At some point during the Proterozoic we think that nutrient supply to the ocean and hence organic carbon burial increased, and this drove an increase in atmospheric oxygen. It didn't need a huge increase, because little net carbon burial is required to explain where the present 21% oxygen in the atmosphere has come from (see Box 10.1). Just a couple of percent net increase in the amount of reduced carbon in continental rocks can explain this. Once the main work of oxidizing the mass of continental rock had been accomplished by hydrogen escape, rather small adjustments to the redox balance—small enough to be compatible with the observed carbon isotope variations—can account for the large build-up of oxygen in the atmosphere. As oxygen increased it would have shut off the last vestiges of hydrogen escape, reducing it to today's very low level.

In principle, a gradual increase in nutrient supply and the net amount of organic carbon buried could have occurred slowly, distributed throughout Proterozoic time, leading to a slow upwards drift in atmospheric oxygen. However, the evidence in support of the 'Canfield Ocean' suggests instead that a step change (or several steps) in organic carbon burial, atmospheric oxygen, and ocean oxygenation occurred in the Neoproterozoic, contemporary with the wild swings of the carbon cycle and climate occurring at that transition. If so, we must explain what held organic carbon burial in check for most of the Proterozoic, and what caused it to increase near the end. The key to this, we think, is the supply of the nutrient phosphorus to the ocean, weathered from rocks on the land surface.

13.3 A nutrient-starved ocean?

As we saw from the biomarker record, the Proterozoic ocean had representatives of all the fundamentally different types of photosynthesizers. Green and purple photosynthetic bacteria apparently thriving in sunlit, anoxic water, fixed carbon without producing oxygen. Oxygenic photosynthesis was performed by cyanobacteria and, by the mid-Proterozoic, by some of the groups of eukaryote algae that are important in today's oceans. But nutrient cycling was very different then, and nutrients were harder to come by.

13.3.1 Phosphorus

Phosphorus (P for short) is, among the big six elements for life, geochemically the toughest to cycle. P is an absolute requirement for all life (being, for example, a constituent of DNA) but it is hard to mobilize: there are no common gases that contain it, so it cannot be transported as a gas through the atmosphere. Furthermore in its normal mineral forms it is very insoluble. If you take typical phosphorus-bearing rock, grind it up and try to dissolve it in seawater, you find that only parts-per-million quantities of it will dissolve (*10*), though its solubility can be substantially increased using acid solutions rather than water.

On the modern Earth, P reaches the ocean by being weathered out of the continental rocks and washed down rivers, or transported there by atmospheric dust. Land plants and fungi have specialized in techniques to extract it from the rocks, as we will discuss in Chapter 16, and help to solubilize it using organic acids and other tricks. Nevertheless, even in the modern world, a great deal of the P in weathered rocks is 'detrital'. It is never dissolved or mobilized from the mineral phase, and simply bypasses biology. When the rock is weathered, it remains bound in grains of sediment or dust. Washed to the sea by wind or water, these are deposited in dispersed form in the surface sediments around the margins of the continents, and the phosphorus in them is never accessed by life. It's quite hard to get an estimate for the proportion of the total P delivered to the ocean in this way. Froelich (*11*) estimated that only 25% of the phosphorus weathered from the continents enters the ocean's biological cycle, 75% therefore is lost to sediments before it dissolves in the ocean. Recent discussions suggest the detrital fraction is between 20% and 40% of the total (*12, 13*).

So P is not easy to get hold of even today, but in the Proterozoic we'd expect things to have been much more difficult still. The land had not been colonized, and there were no land plants to mine the nutrients from continental rocks. Consequently the availability of P would have been lower, and the fraction that bypassed biological cycling higher.

The majority of phosphorus in continental rocks is in dispersed form, phosphates typically forming 0.1% of a sedimentary rock for example. However, some is found in phosphate-rich deposits, usually in minerals called apatites, which may contain 20% or more phosphate. The P in such phosphatic rocks is thought to have been predominantly biologically derived, and biological processes are responsible for its concentration prior to mineralization. For example, phosphate can be removed from solution when iron is precipitated out by the formation of oxides and hydroxides, so P that was previously biologically available can be trapped in this way during formation of BIFs. Biologically utilized P is also transported to the sediments bound in organic matter or on calcium minerals formed by organisms. In either case it may then be concentrated by the action of bacteria within the near-surface sediments (*13*) to eventually form apatites. So the abundance of phosphatic rocks can be taken as a rough indicator of how active the biological part of the phosphorus cycle is.

Phosphatic rocks are extremely rare before the Neoproterozoic. There's some evidence of them starting about 750 million years ago, but they suddenly become abundant at around 600 million years ago, and continue up to and beyond the boundary with the Cambrian (*14*), at which there is a 'phosphogenic event' (*15*). They are then found in varying quantities, through the Phanerozoic up until the present day. We could argue that their near-absence through much of the Proterozoic is due to poor preservation, or that their appearance is a secondary effect connected with the disappearance of the sulphidic oceans or increase of oxygen in the atmosphere. However, the simplest explanation is that phosphorus becomes much more available to the biota starting in the late Proterozoic. Instead of being transferred largely untouched from the continents to sediments, it began to be increasingly mobilized.

The extensive phosphatic rocks of the late Neoproterozoic are important then, because they attest to a fundamental change in the phosphorus cycle on the planet. Those of the Doushantuo formation in Southern China (spanning about 635 to 550 million years ago) also have another important part to play in telling the story of the time, because they are a site of exceptional preservation of microfossils. Preserved in the phosphate-rich rock are fossils which reveal the development of fungi, algae and the very earliest animals—in the form of early stage embryos of a few cells (we will return to these in Chapter 14).

13.3.2 Nitrogen

An alternative, but we think complementary, idea is that lack of nitrogen held back the productivity of the oceans (and hence carbon burial and atmospheric oxygen), during the Proterozoic. This idea (as far as we are aware) is due to Paul Falkowski of Rutgers University—known to many of his colleagues as 'Falko'.

Falko's reasoning goes like this: In completely anoxic environments (the deep Black Sea for example), fixed nitrogen, the kind readily taken up by organisms, exists mostly as ammonium ions, (NH_4^+). In fully oxidized environments, (most of the modern oceans), the stable form of fixed nitrogen is nitrate ions, NO_3^-. Either form can be taken up by plankton to fuel carbon fixation. In between, at low but non-zero oxygen concentrations, neither form is stable. Ammonium is quickly reacted to nitrate. Nitrate in these conditions is utilized by denitrifying bacteria, as an electron acceptor—a source of oxidizing power. In the process they reduce it to nitrous oxide and to nitrogen gas, N_2. In either of these forms, it is biologically nearly useless: the two nitrogen atoms bond strongly to each other, and in N_2 the N-N triple bond is very hard to break. Unavailable to all life except the specialist nitrogen fixers, the fate of these gases is to be returned to the large pool of nitrogen gas in the atmosphere, and it is lost to biology in the oceans.

This would not matter if sufficient nitrogen fixation were occurring elsewhere to make up for the loss. Nitrogen fixation, as we discussed in Chapters 4 and 7, is practiced by several groups of bacteria including cyanobacteria, but it is a metabolically expensive process. In an ocean with a very large volume of low-oxygen waters, such as would have occurred through most of the Proterozoic and which favour denitrification, the nitrogen fixers may have simply been unable to keep up with the denitrifiers, so that the ocean became low in nitrogen and this limited its productivity.

Even in the modern ocean, when deep, nutrient-rich water upwells to the surface and is used by plankton to fix carbon, it is usually fixed nitrogen that is the first nutrient to run out: oceanographers refer to nitrate as the 'proximate' limiting nutrient. However, it only *just* runs out before phosphate does. It's a remarkable fact that the concentrations of phosphate and nitrate in deep water are almost exactly in the ratio that most plankton need to grow, with just a slight excess of phosphate. This was first pointed out by the oceanographer Alfred Redfield in the 1930s, and it's commonly referred to as the Redfield paradox (or puzzle).

It's now generally accepted that this situation results from an elegant negative feedback, that keeps oceanic phosphorus and nitrogen closely tied to one another. The slight fixed nitrogen deficit creates a niche in the surface ocean that enables nitrogen fixers, also known as 'diazotrophs', to flourish there. Because fixed nitrogen runs out just before phosphorus, it's worth their while to fix nitrogen, expensive though it is in terms of energy: all the other nutrients they need are in adequate supply, so by fixing nitrogen the diazotrophs have all they need. Furthermore, because only they are able to make a living in an environment that lacks fixed nitrogen, their competitors are kept at bay.

The feedback mechanism works like this (Fig. 13.5a): suppose the diazotrophs are fixing nitrogen faster than the denitrifiers, (who are deeper in the water column, and may be thousands of miles away) are removing it. As the fixers die, their fixed nitrogen is released into the ocean, and eventually the nitrate concentration of the ocean as a whole increases. Then, nitrate no longer runs out before phosphate when water upwells. At that point the diazotrophs find they have put themselves out of a job: they no longer have a competitive advantage because nitrogen is no longer in shorter supply than other nutrients. Other plankton, that don't use energy to fix nitrogen, are now at an advantage. The nitrogen-fixers decline and the rate of nitrogen fixation declines with them. This feedback system eventually comes to a steady state when the N fixation rate equals the denitrification rate, and when nitrate is just a little scarcer than phosphate. The result is that, though nitrate is the 'proximate' limiting nutrient, phosphate is the really important one, which sets the overall amount of productivity.

(a)

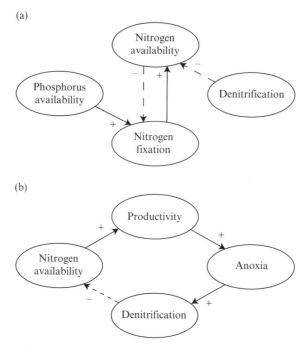

(b)

Fig. 13.5 Negative feedbacks in the ocean involving: (a) control of nitrogen fixation by phosphorus and nitrogen availability, and (b) self-limitation of denitrification

Falko's idea then is that, in the Proterozoic ocean, this feedback mechanism is reset. Because conditions then were much more conducive to denitrification, fixed nitrogen fell below 'parity' with phosphate. However, our own modelling, and also that of a recent paper which includes Falko as an author (*16*), suggests that things are not quite so straightforward. A key reason is that denitrification is ultimately a self-limiting process (Fig. 13.5b): Denitrifiers depend on anoxia—they only gain an advantage over organisms performing aerobic respiration when oxygen has virtually run out in the deep waters. However, by removing available nitrogen from the water, denitrifiers reduce productivity in the surface ocean, and it is this productivity that fuels oxygen consumption and hence anoxia at depth. The result is another negative feedback mechanism, which keeps the activity of the denitrifiers in check.

The overall result is a strong interplay between oxygen, phosphate and nitrogen concentrations, and while there are conditions in which nitrogen limitation can hold back the system, it usually follows phosphate: if phosphate concentrations are high enough, fixed nitrogen concentrations tend to increase. We interpret this to mean that usually, it remains phosphorus rather than nitrogen which is the fundamental limiting factor.

13.3.3 Trace metals

Related to the idea of nitrogen starvation is a further suggestion for limitation of the Proterozoic oceans. It has been argued that the most important limitation might not be a major nutrient such as phosphorus or nitrogen, but trace metals, in particular molybdenum and iron (*17*). As we've seen, the concentrations of both these metals would have been low in the mid-Proterozoic ocean, compared to the Archean. It's an interesting fact that nitrogenases, the enzymes required to fix nitrogen, all require trace metals. The most common ones utilize, guess which two? Iron and molybdenum. There are other alternatives, with some organisms using vanadium instead of molybdenum, and some iron alone. The lack of iron and molybdenum would certainly not have helped the mid-Proterozoic oceans with their nitrogen supply problem. It would likely have further added to the set of circumstances which tended to restrict the supply of nutrients to the ocean, and keep its productivity in check.

Indeed, it should have led to an additional negative feedback that tended to stabilize anoxic and sulphidic ocean conditions (Fig. 13.6). Imagine if some factor tended to increase nitrogen availability, this would fuel increased productivity and promote anoxic and sulphidic conditions in the deep ocean. This in turn would reduce the availability of molybdenum and iron, suppressing nitrogen fixation, and counteracting the initial increase in nitrogen availability. (Equally if some factor tended to reduce nitrogen availability, this would suppress productivity, reducing anoxia and sulphidic conditions, increasing molybdenum and iron availability, increasing nitrogen fixation, and counteracting the initial decrease in nitrogen availability.) Together with the other negative feedbacks we described above (Fig. 13.5), we have a suite of mechanisms that can help explain the maintenance of anoxic and sulphidic ocean conditions through much of the Proterozoic.

Interestingly, the availability of trace metals in the mid-Proterozoic oceans also provides some clues regarding the evolution of eukaryotes. Careful analysis of the genetic encoding of biochemical pathways shows that some key enzymes for eukaryote metabolism have at their active sites different metal atoms to the equivalent enzymes in prokaryotes (*18*). Zinc in particular is widely used in the proteins of eukaryotes but not prokaryotes. Zinc would have been vanishingly scarce in either an anoxic or a euxinic (sulphidic) deep ocean, but would have been readily available in oxygenated waters. The widespread use of copper and molybdenum in eukaryote metabolism also suggest an origin in oxygenated waters. This lends some support to the traditional view that eukaryotes evolved in an oxygenated environment. Indeed if the deep

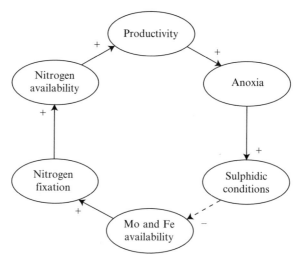

Fig. 13.6 Negative feedback in the Proterozoic ocean involving limitation of nitrogen fixation by molybdenum (Mo) and iron (Fe)

ocean remained anoxic and sulphidic for much of the Proterozoic, it makes it seem unlikely that eukaryotes evolved either in the deep ocean or in the oxygenated surface waters that were mixing with it. Instead the most likely location for eukaryote evolution would have been in oxygenated coastal seas where water running off from the land would have been rich in zinc, copper and molybdenum.

13.4 Summary

We have several possible nutrient-related ideas then, for why the Proterozoic oceans may have been less productive than the modern ones. The ideas aren't mutually exclusive—in fact they are mutually supportive. Of them, nitrogen, iron and molybdenum limitation were *consequences of* a risen but still low concentration of oxygen in the ocean—they may also have been *contributors to* the low atmospheric oxygen, if the low ocean productivity that they promoted helped restrict the amount of carbon being buried. In other words, they were positive feedbacks on atmospheric oxygen (Fig. 13.7).

In the late Proterozoic, if some process tended to increase atmospheric oxygen to the point that it began to remove anoxic conditions from the deep ocean then these feedbacks would have kicked in: The loss of sulphidic conditions (which would have happened first) would increase molybdenum and iron availability, fuelling nitrogen fixation, increasing nitrogen availability, productivity and organic carbon bur-

(a)

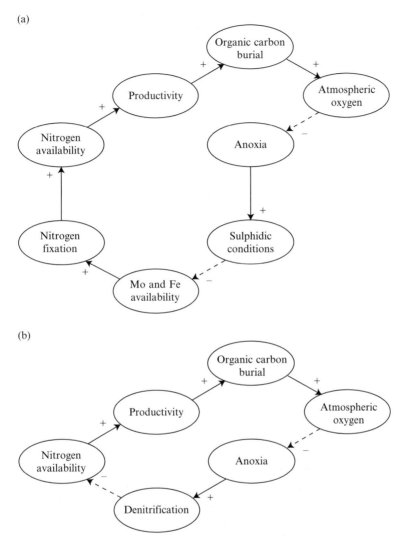

(b)

Fig. 13.7 Positive feedbacks on atmospheric oxygen in the Proterozoic involving (a) molybdenum (Mo) and iron (Fe) limitation of nitrogen fixation and (b) denitrification

ial, and thus amplifying the rise in atmospheric oxygen (Fig. 13.7a). If anoxia began to be removed from the deep waters altogether, then denitrification would be suppressed, increasing nitrogen availability, productivity and organic carbon burial, thus further amplifying the rise in atmospheric oxygen (Fig. 13.7b). These positive feedbacks may help to explain the rapid transition to higher atmospheric oxygen in the late Proterozoic. However, neither represents a primary driver of change.

The first mechanism, low phosphorus supply, increasing, perhaps by an order of magnitude, in the late Proterozoic, can be called upon to play a more primary role in driving up oxygen concentration at the time. Phosphorus became far more available in the oceans in the late Proterozoic, and the extra P must have come from the land. A good reason for this would be if biology was beginning to get a foothold on the land surface, and take control of continental weathering. In fact we argue that P supply and atmospheric oxygen would have increased at the right time if (and only if) the land surface was beginning to be colonized by the late Proterozoic (*19*). Geological drivers can be invoked as well, but the rise in oxygen was permanent, whereas geological changes tend to be cyclical. This is well before most textbooks describe the colonization of the land, so now we had better examine this idea and see what other evidence there is for it.

References

1. C. Scott *et al.*, *Tracing the stepwise oxygenation of the Proterozoic ocean. Nature* **452**, 456 (2008).

2. M. Mentel, W. Martin, *Energy metabolism among eukaryotic anaerobes in the light of Proterozoic ocean chemistry. Philosophical Transactions of the Royal Society B-Biological Sciences* **363**, 2717 (2008).

3. H. D. Holland, *The Chemical Evolution of the Atmosphere and Oceans.* (Princeton University Press, Princeton, 1984).

4. D. E. Canfield, *A new model for Proterozoic ocean chemistry. Nature* **396**, 450 (1998).

5. S. W. Poulton, P. W. Fralick, D. E. Canfield, *The transition to a sulphidic ocean similar to 1.84 billion years ago. Nature* **431**, 173 (2004).

6. J. J. Brocks *et al.*, *Biomarker evidence for green and purple sulphur bacteria in a stratified Palaeoproterozoic sea. Nature* **437**, 866 (2005).

7. G. J. H. McCall, *The Vendian (Ediacaran) in the geological record: Enigmas in geology's prelude to the Cambrian explosion. Earth-Science Reviews* **77**, 1 (2006).

8. D. J. Frost, C. A. McCammon, *The redox state of Earth's mantle. Annual Review of Earth and Planetary Sciences* **36**, 389 (2008).

9. J. M. Hayes, J. R. Waldbauer, *The carbon cycle and associated redox processes through time. Philosophical Transactions of the Royal Society B-Biological Sciences* **361**, 931 (2006).

10. E. Atlas, R. M. Pytkowicz, *Solubility behavior of apatites in seawater. Limnology and Oceanography* **22**, 290 (1977).

11. P. N. Froelich, *Kinetic control of dissolved phosphate in natural rivers and estuaries—A primer on the phosphate buffer mechanism. Limnology and Oceanography* **33**, 649 (1988).

12. C. R. Benitez-Nelson, *The biogeochemical cycling of phosphorus in marine systems. Earth-Science Reviews* **51**, 109 (2000).

13. K. B. Follmi, *The phosphorus cycle, phosphogenesis and marine phosphate-rich deposits. Earth-Science Reviews* **40**, 55 (1996).

14. P. J. Cook, J. H. Shergold, *Phosphorus, phosphorites and skeletal evolution at the Precambrian Cambrian boundary. Nature* **308**, 231 (1984).

15. Y. Shen, M. Schidlowski, X. L. Chu, *Biogeochemical approach to understanding phosphogenic events of the terminal Proterozoic to Cambrian. Palaeogeography Palaeoclimatology Palaeoecology* **158**, 99 (2000).

16. K. Fennel, M. Follows, P. G. Falkowski, *The co-evolution of the nitrogen, carbon and oxygen cycles in the Proterozoic ocean. American Journal of Science.* **305**, 526 (2005).

17. A. D. Anbar, A. H. Knoll, *Proterozoic ocean chemistry and evolution: A bioinorganic bridge? Science* **297**, 1137 (2002).

18. C. L. Dupont, S. Yang, B. Palenik, P. E. Bourne, *Modern proteomes contain putative imprints of ancient shifts in trace metal geochemistry. Proceedings of the National Academy of Sciences USA* **103**, 17822 (2006).

19. T. M. Lenton, A. J. Watson, *Biotic enhancement of weathering, atmospheric oxygen and carbon dioxide in the Neoproterozoic. Geophysical Research Letters* **31**, L05202 (2004) pp. 1–5.

14

The Neoproterozoic

We hypothesize that photosynthesizing life colonized the late Proterozoic land surface, liberating phosphorus from rocks, and driving up atmospheric oxygen, but what could have been the responsible organisms? Plants do not appear on the land surface until much later, around 470 million years ago (as we will discuss in Chapter 16). These first plants were bryophytes, a classification which contains modern mosses and liverworts—and it is their characteristic spores that record their first occurrence. However, we can be pretty sure that bryophytes were not the first organisms, or even the first eukaryotes, to venture on to the land. We can get some clues to the nature of the first colonizers from what happens when bare rock is colonized today. The first organisms to get a foothold are usually not bryophytes, but lichens. You'll see lichens encrusting gravestones in our older churchyards for example, and growing on stone buildings that are more than a decade or so old, where they can be responsible for speeding up the weathering and deterioration of the buildings, especially if they are built of limestone. Lichens are remarkable not only for their ability to colonize bare rock but because they are a partnership between quite different types of life.

14.1 Lichens—a marriage made in heaven

The Swiss botanist Simon Schwendener was the first to realize that lichens are not single organisms, but a symbiosis between fungi and algae, across two of the fundamental divisions of life. Schwendener was a farmer's son, marked out to take over the family farm, but his talents and inclination lay in science rather than farming. By dint of his talent for botany, and a mastery of techniques of light microscopy, he rose steadily to become Professor at the University of Basel (1). In 1867, shortly after that appointment, he put forward his 'dual lichen' hypothesis at the annual meeting of the Swiss Natural History society. His idea was met with great scepticism then, and until the end of the nineteenth century continued to be vigorously, and sometimes aggressively, rejected by the lichenoligists of the day (*1*). Attempts to disprove the dual nature of lichens continued until the 1950s, by which time however, they had been successfully recreated in the laboratory by combination of separately cultured fungi and algae.

Lichen consists of a photosynthesizer, which can be either a eukaryote alga or a cyanobacterium, in close association with a fungus. Usually the fungus wraps itself around the algal cells, but sometimes it actually penetrates them. There are many different fungal-algal pairs that form lichens, and the association has clearly formed repeatedly during the course of evolution. (Indeed sometimes different parts of the same lichen will contain prokaryotic cyanobacteria and eukaryotic algae.) Any of dozens of different types of algae and cyanobacteria can form them, but most of these can also be found free-living. However, the fungi that form lichens are rarely found apart from their algae. We can deduce from this that there is a sense in which the fungus is the controlling partner of the two, and Schwendener himself described lichens as a parasitism of algae by fungi. The fungus takes up metabolites such as polysaccharides that the algae produce. Effectively it is using the algae as a source of carbon, and benefiting thereby from their ability to perform photosynthesis. But it is clear that the algae also gain benefit from the association, most especially by getting increased access to nutrients and minerals.

The fungi put substantial resources into producing organic acids (oxalic acid is the most prevalent) which chemically dissolve the rock, helping them to get access to the elements they need. They also promote dissolution of the rock by releasing CO_2, forming carbonic acid and promoting weathering. They grow long hyphae: these are hair-like extensions of their cells which are unique to fungi. These are made possible only by the extensive intra-cellular cytoskeleton that the eukaryotes evolved during the Proterozoic. The hyphae of lichens penetrate the underlying rock (Fig. 14.1), often along cracks and fissures, sometimes to a depth of several millimetres (*2*). They also can take up water, which penetrates into the rocks along the lichen. This assists in the dissolution, and may also freeze, helping to further break up the rock. By all of these techniques, the lichens contribute to the weathering of rock surfaces (*3*). The combination of penetrative hyphae and organic acids are used to dissolve, mobilize and take up the elements that the two organisms need, in particular phosphorus. When the hyphae hit on a vein of phosphorus rich rock such as apatite they spread out and multiply along the vein, dissolving out the nutrient. Lichens can literally hollow out rock and penetrate it, and some, 'endolithic' lichens live inside it rather than on its surface, with their photosynthetic algae so close to the interface that they still receive light.

The fundamental reason that fungal-photosynthesizer associations are so successful (both lichens and mycorrhizal—plant associations,

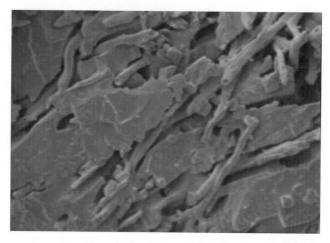

Fig. 14.1 Fungal hyphae associated with a crustose lichen boring into marble (from Swartbankberg, Namibia). This scanning electron microscope image shows an area approximately 75 by 50 microns. [Photograph kindly provided by Heather Viles of the University of Oxford.]

which today are fundamental to the presence of plants on land) is that the needs of the two partners are so well matched. Photosynthesizers can fix carbon and derive energy out of water and carbon dioxide, providing they also have access to nutrients. Fungi can access nutrients out of rock, but the trick of photosynthesis has eluded them. Together they formed (and still form) a partnership which could prosper in environments where neither by itself could survive.

Typically today, ecological succession on bare rock involves the slow establishment of lichens, growing at a millimetre or two a year. They can grow here because of their hyphae and organic acids which enable them to penetrate and extract minerals from the rock, their adaptation to slow growth, and their ability survive freezing and dessication. After sufficient time has elapsed (it may be centuries), a part mineral, part organic dust begins to be trapped around and beneath the lichen. Mosses can begin to grow in this medium, and if it is sufficiently undisturbed, they will eventually grow over the lichens and supplant them. The first step to what will eventually become soil has been taken. In the modern world then, colonization follows a sequence starting with lichens, after which bryophytes appear, and only after that do larger, more recently evolved plants begin to establish themselves. This is suggestive that a similar sequence of events occurred during the first colonization of the land surface. As it happens, most modern lichens probably are relatively late evolved and were not around during the Proterozoic. However, there is good evidence that the lichen style of symbiosis, between fungi and photosynthesizers, evolved early, and played a crucial role.

14.2 Fossil and molecular evidence for early 'proto-lichens'

Unfortunately, the fossil record for lichens is very sparse. They are a poor candidate for fossilization, since they have no mineral parts to resist degradation, and they flourish most often on bare rock and dry land, whereas fossils are only laid down in sediments, under water. Fossil evidence for lichens and lichen-like symbioses comes therefore almost exclusively from the very rare environments where exceptional preservation has occurred. For instance they are found preserved in amber—but not of course until the 'recent' past when trees produced amber. The earliest examples of structures clearly like modern lichens (4) come from the early Devonian Rhynie Chert in Scotland, which is another location of exceptional preservation. However, the earliest 'lichen-like symbiosis' so far documented comes from the Doushantuo phosphorite formation, already mentioned above. Here, about 600 million years ago, precipitation of phosphates preserved a window on life in the late Proterozoic shallow seas, which includes a close association between algal cells and fungi with hyphae that seem to be enveloping them (5). However, these fossil organisms don't resemble in detail modern lichens, not least because they lived underwater rather than on exposed rocks (as do many fungi forming lichen-like associations today).

It would be satisfying at this point to be able to report that modern lichens have clearly survived unchanged from the Proterozoic, and that we can take them as close models of the organisms that first colonized the land. Unfortunately we can't, at least for the dominant lichens to be found today. Molecular analysis of the fungal evolutionary tree indicates that most of the modern lichen-forming fungi are 'higher' fungi (Ascomycota), arising in the 'branches' of the tree (6), and not particularly close to the presumed root, where the fungi first emerge. However, this does not necessarily mean that lichens are recent in origin. Recent work has shown that many higher fungi that do not form lichens have lichen-symbiotic ancestors (7), suggesting that the lichen habit may have been the ancestral state for the first higher fungi.

Furthermore, it now appears that the fungal lineage could be very ancient, with much of its diversification having happened back in the Proterozoic. The fossil record of fungi potentially extends as far back as 1.4 billion years ago (as discussed in Chapter 12). Some molecular clock estimates also give an origin for fungi over 1.4 billion years ago (8), and molecular evolutionist Blair Hedges has been arguing for some time for an early date for the colonization of the land. As discussed in

Chapter 12, we remain wary of the earliest molecular clock dates without corroboration from other sources. However, more recent 'relaxed' molecular clocks (see Fig. 12.2) still suggest an early divergence of the ancestors of fungi and animals around 960 million years ago, and the divergence of major fungal lineages (including the start of the Ascomycota) around 730 million years ago, right around the time of the first extreme Neoproterozoic glaciations (see Fig. 14.2).

Considerable evolution has occurred in many of the lichen-forming fungi during the last 150 million years, so we think it probable that most modern lichens are not very closely related to our postulated Neoproterozoic early colonizers. On reflection, this is not so surprising. The environment of the land surface has changed dramatically since that time, with the coming of plant competitors, and especially animal grazers. If the early proto-lichens were at all nutritious, they would have been a prime target for being eaten, and since they must have been constrained to grow only slowly given that their prime source of nutrient was bare rock, they might have gone rapidly extinct once land animals evolved. Nonetheless, we can hope that some of their ancestors would have survived relatively unchanged. There remains a possibility therefore that there still exist today among the more primitive fungi, examples of the proto-lichen symbiotic relationships we postulate for the late Proterozoic.

Indeed such relationships do exist, and among fungi that are closer to the root of the fungal genetic tree. *Geosiphon pyriformis* for example, is the only known fungus that 'captures' cyanobacteria as endosymbionts. It lives at the surface of nutrient-poor soils. Today it is rare, known in the wild at only one site in Germany. The tips of its hyphae seek out cyanobacteria of a particular genus, and when they contact

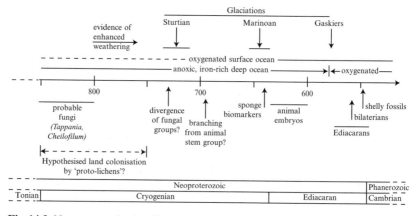

Fig. 14.2 Neoproterozoic timeline concentrating on the Cryogenian and Ediacaran periods. The precise timing, especially of the Sturtian and Marinoan glaciations is uncertain

them, engulf them. Over time the tip then swells into a bladder a millimetre or two in size, within which the cyanobacteria continue to thrive and multiply (9). Taxonomically, *Geosiphon* is closely related to fungi that form mycorrhizae with (higher) plants, about which we'll have much more to say in Chapter 16. The original 'proto-lichens' which we postulate first colonized the land in the Neoproterozoic are probably therefore better described as proto-mycorrhiza. They would at first have formed symbioses with earlier-evolved algae rather than with the relatively advanced land plants with which they are mostly associated today. Most of these early relationships have now been supplanted by symbioses with more successful and later-evolved plants. In this scenario, *Geosiphon* is a lonely representative of an earlier, now largely outmoded, way of life.

14.3 Geochemical evidence of biotic weathering

We are hypothesizing a colonization of the land starting around 800 million years ago in the Neoproterozoic (see Fig. 14.2) more than 300 million years before the Ordivician date that is normally accepted for the first land plants. We've put this forward as a reason for the increased abundance of phosphorites as we approach the Proterozoic–Phanerozoic boundary. But photosynthetic activity on land, and the increased chemical weathering caused by land colonization should have left other geochemical traces. Surely changing the whole nature of chemical weathering on the continents would have more general effects, creating soils and altering sediments, and some other evidence should document it?

Such effects are indeed visible, and are well known to geologists, though an explicit link to the invasion of the land has only recently been pointed out by Martin Kennedy and colleagues (*10*). The products of extensive chemical weathering of continental rocks are clay minerals, fine-grained and of entirely different mineralogy to sediments derived from mostly physical erosion. Physically weathered sediments that have not gone through a soil phase retain more of the original, metamorphic and igneous minerals, such as feldspar and quartz. Globally there is a transition away from these and towards clay minerals, associated with the Phanerozoic. It occurs over a very long time base—indeed it might be considered to be going on still today. The important question for the present discussion though is; when did it begin? This would mark the first influence of increased chemical weathering on the continents. Kennedy *et al.* present data to suggest that the shift gets under way in the Neoproterozoic, perhaps around 800 million years ago.

We should be clear that there were probably prokaryotic communities on land before our postulated proto-lichens. There is evidence of the characteristic carbon isotope signature of photosynthesis recorded in rock crusts and ancient soils from the middle of the Proterozoic (*11*, *12*). Some accompanying microfossils and other much earlier evidence suggests microbial mats, fuelled by cyanobacteria (*13*). Some of these microbial communities lived under rock surfaces, others bound sand together (*14*), and some apparently sat atop soils. Remarkably there are ancient, clay rich formations called 'laterites' from the early Proterozoic, which are similar to those found today in tropical settings under rich vegetation cover (*15*). Laterites typically have iron and phosphorus leached out of them by the action of organic acids, and this same signature is seen in Proterozoic examples (*16*). Once fungi were present they could have contributed to leaching as they can be prolific producers of organic acids, especially when in symbioses as lichens. Intriguingly the carbon isotope shift of these scattered terrestrial samples is stronger by 800 million years ago than at 1.2 billion years ago, suggesting some development of the land community.

A final source of evidence for early land colonization comes from a recent reinterpretation of the carbon isotope record of marine carbonates by Paul Knauth and Martin Kennedy (*17*). As we showed in Fig. 2.1, this swings around wildly in the Neoproterozoic, including some very negative values which have spurred complex explanations including the oxidation of a massive pool of organic carbon in the ocean (*18*). However, such negative values are also seen in Phanerozoic carbonates from coastal sediments, where they are readily explained by an influx of isotopically light plant biomass from the land surface, transported via groundwater. Knauth and Kennedy show that the combined variations of carbon and oxygen isotopes in Neoproterozoic samples cover the same range as Phanerozoic ones, and are consistent with the same cause. The inference is that there was considerable photosynthetic productivity on the Neoproterozoic land surface, supplying carbon via groundwaters to coastal sediments. The signal is not apparent before about 850 million years ago, consistent with the other sources of evidence for greening of the land around this time.

14.4 Consequences for the atmosphere and climate: rising oxygen, falling carbon dioxide

Let us assume then that land colonization by eukaryotes began at some point around 800 million years ago. We can't as yet call the organisms responsible 'plants', but we assume they were associations of fungi with algae and/or cyanobacteria—proto-lichens. They began to accelerate the chemical weathering of continental rocks, and this would have had two fundamental effects on the atmosphere: increasing oxygen, and reducing carbon dioxide concentrations.

Why should oxygen go up? It's not quite as simple as, 'more plants equals more biological productivity, equals more oxygen'. A rise in oxygen occurs because overall, there is burial of more reduced carbon. (We'll assume for a moment that weathering of reduced products on the land, which is the main sink, remains constant.). More carbon would be buried because the land biota make more nutrients biologically available. Among these nutrients is phosphorus, more of which is washed in dissolved form to the ocean. This increases the biological production there. In fact, the dominant way that P, once dissolved in the oceans, is removed to the sediments is in the bodies of dead organisms, and this process buries carbon as well as phosphorus. This represents a net source of oxygen to the atmosphere.

It is really the availability of the extra nutrients that enables a sustained increase in carbon burial. A 'pulse' of increased carbon burial is possible without such an increase in supply of nutrients, but because phosphorus in particular is removed from the ocean and buried along with carbon, the greater burial rate would very quickly deplete P. This would throttle down the carbon burial. Our modelling indicates that you can't bury enough extra carbon to noticeably affect the atmospheric oxygen content, unless there is a sustained increase in the supply of limiting nutrient.

Of course, if other nutrients were limiting, such as nitrogen or molybdenum, then it would need addition of those also, to produce the increase in oxygen. There is a difference however. Removal of organic carbon by burial also usually removes phosphorus, whereas nitrogen and molybdenum removal are not so inevitably tied to carbon burial. For example, most of the nitrogen that follows carbon into the sediments is returned to the overlying water and atmosphere in gaseous form by denitrifiers.

Enhanced chemical weathering of the land surface by proto-lichens would also have an effect that might *decrease* oxygen, since it would expose more reducing crustal rocks to oxidation by the atmosphere. We investigated the balance of these processes when we first suggested this mechanism, and found however that this was the smaller effect (*19*). It is the oxygen source that dominates, especially if the proportion of phosphorus that is brought into the biological cycle is increased.

By increasing overall weathering rates, colonization of the land would have had a second major effect on the composition of the atmosphere; tending to remove carbon dioxide and thus cool the climate. The reasons why land colonization would lead to reduced carbon dioxide are easy to understand. Remember (see Chapter 7) that the weathering of silicates in rocks by carbonic acid, derived from the atmosphere, is the ultimate sink for carbon dioxide. Over most of Earth history, the tendency for the weathering rate to slow down as temperature decreases has been an important negative feedback. It has helped keep the surface temperatures fairly comfortable for example, despite a steadily increasing output from the Sun. If proto-lichens began to enhance the chemical weathering of the continental surface, they would have increased silicate weathering and this would have caused reduction in atmospheric CO_2 concentrations, upsetting the steady state that had maintained comfortable temperatures throughout the Proterozoic. If the acceleration of weathering was great enough it might have caused a Snowball Earth.

14.5 The Neoproterozoic glaciations

Thus we come to the Neoproterozic glaciations. As we highlighted in Chapter 2, there is much debate over the nature and severity of these glaciations. Whilst most researchers would agree with the evidence that continental ice sheets reached sea level in low latitudes, they differ over precisely what happened to the global climate and the carbon cycle. However, almost everyone would agree that the glaciations were far colder and lasted far longer than anything seen in the Phanerozoic. There are (at least) three glacial episodes in the record between about 750 million years ago and 570 million years ago. The first two, the Sturtian (ending around or before 700 million years ago) and Marinoan (also called the Varanger, ending 635 million years ago), were lengthy and deep, covering very extensive parts of the planet, and making more recent ice ages seem just mildly chilly by comparison. The third, the Gaskiers (occurring about 580 million years ago), was less severe and shorter. (Dates for many of the Neoproterozoic events are not firmly established as yet (20).) Proponents assert that the Sturtian and Marinoan glaciations were true global Snowball Earth events, lasting millions and probably tens of millions of years each. Box 14.1 gives more detail on the geochemical evidence used to support the Snowball Earth hypothesis, and some of the disagreements over its interpretation.

A Snowball Earth, as we discussed in Chapter 7, is a condition where ice-albedo positive feedback 'runs away' and the planet becomes covered in snow and ice. It might come about as follows. Some process—let's assume proto-lichens weathering the continents, reduces atmospheric carbon dioxide, and thereby global temperatures, so that ice caps begin to form. The ice caps reflect some sunlight, cooling the planet even further. Providing there is a reasonable area of the land surface that is not close to the poles however, the proto-lichens continue to flourish in the warmer latitudes, removing more CO_2. The ice caps grow yet bigger, reflecting more of the Sun's energy. Eventually, when the ice caps reach to about 30 degrees from the equator, a tipping point is passed. The ice now reflects so much sunlight that the process runs away and the rest of the planet quickly becomes ice-bound (sea-ice

Box 14.1 Snowball Earth: the geochemical evidence

In its 'hard' form, the Snowball Earth hypothesis postulates an inhospitable planet completely covered in ice for millions of years, in which primary production crashed and carbon dioxide steadily built up in the atmosphere, followed by a 'super-greenhouse' state with extremely rapid rates of weathering. Three main sources of evidence have been drawn on to support this scenario:

1. **Cap carbonates**. After each glaciation there is a massive, widespread deposit of carbonate rocks. These are argued to represent the chemical combination of all the acidic carbon dioxide that built up in the atmosphere and ocean during the snowball, with alkaline calcium and magnesium, which would have weathered extremely rapidly to the ocean in the super-greenhouse after the snowball.

2. **The $\delta^{13}C$ composition of marine carbonates** (Fig. 14.3). This shows extremely low values both before and straight after each postulated snowball. A value of around -5 ‰ has been argued to be consistent with an almost total collapse of marine productivity during the glaciations, which left the atmosphere and ocean reservoirs close to the isotopic composition of carbon coming in from the mantle. The nosedive of $\delta^{13}C$ before glaciations was also initially interpreted as productivity collapsing in a cooling world.

3. **BIFs**. Banded iron formations reappear in the rock record in the Neoproterozoic after a gap of around a billion years. With the ocean largely isolated from the atmosphere in a snowball, the ocean would become anoxic because there would be a meagre supply of oxygen to oxidize reduced material entering via hydrothermal vents. Wherever photosynthesizers could grow, perhaps under thin ice or in cracks (known as 'leads') in the sea-ice, the iron would be oxidized to form BIFs.

Unsurprisingly, some controversy surrounds the interpretation of each of these sources of evidence. Recently it has been found that the deep ocean was anoxic and iron-rich in intervals before and between the glaciations (*21*). So, the conditions for depositing BIFs already appear to have been in place, and it is unclear that the unusual conditions of a snowball need to be invoked to explain their presence. The greatest leeway in interpretation probably surrounds the carbon isotope record (Fig. 14.3) (*22*). In recent years, it has been fashionable to invoke isotopically-light methane as responsible for very negative $\delta^{13}C$. Thus, the oxidation of a methane greenhouse (contributing to cooling) has been invoked to explain the nosedive of $\delta^{13}C$ before glaciations. At the end of glaciations, the oxidation of methane out-gassing from melting frozen clathrates (adding to warming) has been invoked (*23*). However, most recently it has been pointed out that extremely negative $\delta^{13}C$ values could be due to the input via groundwater of isotopically-light carbon from productivity on land (*17*). This would fit neatly with our scenario; the nosedive in $\delta^{13}C$ before glaciation could be associated with increased productivity and bicarbonate weathering on the land surface, the low values afterward could be partly associated with recolonization of the land surface. (The interpretation of the cap carbonates is discussed in the main text.)

can spread globally within decades, whilst land ice sheets grow more slowly, over thousands of years). During that final phase, the sub-zero temperatures would at last stop the proto-lichens from removing more CO_2, but by then it would be too late to stop the runaway.

According to the hypothesis, what saved the Earth from being locked forever in the Snowball state was the fact that on the continents, weathering was brought to a stop by the sub-zero temperatures, so that carbon dioxide was no longer removed from the atmosphere. The CO_2 continually exhaled from volcanoes could then build up in the atmosphere, increasing the greenhouse effect. Eventually, after perhaps ten million years, the CO_2 blanket was thick enough that the planet began to unfreeze again. The concentrations of CO_2 needed to do this were very high— much higher than had been required to keep the planet warm in the absence of ice, because now the white covering reflected away so much of the Sun's energy. However, once the unfreezing started, it too would be a runaway process, with decreased ice cover leading to less reflection, further heating, yet more melting, and so on. In an instant of geological time—again, no more than a few thousand years, the planet would go from being frozen from poles to equator, to being very hot. The melting of the ice would occur much faster than weathering processes could remove carbon dioxide, with the result that the planet would go into a 'super-greenhouse' state.

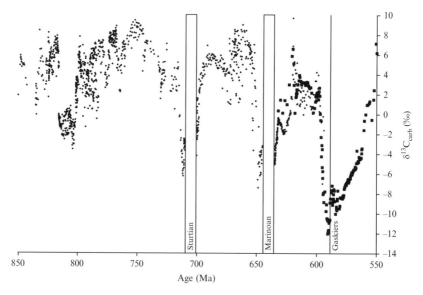

Fig. 14.3 Carbon isotope composition of marine carbonates through the Neoproterozoic. Note that the age scale, particularly the first half, is highly uncertain. [Data kindly provided by Galen Halverson of the University of Adelaide (22).]

One of the lines of evidence and controversy in the Snowball Earth debate focuses on this super-greenhouse. Eventually it would dissipate as, over time, weathering removed the CO_2 from the atmosphere. It is characteristic of the Neoproterozoic glaciations that the glacial deposits are overlain by distinctive carbonate rocks, called cap carbonates (see Box 14.1). Carbonate deposition is usually associated with warm climates, conditions quite different to those under which glacial deposits are laid down, so the presence of the cap carbonates is something of a puzzle. Proponents of Snowball Earth explain them as the result of the super-greenhouse state that followed once the ice had melted: The extreme enhancement of rock weathering, due to high carbon dioxide concentrations and high planetary temperatures, led to their formation. The carbon dioxide that had formerly been in the atmosphere ended up in these carbonate rocks. Argument swirls around almost every aspect of this theory however: did the cap carbonates form quickly, as the original theory proposed, or more slowly? Were they really laid down immediately after the glacial episodes? And should carbonate precipitate at all in the aftermath of the snowball, given the very high concentrations of carbon dioxide in the atmosphere being postulated, which would acidify the oceans and tend to dissolve carbonate rock?

A further key area of disagreement is over whether the glaciations were 'hard' snowballs, with the oceans entirely frozen, or 'soft' 'slushballs', with unfrozen regions of ocean at least near the equator. This has potentially important implications for the survival and evolution of life. Some climate models for example, stop short of complete runaway, and allow a band of tropical ocean to remain unfrozen (*24*), thus providing ample habitat for eukaryotic life to survive (and indeed thrive). However, the same climate models that can produce a slushball often, if pushed harder, produce a hard snowball. Some slushball proponents go further and claim that exotic feedback mechanisms would have prevented the snowball state from being entered (*25*). The argument goes that as the planet cooled, more oxygen would have dissolved in the ocean and this in turn would have led to the oxidation of a large reservoir of organic carbon, returning it to the atmosphere as CO_2. However, there are so many questionable assumptions and inconsistencies in this particular scenario that even one of its co-authors has rejected it (*26*). Furthermore, proponents of the hard snowball scenario argue that a slushball cannot explain the geological evidence, especially the formation of cap carbonates and the deposition of BIFs (Box 14.1).

But how can a hard snowball be reconciled with biological and biomarker evidence that communities of eukaryotes and prokaryotes including photosynthesizers, survived, apparently relatively unscathed, through the glaciations (*27, 28*)? To help life through a long, hard snowball, a number of models suggest that the sea-ice cover would be relatively thin near the equator, allowing enough sunlight to penetrate through the ice to support photosynthetic eukaryotes (algae) underneath. Also, there would likely be cracks ('leads') in the ice, exposing some ocean water, and meltwater ponds would probably form on the ice in the tropical 'summer'. Still it would be bitterly cold, and the only warm places on the planet would be near volcanic sources of heat. Large eruptions could melt their way through continental ice sheets (as they do today) and hot springs on land would provide oases of warm water in an otherwise frozen desert.

To adjudicate between the two scenarios we should identify where they make quite different predictions and seek the corresponding evidence. Perhaps the clearest prediction is that slushball events would be much shorter, because they would require much less CO_2 build up to melt them. On this point, recent evidence has fallen against the slushball. First, estimates put the duration of the Marinoan glaciation as at least 3 million years and most likely 12 million years (*29*). Second, a new proxy for estimating CO_2 levels in the atmosphere gives roughly 12 000 ppm (1.2 % of the atmosphere) in the cap carbonates after the Marinoan glaciation (*30*) and up to 80 000 ppm (8 % of the atmosphere) during it (*31*). These numbers are more consistent with a prolonged, hard snowball, which it is estimated would require about 200 000 ppm (20% of the atmosphere) to melt. However, there are lingering problems of getting trapped in a hard snowball, in that even this much CO_2 may be insufficient to melt it (*32*), and the build-up of CO_2 may get ever slower with time (*33*). Perhaps the actual exit mechanism involved the formation of clouds of 'dry ice' (frozen CO_2), which scatter heat radiation and could have been a strong warming agent (*34*).

At present it is fair to say that neither the hard snowball nor the slushball scenarios seem entirely to fit the facts. Some geologists still hotly dispute whether the geological evidence is compatible with global glaciation, and have offered further alternative scenarios (see Box 14.2). Fortunately, we can make good progress without having to definitely answer whether the glaciations were deep but ultimately 'conventional' or if they were true snowballs. Virtually everyone would agree that there were several glaciations. The glacial states were followed by a recovery to warm conditions, and then, after a considerable length of time, a return to extreme glacial conditions. The fact that the glaciations occurred cyclically shows that something changed at that time, to make the planet liable to freezing over. We believe this was the evolution of a land biota, but there have been other, more purely geological suggestions (*35*, *36*). Below we follow through our 'biological' explanation for the Neoproterozoic events. In Box 14.2 we discuss alternative possible geological reasons why the glaciations might have occurred.

Box 14.2 Alternative models for the Neoproterozoic glaciations

Why, after a billion and a half years of the Proterozoic in which there is no evidence for any ice on the planet, should the Earth have been gripped by such a sequence of long and deep glaciations? Whilst we think that the most important factor was the colonization of the land, geologists who have thought about this problem usually invoke plate tectonics in preference to a biological explanation. Two tectonic explanations for low atmospheric carbon dioxide at the time, and hence a cold planet, are:

1) **Low-latitude continents:** If the main continental mass was comparatively close to the equator in the Neoproterozoic, then it would have been warmer than the average temperature of the planet (*35*). The carbonate-silicate weathering 'thermostat' responds to the temperature of continental silicate rock being weathered, so if all the continents are anomalously warm, atmospheric CO_2 is drawn down more. It certainly is the case that most of the land mass was at low latitudes in Neoproterozoic times. However, one could argue that this is not a very unusual situation. 50% of the Earth's surface is between latitude 30°N and 30°S in any case (that is within the tropics), so we might expect that normally a substantial proportion of the continental area will be found here. In other words, the average temperature of the continents is usually closer to the tropical than polar values, because in terms of area, the polar regions are small compared to the tropics.

2) **Break-up of a supercontinent:** The Neoproterozoic glaciations coincide with the time when the supercontinent Rodinia was breaking up. It is believed that supercontinents have formed and broken up repeatedly over the history of the Earth, in a great cycle (sometimes called the Wilson cycle), lasting many hundreds of millions of years (Fig. 14.4). During the break-up phase, the continent rifts and divides, creating an ocean that widens from a central ridge, like the present Atlantic. On the 'passive margins' of such an ocean, material that is weathered from the continents can build up as sedimentary rocks on the sea floor, because it is not being subducted into the mantle. Carbonate rocks can accumulate, without being recycled via subduction and having the CO_2 they contain released by volcanoes. Therefore, during this break-up phase, there is a tendency for the weathering sink of CO_2 to dominate over the tectonic source, leaving less CO_2 in the atmosphere.

Nick Eyles, a Canadian geologist who is a long standing critic of Snowball Earth, argues that this break-up phase of the supercontinent cycle is particularly associated with periods of glaciation (*37*): Fig. 14.4 below is redrawn from his paper, showing his association of glacial epochs with supercontinent break-up throughout Earth history. He goes further than this however, and argues that the Neoproterozoic climate was not particularly unusual.

The 'zipper-rift' model: Eyles and his colleagues reinterpret the Neoproterozoic deposits, arguing that many of them are not glacial in origin at all. Instead their formation is better interpreted as regional and non-synchronous, associated with progressive rifting of the continents like a slow moving zip-fastener (*38*). They argue that the deposits are not compatible with the idea of sudden and global rises and falls in sea level such as the hard Snowball hypothesis would require. Some of the deposits are glacial in origin, but they can be explained by non-global glaciation spread through time, and not greatly different to more recent cold episodes. They name their alternative view the 'zipper-rift' model.

The idea that supercontinent break-up is a particularly favourable time for glaciation to occur is an attractive and plausible one, but it is not a universal explanation for glaciations. Break-up is neither a necessary nor a sufficient condition, as there doesn't seem to be any ice associated with the break-up of continents during the mid-Proterozoic for example, while the Ordovician glaciation and the recent Cenozoic cooling are not associated with break-up. Indeed, other geologists have associated periods of continental collision with enhanced weathering and therefore lower atmospheric CO_2, as mountains are thrust up and then rapidly weathered down. This explanation has been advanced for the 'recent' Cenozoic cooling of the Earth (over the last 35 million years or so), with the Himalaya being the source of enhanced weathering. A final caveat is that our knowledge of the very early continental cycle is very sketchy, so the association of glacial formations with any particular phase of continent-building or break-up is not very secure then.

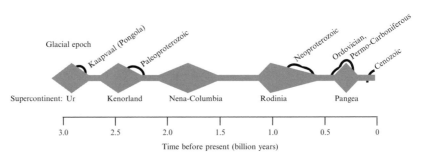

Fig. 14.4 Eyles' association of glacial epochs through Earth history with the break-up phase of the supercontinent cycle. [Adapted from Eyles (*37*).]

14.6 Why did they start and why did they stop?

Any complete proposal to explain the interval of Neoproterozoic glaciations needs to address why they started when they did and why they stopped. The only comparable events in the rock record occurred around the time of the Great Oxidation and nothing like them has happened since. Appealing to the fainter younger Sun is clearly insufficient explanation because extreme glaciations only occupy two discrete intervals of deep time. Equally explanations that appeal to cyclical geological processes such as the break-up of super-continents cannot be the whole answer, because those cycles are more frequent than the intervals of extreme glaciations. Super-volcano eruptions are also more frequent. Others have appealed to astronomical causes including asteroid impacts (also too frequent), or the solar system passing through dense molecular clouds roughly every billion years (precise timing unknown), but these cannot account for the association of extreme glaciations with rises in oxygen. Cyclical geological conditions, especially continental break-up and the associated volcanic outpouring of easy-to-weather basalt, probably primed the Neoproterozoic Earth (*36*). But we appeal to biological evolution because it can explain both the starting of the glaciations and why they stopped.

Our scenario goes something like this: The colonization of the land by proto-lichens, beginning sometime after 800 million years ago, would have decreased atmospheric CO_2 and increased oxygen. Although the absolute fluxes of these two gases would be similar, the effect of decreasing CO_2 would be felt first, because the atmospheric 'residence time' is much shorter for CO_2 (see Box 14.3). The land colonizers would likely have decreased carbon dioxide concentrations sufficiently to affect the climate before they could cause oxygen to rise toward modern levels. The decrease in atmospheric CO_2, triggering the ice-albedo feedback, could have brought on an extreme glaciation (the Sturtian) lasting tens of millions of years perhaps from about 720 to 700 million years ago. The colonization would also have raised oxygen concentrations, before being stopped by the sub-zero conditions.

Some of the organisms responsible must have survived the great ice age, living in localized refugia (near hot springs perhaps). Following the melting of the ice, and the removal by the ice sheets of any chemically worked 'proto-soils' they might have created, these survivors recolonized the continents from bare rock again. Eventually this brought on the second global (Marinoan) glaciation. Following the second glaciation, the land was recolonized yet again by proto-lichens. Oxygen

Box 14.3 Residence times of oxygen and carbon dioxide

The residence time is a measure of how long it takes for a given flux to change the concentration of a gas, and is given by the mass of it in the atmosphere divided by the flux. Gases with higher concentrations tend to have longer residence times, and respond more slowly to changes. Today for example, carbon dioxide is just 0.04% of the atmosphere while oxygen has a much larger concentration of 21%, and this is why we can burn enough fossil fuel to double carbon dioxide while barely affecting the amount of oxygen. Though CO_2 was higher in the Neoproterozoic and oxygen much lower, the CO_2 residence time was still less than that of oxygen.

notched up further, rising to concentrations sufficient for animals to flourish in some of the shallow coastal seas. But what stopped the oxygen level from rising indefinitely and the cycle of glaciations from endlessly repeating?

Our theory is that after the Marinoan glaciation, the rising concentration of oxygen interfered with the process of photosynthesis, making it more difficult for our postulated proto-lichens to fix carbon dioxide at low concentrations (Fig. 14.5). Too much oxygen is a hindrance to photosynthesizers that are trying to fix carbon dioxide using the Calvin cycle, because oxygen competes with CO_2 to react with their Rubisco, in the process called photo-respiration. This is a null cycle that wastes the organisms' resources, so if the oxygen rose too high the production by the land proto-lichens would eventually be shut off. This prevented them from drawing the CO_2 down low enough to bring on further global glaciations. There was a final glaciation (the Gaskiers), but it was short lived and not global. In the aftermath of that, the deep ocean finally became sufficiently oxygenated for animals to flourish there. The atmospheric oxygen level was now such that carbon dioxide was more limiting to the land biota and they could not bring it down to the level that would trigger another glaciation. The planet settled into an ice free and stable state.

Important to this description is what ends the series of glaciations, which we've assumed above is the effect of rising oxygen on carbon

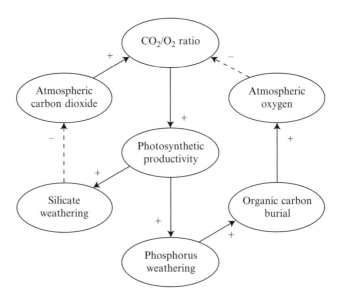

Fig. 14.5 Negative feedbacks involving the ratio of carbon dioxide to oxygen in the atmosphere. Its effect on photosynthesis, could have halted the rise of oxygen and ended the interval of glaciations by suppressing biologically-induced weathering

fixation. Higher oxygen pushes up the lowest concentrations of carbon dioxide that the photosynthesizers can work with. This effect is well known in most 'C3' plants, which includes most of the higher plants of today (39). However, the compensation point at which it becomes effective can be lowered if the organisms evolve 'carbon concentration mechanisms'. Most (but not all) modern algae and cyanobacteria have evolved versions of these (40). By a variety of different mechanisms, these photosynthesizers boost the carbon available to their Rubisco above the concentration that simply diffuses to the site of uptake. The concentration mechanisms are energetically costly, but given the low-CO_2, high-oxygen world of today, some organisms have found this a cost worth paying.

We don't know if the photosynthesizers of the Neoproterozoic had carbon concentration mechanisms or not. Since this was probably the first time that the Earth had experienced such low concentrations of CO_2, they would not previously have been needed. It is unlikely therefore that the necessary biochemistry was pre-evolved and ready for use. Selection pressures may well have caused their evolution in response to the conditions of the time. However, even though some degree of carbon concentration may have evolved, the rise in oxygen would still be expected to decrease the efficiency with which photosynthesizers could take up CO_2. The basic description above should still remain valid, but it is difficult to be prescriptive about the exact levels of oxygen and CO_2 that would result.

In keeping with our 'feedbacks' approach, we should also ask; what (if anything) stopped the rise in oxygen? Why should it not have increased almost indefinitely, given that the theoretical capacity for burying reduced carbon is large, and there is no obvious reason why the system should run short of any of the elements involved? We think part of the answer is that phosphorus weathering also became limited by the combination of high oxygen and low carbon dioxide concentration constraining the proto-lichens, and this in turn restricted the source of oxygen from organic carbon burial (Fig. 14.5). But there was more to it than this: as the ocean became more oxygenated, the removal of phosphorus into sediments would have become more efficient, providing further negative feedback on organic carbon burial and atmospheric oxygen (41) (discussed further in Chapter 15, see Fig. 15.3). Together these mechanisms could have stabilized atmospheric oxygen.

14.7 Snowball Earths as a causal factor in the evolution of animals

The scenario we have just outlined is consistent with a popular view that the radiation of animals after the Marinoan and Gaskiers glaciations was intimately linked to rising oxygen, especially the oxygenation of the surface and then the deep ocean. But there is almost certainly more to the evolution of animals than simply removing an environmental constraint. Since it was first recognized that the late Proterozoic was a time of extensive glaciations, there has been speculation that there may be a causal link between them and the subsequent Cambrian radiation. Did the environmental upheaval somehow encourage an unprecedented diversification into new forms? For the most part these suggestions have been quite vague, with little discussion of the mechanisms that might be involved. Severe glaciations may have led to the extinctions of other life forms and the clearance of ecological niches that the metazoans could radiate into, but the same could be said for any new life form. Might there be any reason why metazoans in particular might be the beneficiaries?

Perhaps: there is something special about animals. They are not the only multi-cellular life forms on Earth—plants, fungi, and various marine algae have independently evolved the multi-cellular habit. However, animals are committed to multi-cellularity in a way that that these other eukaryotes are not. Animal cells usually become *terminally differentiated*. Most animal cells are mortal, and doomed eventually to come to the end of a cell line, when the genes in them cannot be passed on. To survive, the genes rely on the fact that they are identical to those in the reproductive cells of the animal. This isn't in general true of plants, fungi, or multi-cellular algae, in which cells remain 'totipotent', meaning that each cell can potentially give rise to the whole organism, and may do so if it is separated, as for example when a gardener takes a cutting from a plant. Probably because of this, higher animals are more differentiated, with a much greater number of different cell types than the other groups have. A typical animal today has about a hundred such cell types, whereas for plants, the number is only a dozen or so.

The individual cells of multi-cellular organisms display a form of altruism. They do not reproduce indefinitely without regard to their neighbour cells, consuming as much resource as they can, but are constrained by their shared genes to cooperate. While this is as true for a tree or a giant kelp as it is for an animal, the totipotent cells of algae, fungi and plants do have a sort of get-out clause. If they happen to find themselves detached from the rest of the organism, but are otherwise able to grow, they will go ahead and produce a new organism, including all its reproductive structures. As a general rule animal cells cannot do this. Without their community—the animal—their genes have no future. Among animals, only the sponges have a degree of totipotency similar to plants or fungi, which is interesting because sponges were among the first of all the still-existing phyla to split off from the rest of the metazoa. The origin of terminal differentiation in animals might therefore be dated to a time after sponges evolved but before other types had diverged.

The reason that this might be particularly relevant has been discussed by our former student Richard Boyle (*42*). Rich's idea is that extreme glaciations would produce just the conditions in which altruistic traits would most readily evolve. This was an environment in which many small populations of proto-animals would have been isolated from one another for very long periods of time.

There is a difficulty that altruistic genes must overcome, even if cooperation would be to their and their cell's long-term advantage because it would help the entire community to survive, including themselves. The difficulty is that altruism is assumed to carry a fitness cost and unless all the cells in the community share the altruistic genes, the altruists will lose out in the short term to 'cheaters' who save this cost and multiply more rapidly, at the expense of the altruists. In a large population, altruistic behaviour is difficult to evolve: it tends to be disrupted, before it comes to dominate. For it to prosper, the altruistic gene has to have a high frequency in the population, and it must be linked to mechanisms for keeping out invaders and suppressing cheaters. Animals have developed sophisticated methods to perform these functions. They have genes for example that will kill cells that might otherwise pose a threat, while their immune systems keep foreign invaders at bay. The difficulty is to understand how the altruistic genes came to dominance in the first place.

Picture the scene in a snowball Earth: Everywhere is like Antarctica today, except that is a good deal colder in the winter hemisphere. Eukaryote life is restricted to scattered oases, including hot springs on land, possible meltwater ponds on sea-ice, and patches of open water exposed by cracks in the sea-ice. In the midst of such an extreme glaciation, populations of eukaryote cells with the potential to form 'proto-animals' would have survived only in these refuges, isolated from one another. Such small populations tend to contain less genetic variation than large ones, which is another way of saying that their members tend to be more closely related: this is called the 'founder effect'. Founder effects give the chance for genes that would otherwise lose out to competition to be given an initial boost. Meanwhile, the variation between groups would be large. By chance there would be some groups with a predominance of altruistic genes, and some with a predominance of cheats. It is within the groups dominated by altruists that Rich hypothesizes the first animals evolved. His scenario relies on near isolation of the populations, because otherwise the influx of cheats from elsewhere would disrupt any nascent cooperation. For it to work really well, it also requires that cooperative populations survive better in the face of extremely adverse environmental conditions, whereas populations dominated by cheats tend to go extinct. Then although the number of groups tends to dwindle, those that remain tend to be cooperative. Eventually, after millions of years, when the ice melted, the new ecological niches revealed by the receding glaciers would be inherited by populations of altruistic cells, dependent on their community (the animal) to survive, but free to specialize in hundreds of different ways.

The hypothesis that extreme conditions, in this case extreme cold, leads to cooperative behaviour is based on the idea that it helps to pool resources, and cooperative groups (multi-cellular organisms), especially those with division of labour (cell differentiation), can do this much better than isolated individuals (the cells of which they are composed). This idea can be tested by going to one of the most extreme environments on today's Earth; Antarctica, and examining how organisms survive there. Often it is by teamwork. Consider the human pioneers that first explored this great wilderness—Amundsen, Scott, Shackleton—they didn't go alone—their only hope for survival was as part of a team, and even then individuals, and in Scott's case a whole team, perished. The first snowball explorers were faced with an even bleaker environment, for which one analogue is the McMurdo dry valleys of Antarctica today. There one finds some striking examples of cooperation to cope with a severe environment. Despite the severe conditions there are animals living there—three species of nematode worm—which play a disproportionately large role in recycling carbon for the ecosystem.

Rich's hypothesis leads to a clear prediction; that evidence of the first ancestors of modern animals should appear during (or straight after) the interval of global glaciations. Thus it was with excitement that we read about some recent biomarker evidence which suggests that sponges were present before the end of the second great glaciation (the Marinoan), 635 million years ago, and these biomarkers become abundant in its aftermath (*43*). Whether the biomarkers record actual sponges or their predecessors is unclear, as is whether sponges were the first group of animals to branch off from the stem-group leading to all animals (*44*). But these details matter less for the hypothesis. An important caveat is that sponge cells are 'totipotent'—so sponges may represent an intermediate step on the path to the most altruistic forms of animal multi-cellularity—but it is just such an intermediate step that we expect to appear first.

So, there is now fossilized evidence that the origins of animals were in the interval of global glaciations, and well before the proliferation of Ediacaran and then Cambrian fossils. This is starting to converge with estimates arising from the latest 'relaxed' molecular clocks, which put the first branching from the stem group of animals during the interval of global glaciations (e.g. 695 million years ago in Fig. 12.2). We can hope for further discoveries that will help tie the chronology down. For now we turn our attention to what happened in the aftermath of the glaciations.

14.8 Life in the aftermath

Whilst abundant oxygen cannot in itself make animals evolve, it is necessary for them to become large or highly mobile. During and between the Sturtian, Marinoan and Gaskiers glaciations, the deep ocean was rich in iron indicating an anoxic state, in which sulphur was relatively scarce (*21*). Anoxia would have made the deep ocean uninhabitable to animals. However, shallow shelf sea waters were probably already oxygenated before the global glaciations, and they certainly were in the aftermath of the Marinoan glaciation, 630 million years ago (*21, 45 46–47*). Some 50 million years later, in the aftermath of the Gaskiers glaciation, reactive iron levels in the deep ocean fell, indicating that at least parts of the deep ocean finally became fully oxygenated then, and (in principle) animals could live anywhere (*21, 48*). Estimates place the resulting atmospheric oxygen levels as at least 15% of the present level (i.e. over 3% of the atmosphere).

The first fossilized animal embryos appear in the aftermath of the second, Marinoan glaciation in the oxygenated waters of a shallow shelf sea, alongside multi-cellular algae (seaweed). They are beautifully preserved in the Doushantuo phosphorite formation found today in Southern China (*49, 50*). The embryos are caught in the act of cell division and some of the resulting clusters of cells are found within a container or 'egg cyst' (*51*). The presence of this outer shell and of many cell divisions seems to rule out an alternative explanation of the fossils as giant bacteria (*52*). There is some rumbling debate about the precise age of the fossils due to issues with aligning different sections of rock, only some of which have radiometric dates (*53*). However, even the more sceptical commentators now concede that 'embryo-forming animals of some sort had evolved by just after Marinoan time' (*54*). What they are the embryos of is not so clear, with the earliest ones currently interpreted as representing stem-group animals (rather than a particular derived group such as sponges) (*51, 55*). Elsewhere in the Doushantuo formation there is separate evidence of sponges and jellyfish (cnidarians) (*50, 56*).

In the aftermath of the Gaskiers glaciation, with the oxygenation of (at least parts of) the deep ocean, fossils of large eukaryotes first appear in abundance. These strange, multi-cellular creatures are perhaps the most mysterious in all the fossil record. First described from the Ediacara hills in South Australia, in the mid twentieth century (*57*) (but wrongly dated at the time), they are now known from many localities, worldwide. The final period of the Precambrian from 635 to 542 million years ago is now officially known as the Ediacaran, in their honour. Actually, the first definitely Precambrian fossil to be recognised was found in Charnwood Forest in Leicestershire; *Charnia masoni* (Fig. 14.6) is named after its location and Roger Mason, one of three schoolboy discoverers who in 1957 brought it to scientific attention (*58*) (actually

Fig. 14.6 The Ediacaran fossil *Charnia masoni* (with a ruler graduated in mm for scale). [Image from Wikipedia.]

schoolgirl Tina Negus had seen and recorded the fossil a year earlier, but her teacher flatly denied that Precambrian fossils could exist).

The currently oldest dated Ediacarans appear within 5 million years of the end of the Gaskiers glaciations, at around 575 million years ago. They lived on the sea floor in deep, dark and newly oxygenated waters several hundred metres below the surface. Their fossilized imprints in the sediments are exposed today on the Avalon Peninsula of Newfoundland (*59*). The community was relatively low in diversity including organisms that look like 'pizza disks' (*Ivesheadia*) and long frond-like structures, closely related to the original *Charnia* (Fig. 14.6) but up to 2 metres in length and hence assigned to a different species (*Charnia wardi*).

Ten million years later, a much more diverse community had evolved. This radiation of different Ediacaran body types has been described as the 'Avalon explosion' and likened to the later Cambrian explosion (*60*). However, for the most part the Ediacarans don't resemble living animals, and for many it is not clear that they are animals at all—there have been attempts to ally them with fungi (*61*), or underwater lichens (*62*). They frequently are frond-like in shape, like exuberant tropical leafs, but there are other body plans; fans, and disks, perhaps originally cup or cone-shaped, sometimes with three-fold or five-fold symmetric marking. They probably lived on, perhaps attached to, sediment on the sea floor, but precisely how they obtained nourishment is unclear—no mouth or body cavity is visible.

An influential argument has been made that the Ediacarans are simply too weird to be classified as animals, and they should be assigned to their own kingdom, the Vendobionta, after the alternative name of Vendian for this period (*63, 64*). Some have envisaged them as living in an Eden-like peaceful world, the 'Garden of Ediacara' (*65*) where, before predators with eyes and jaws had evolved, they had no need to move rapidly or protect themselves with hard shells.

However, not all the Ediacarans are unclassifiable, some at least are definitely animals, and can be interpreted in terms of known phyla. Many of the forms can be interpreted as ancestral cnidarians (jellyfish), while some are possible ancestral arthropods—for example 'soft bodied trilobites' (*66*), molluscs (*67*) and echinoderms (*68*). A recent report of a new specimen of *Dickinsonia*, an archetypal frond-like Ediacaran, finds evidence for internal structures that could place it with the ctenophores, or comb-jellies (*69*). Another paper describes an early Cambrian 'vendobiont' that looks very much like the Ediacarans, but that also shows structures similar to the comb rows of the comb-jellies (*70*), so could be an intermediate form.

Jellyfish (cnidarians) and comb-jellies (ctenophores) possess sensory organs and a simple nervous system (which sponges do not), but they lack digestive and circulatory organs. They typically have radial symmetry—meaning they have a top and bottom but no front and back.

In contrast, most modern animals are 'bilaterians'—meaning they have a front and a back as well as a top and a bottom—and most bilaterians have a digestive tract including a separate mouth and anus. Although the Cambrian explosion is still seen as the great radiation of bilaterians, both fossils and molecular clocks now suggest that they emerged in the Ediacaran period (*71*). The first widely accepted bilaterian fossils are numerous *Kimberella*, dating from 558 million years ago. These organisms had a shell on their backs and a frilly 'skirt' extending beyond the edge of the shell. They apparently trundled around on the seafloor in shallow waters only a few metres deep, perhaps grazing on microbial mats.

The evolution of animals with the capacity (i.e. a mouth, gut and anus) to eat other creatures has been described as the defining moment for the ecology of the planet. Beforehand most evolution proceeded at a sedate pace steered by the environment, and ecosystems lacked complex webs of feeding relationships. Afterwards there were co-evolutionary 'arms races' between the eater and the eaten: Prey evolved new ways to avoid predators and predators evolved new ways to eat prey. The result was a much more rapid turnover of types of life in the fossil record. Traditionally this transition was viewed as happening in the Cambrian, but recent work now places the major increase in the rate of turnover of fossil lineages during the Ediacaran interval (*72*), consistent with the appearance of bilaterians.

Direct evidence of predation is scarce, but it does appear in the first mineralized skeletons made by animals that could calcify (precipitate calcium carbonate). Around 548 million years ago, at a time of increased organic carbon burial producing rising oxygen, the first small shelly fossils appear in the fossil record. The most famous examples are in the genus *Cloudina*, which looked a bit like a stack of ice-cream cones. They occurred alongside *Namacalathus*, which looks like something out of Dr Suess, with a curved stalk ending in a sphere with holes in it. These organisms apparently lived on microbial reefs in shallow waters, and are never found alongside the soft-bodied Ediacarans. Remarkably, many *Cloudina* specimens show holes that have been bored through their shells by predators (*73*). A popular view is that the hard shells themselves evolved as a means of defence, which the predators then evolved a way to drill through, suggesting that a co-evolutionary arms race had begun.

In summary, most, if not all, of the ingredients for the Cambrian explosion appeared or evolved in the Ediacaran period. In one view, many of the Ediacarans did not go extinct, but rather they evolved under the selection pressures of the Cambrian world, into some of the more-or-less familiar animals. The descendants of the Ediacarans are in that case still with us. This view helps us to understand why the modern animal phyla seem to emerge so suddenly in the Cambrian. Perhaps many of them were already established in the Ediacaran, and the

Cambrian 'explosion' was not therefore quite as explosive as it once appeared.

In the last few million years before the Precambrian-Cambrian boundary, the fossil assemblage becomes richer and more varied. The official boundary that ends the Precambrian is dated at 542 million years ago, when the first complex trace fossil burrows (*Trichophycus pedum*) become widespread (although these too have now been found below the boundary, back in the Ediacaran). At the boundary there is also a pronounced negative carbon isotope excursion, suggesting a crash in ocean productivity, and some evidence for an extinction event, including *Cloudina* and *Namacalathus* among its victims. Leading up to the boundary, conditions in the deep ocean were once again anoxic and sulphidic, and one theory is that extinction in shallower waters was caused by a final belch of hydrogen sulphide from the deep ocean. The Precambrian-Cambrian boundary is also accompanied by the deposition of massive amounts of phosphorus-rich rocks, which in our scenario could be due to a further pulse of colonization of the land surface, supplying phosphorus to the ocean and pushing up atmospheric oxygen. By 530 million years ago trilobites, echinoderms and molluscs are unequivocally present. By 520 million years ago, there has been a radiation of the body plans of the major animal phyla, and we have arrived in the modern world.

References

1. R. Honegger, *Great Discoveries in bryology and lichenology—Simon Schwendener (1829–1919) and the Dual Hypothesis of Lichens. Bryologist* **103**, 307 (2000).
2. R. C. Stretch, H. A. Viles, *The nature and rate of weathering by lichens on lava flows on Lanzarote. Geomorphology* **47**, 87 (2002).
3. J. Chen, H. P. Blume, L. Beyer, *Weathering of rocks induced by lichen colonization—a review. Catena* **39**, 121 (2000).
4. T. N. Taylor, H. Hass, W. Remy, H. Kerp, *The oldest fossil lichen. Nature* **378**, 244 (1995).
5. X. L. Yuan, S. H. Xiao, T. N. Taylor, *Lichen-like symbiosis 600 million years ago. Science* **308**, 1017 (2005).
6. T. Y. James *et al.*, *Reconstructing the early evolution of Fungi using a six-gene phylogeny. Nature* **443**, 818 (2006).
7. F. Lutzoni, M. Pagel, V. Reeb, *Major fungal lineages are derived from lichen symbiotic ancestors. Nature* **411**, 937 (2001).
8. D. S. Heckman *et al.*, *Molecular evidence for the early colonization of land by fungi and plants. Science* **293**, 1129 (2001).
9. A. Schussler, *Molecular phylogeny, taxonomy, and evolution of Geosiphon pyriformis and arbuscular mycorrhizal fungi. Plant Soil* **244**, 75 (2002).

10. M. Kennedy, M. Droser, L. M. Mayer, D. Pevear, D. Mrofka, *Late Precambrian oxygenation; Inception of the clay mineral factory. Science* **311**, 1446 (2006).

11. R. J. Horodyski, L. P. Knauth, *Life on land in the Precambrian. Science* **263**, 494 (1994).

12. G. J. Retallack, A. Mindszenty, *Well preserved late Precambrian paleosols from Northwest Scotland. Journal of Sedimentary Research* **A64**, 264 (1994).

13. Y. Watanabe, J. E. J. Martini, H. Ohmoto, *Geochemical evidence for terrestrial ecosystems 2.6 billion years ago. Nature* **408**, 574 (2000).

14. A. R. Prave, *Life on land in the Proterozoic: Evidence from the Torridonian rocks of northwest Scotland. Geology* **30**, 811 (2002).

15. J. Gutzmer, N. J. Beukes, *Earliest laterites and possible evidence for terrestrial vegetation in the Early Proterozoic. Geology* **26**, 263 (1998).

16. A. Neaman, J. Chorover, S. L. Brantley, *Element mobility patterns record organic ligands in soils on early Earth. Geology* **33**, 117 (2005).

17. L. P. Knauth, M. J. Kennedy, *The late Precambrian greening of the Earth. Nature*, **460**, 728 (2009).

18. D. H. Rothman, J. M. Hayes, R. E. Summons, *Dynamics of the Neoproterozoic carbon cycle. Proceedings of the National Academy of Sciences USA* **100**, 8124 (2003).

19. T. M. Lenton, A. J. Watson, *Biotic enhancement of weathering, atmospheric oxygen and carbon dioxide in the Neoproterozoic. Geophysical Research Letters* **31**, L05202 (2004).

20. G. J. H. McCall, *The Vendian (Ediacaran) in the geological record: Enigmas in geology's prelude to the Cambrian explosion. Earth-Science Reviews* **77**, 1 (2006).

21. D. E. Canfield *et al.*, *Ferruginous conditions dominated later Neoproterozoic deep-water chemistry. Science* **321**, 949 (2008).

22. G. P. Halverson, P. F. Hoffman, D. P. Schrag, A. C. Maloof, A. H. N. Rice, *Toward a Neoproterozoic composite carbon-isotope record. Geological Society of America Bulletin* **117**, 1181 (2005).

23. M. Kennedy, D. Mrofka, C. von der Borch, *Snowball Earth termination by destabilization of equatorial permafrost methane clathrate. Nature* **453**, 642 (2008).

24. W. T. Hyde, T. J. Crowley, S. K. Baum, W. R. Peltier, *Neoproterozoic 'snowball Earth' simulations with a coupled climate/ice-sheet model. Nature* **405**, 425 (2000).

25. W. R. Peltier, Y. G. Liu, J. W. Crowley, *Snowball Earth prevention by dissolved organic carbon remineralization. Nature* **450**, 813 (2007).

26. P. F. Hoffman, J. W. Crowley, D. T. Johnston, D. S. Jones, D. P. Schrag, *Snowball prevention questioned. Nature* **456**, E7 (2008).

27. F. A. Corsetti, S. M. Awramik, D. Pierce, *A complex microbiota from snowball Earth times: Microfossils from the Neoproterozoic Kingston*

Peak Formation, Death Valley, USA. Proceedings of the National Academy of Sciences USA **100**, 4399 (2003).

28. A. N. Olcott, A. L. Sessions, F. A. Corsetti, A. J. Kaufman, T. Flavio de Oliviera, *Biomarker evidence for photosynthesis during Neoproterozoic glaciation. Science* **310**, 471 (2005).

29. B. Bodiselitsch, C. Koeberl, S. Master, W. U. Reimold, *Estimating duration and intensity of Neoproterozoic snowball glaciations from Ir anomalies. Science* **308**, 239 (2005).

30. H. M. Bao, J. R. Lyons, C. M. Zhou, *Triple oxygen isotope evidence for elevated CO_2 levels after a Neoproterozoic glaciation. Nature* **453**, 504 (2008).

31. H. Bao, I. J. Fairchild, P. M. Wynn, C. Spotl, *Stretching the envelope of past surface environments: Neoproterozoic glacial lakes from Svalbard. Science* **323**, 119 (2009).

32. R. T. Pierrehumbert, *High levels of atmospheric carbon dioxide necessary for the termination of global glaciation. Nature* **429**, 646 (2004).

33. G. Le Hir, G. Ramstein, Y. Donnadieu, Y. Godderis, *Scenario for the evolution of atmospheric pCO_2 during a snowball Earth. Geology* **36**, 47 (2008).

34. F. Forget, R. T. Pierrehumbert, *Warming early Mars with carbon dioxide clouds that scatter infrared radiation. Science* **278**, 1273 (1997).

35. D. P. Schrag, R. A. Berner, P. F. Hoffman, G. P. Halverson, *On the initiation of a snowball Earth. Geochemistry Geophysics Geosystems* **3**, 1036 (2002).

36. Y. Donnadieu, Y. Godderis, G. Ramstein, A. Nedelec, J. Meert, *A 'snowball Earth' climate triggered by continental break-up through changes in runoff. Nature* **428**, 303 (2004).

37. N. Eyles, *Glacio-epochs and the supercontinent cycle after similar to 3.0 Ga: Tectonic boundary conditions for glaciation. Palaeogeography Palaeoclimatology Palaeoecology* **258**, 89 (2008).

38. N. Eyles, N. Januszczak, *'Zipper-rift': a tectonic model for Neoproterozoic glaciations during the breakup of Rodinia after 750 Ma. Earth-Science Reviews* **65**, 1 (2004).

39. N. E. Tolbert, C. Benker, E. Beck, *The oxygen and carbon-dioxide compensation points of C-3 plants—possible role in regulating atmospheric oxygen. Proceedings of the National Academy of Sciences USA* **92**, 11230 (1995).

40. J. A. Raven, C. S. Cockell, C. L. La Rocha, *The evolution of inorganic carbon concentrating mechanisms in photosynthesis. Philosophical Transactions of the Royal Society B-Biological Sciences* **363**, 2641 (Aug, 2008).

41. P. Van Cappellen, E. D. Ingall, *Benthic phosphorus regeneration, net primary production, and ocean anoxia: A model of the coupled marine biogeochemical cycles of carbon and phosphorus. Paleoceanography* **9**, 677 (1994).

42. R. A. Boyle, T. M. Lenton, H. T. P. Williams, *Neoproterozoic 'snowball Earth' glaciations and the evolution of altruism. Geobiology* **5**, 337 (2007).

43. G. D. Love *et al.*, *Fossil steroids record the appearance of Demospongiae during the Cryogenian period. Nature* **457**, 718 (2009).

44. J. J. Brocks, N. J. Butterfield, *Early animals out in the cold. Nature* **457**, 672 (2009).

45. D. A. Fike, J. P. Grotzinger, L. M. Pratt, R. E. Summons, *Oxidation of the Ediacaran Ocean. Nature* **444**, 744 (2006).

46. K. A. McFadden *et al.*, *Pulsed oxidation and biological evolution in the Ediacaran Doushantuo Formation. Proceedings of the National Academy of Sciences USA* **105**, 3197 (2008).

47. Y. Shen, T. Zhang, P. F. Hoffman, *On the coevolution of Ediacaran oceans and animals. Proceedings of the National Academy of Sciences USA* **105**, 7376 (2008).

48. D. E. Canfield, S. W. Poulton, G. M. Narbonne, *Late-Neoproterozoic deep-ocean oxygenation and the rise of animal life. Science* **315**, 92 (2007).

49. S. H. Xiao, Y. Zhang, A. H. Knoll, *Three-dimensional preservation of algae and animal embryos in a Neoproterozoic phosphorite. Nature* **391**, 553 (1998).

50. S. H. Xiao, X. L. Yuan, A. H. Knoll, *Eumetazoan fossils in terminal Proterozoic phosphorites? Proceedings of the National Academy of Sciences USA* **97**, 13684 (2000).

51. L. Yin *et al.*, *Doushantuo embryos preserved inside diapause egg cysts. Nature* **446**, 661 (2007).

52. J. V. Bailey, S. M. Joye, K. M. Kalanetra, B. E. Flood, F. A. Corsetti, *Evidence of giant sulphur bacteria in Neoproterozoic phosphorites. Nature* **445**, 198 (2007).

53. D. Condon *et al.*, *U-Pb ages from the Neoproterozoic Doushantuo formation, China. Science* **308**, 95 (2005).

54. G. E. Budd, *The earliest fossil record of the animals and its significance. Philosophical Transactions of the Royal Society B-Biological Sciences* **363**, 1425 (2008).

55. J. W. Hagadorn *et al.*,*Cellular and subcellular structure of Neoproterozoic animal embryos. Science* **314**, 291 (2006).

56. C. W. Li, J.-Y. Chen, T.-E. Hua, *Precambrian sponges with cellular structures. Science* **279**, 879 (1998).

57. R. C. Sprigg, *Early Cambrian (?) jellyfish from the Flinders Ranges, South Australia. Transactions of the Royal Society of South Australia* **71**, 212 (1947).

58. T. D. Ford, *Precambrian fossils from Charnwood Forest. Yorkshire Geological Society Proceedings* **31**, 211 (1958).

59. G. M. Narbonne, J. G. Gehling, *Life after snowball: The oldest complex Ediacaran fossils. Geology* **31**, 27 (2003).

60. B. Shen, L. Dong, S. Xiao, M. Kowaleski, *The Avalon explosion: evolution of Ediacra morphospace. Science* **319**, 81 (2008).

61. K. J. Peterson, B. Waggoner, J. W. Hagadorn, *A fungal analog for Newfoundland Ediacaran fossils? Integrative and Comparative Biology* **43**, 127 (2003).

62. G. J. Retallack, *Were the Ediacaran fossils lichens? Paleobiology* **20**, 523 (1994).

63. A. Seilacher, *Vendozoa—organismic construction in the Proterozoic biosphere. Lethaia* **22**, 229 (1989).

64. A. Seilacher, *Vendobionta and Psammocorallia—lost constructions of Precambrian evolution. J. Geol. Soc.* **149**, 607 (1992).

65. M. McMenamin, *The Garden of Ediacara*. (Columbia University Press, New York, 1998), pp. 295.

66. R. Jenkins, in *Origin and Early Evolution of the Metazoa*, J. H. Lipps, and Signor, PW, Ed. (Plenum Press, New York, 1992), pp. 131–76.

67. M. A. Fedonkin, B. M. Waggoner, *The Late Precambrian fossil Kimberella is a mollusc-like bilaterian organism. Nature* **388**, 868 (1997).

68. J. G. Gehling, *Earliest known echinoderm—a new Ediacaran fossil from the Pound Subgroup of South-Australia. Alcheringa* **11**, 337 (1987).

69. X. L. Zhang, J. Reitner, *A fresh look at Dickinsonia: Removing it from Vendobionta. Acta Geologica Sinica-English Edition* **80**, 636 (2006).

70. D. G. Shu et al., *Lower Cambrian vendobionts from China and early diploblast evolution. Science* **312**, 731 (2006).

71. K. J. Peterson, J. A. Cotton, J. G. Gehling, D. Pisani, *The Ediacaran emergence of bilaterians: congruence between the genetic and the geological fossil records. Philosophical Transactions of the Royal Society B-Biological Sciences* **363**, 1435 (2008).

72. N. J. Butterfield, *Macroevolution and macroecology through deep time. Paleontology* **50**, 41 (2007).

73. S. Bengtson, Y. Zhao, *Predatorial Borings in Late Precambrian Mineralized Exoskeletons. Science* **257**, 367 (1992).

PART V
INTERLUDE

15
Animals and oxygen

The Cambrian explosion marks the start of the Phanerozoic—the Eon of visible, mobile life—beginning 542 million years ago, and still ongoing. Many histories of life on Earth start at this point having skipped over the first 3 billion years or so. Certainly we know more about what has happened in the last 542 million years than the preceding 3 billion. Many books have been written on the evolution of complex life during the Phanerozoic, or just on a single fascinating group of animals within it. We will not attempt to cram in all this history. Instead, in this and the next two chapters, we want to briefly focus on some of the key environmental changes and evolutionary events during the Phanerozoic, which we think are pertinent to understanding the new revolution that the Earth may now be entering. A timeline highlighting some of these major events is given in Fig. 15.1.

The Phanerozoic began with an evolutionary bang—the rate of appearance of new species in the Cambrian explosion, and in the preceding Ediacaran period, is well above the levels seen in the preceding three billion years of life on Earth. Origination rates were also far greater than at any time since (1). As well as a plethora of hard shelled marine invertebrates, the first marine vertebrates evolved early in the Cambrian (2). These early 'chordates' were the start of the branch of the tree of life that would eventually lead to us.

Looking at the rapid evolution of life in the Cambrian, one might wonder if there is any reason why intelligent life couldn't have emerged in Cambrian times: Why did it take another half billion years before an 'observer' species evolved that could look back at Earth history and marvel at it? Was evolution held back by some environmental constraint? Or was it more of an ecological one? Or was there an additional, extremely improbable evolutionary step on the way to our type of intelligence? In the following chapters we will explore each of these options, but we start by addressing the possibility that the appearance of intelligent life was held back by a simple global environmental constraint; lack of oxygen in the atmosphere.

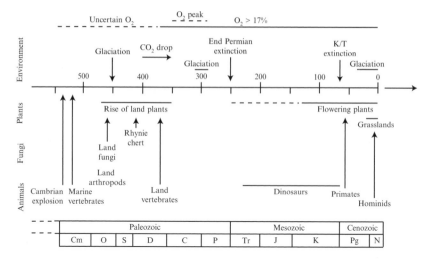

Fig. 15.1 A timeline of the Phanerozoic (ages in millions of years ago), highlighting key periods of environmental change and key events in the evolution of complex life Geologists divide the Phanerozoic Eon up into three Eras; the Paleozoic (542–251 million years ago, or Ma for short), Mesozoic (251–65.5 Ma) and Cenozoic (65–0 Ma). The Eras are further divided into a total of eleven Periods: the Cambrian (Cm, 542–488 Ma), Ordovician (O, 488–443 Ma), Silurian (S, 443–416 Ma), Devonian (D, 416–359 Ma), Carboniferous (C, 359–299 Ma), Permian (P, 299–251 Ma), Triassic (Tr, 251–200 Ma), Jurassic (J, 200–146 Ma), Cretaceous (K, 146–65 Ma), Paleogene (Pg, 65–23 Ma) and Neogene (N, 23–0 Ma). Most of the divisions between Periods within the Phanerozoic are marked by extinction events and striking changes, mostly in marine fossil life

15.1 Did a lack of oxygen hold back intelligent life?

Oxygen is essential for thinking. Intelligence is all about information processing, and information processing requires energy. Also, to support a large brain requires a fairly large body and to maintain a large body, especially one that moves around, requires a lot of energy. Even at rest, our bodies dissipate 100 watts of power, enough to boil the water for a cup of tea in 10 minutes, and 20% of this power is dissipated in our brains (*3*). We get the necessary energy from 'burning' food with oxygen, and if there's not enough oxygen, brain function is among the first things to suffer. So, sufficient oxygen in the atmosphere is a necessary requirement for intelligent life. But just how much oxygen do we need?

The inhabitants of Biosphere 2, in the Arizona desert, had a stark reminder of our dependence on oxygen when levels within their sealed home dropped from 20.9% to 14.5% over 16 months (*4*). The crew became lethargic, irritable, and toiled in their efforts to grow food. On the advice of the doctor in the crew, they eventually had to be 'rescued' by an injection of pure oxygen into the facility. One of the biospherians,

Linda Leigh, described the resulting feeling of euphoria: 'I went from an atmosphere of 14.2% O_2 to an atmosphere of 26% in less than five minutes... it was a sense of intense well being to breathe deeply and not feel like I needed another deep breath immediately thereafter... strangely hyperactive.... I got a sudden impulse to run for no conscious reason, just an impulse which drove my legs.... I felt like a 'born againer' praising the virtues of oxygen...' (*4*).

Given time, the biospherians could probably have acclimatized and even reproduced. A mining village in South Peru is the highest permanent human settlement at just over 5000 metres. There the partial pressure of oxygen is nearly halved, equivalent to about 11% of the atmosphere at sea level. The village has been inhabited for over 40 years and a number of children have been born in this time. This sets a lower limit on oxygen for human persistence, but living with such low oxygen levels clearly involves special adaptations. The body adjusts to increase the amount of oxygen available in the arteries, achieving 85% of the sea level value, but the ability to do work is still reduced by about 25% and mental abilities by 10–30% (*5*). These figures come from health assessments for the Atacama Large Millimetre Array astronomical site in Northern Chile, also at 5000 metres altitude. The assessments recommended increasing the partial pressure of oxygen in the buildings to the equivalent of about 14% of the atmosphere at sea level. That will be like working at about 3500 metres, which is the point at which altitude sickness kicks in for many people. Oxygen presumably needed to be higher than this for humans to evolve.

As a ball park figure, let us take about 15% oxygen in the atmosphere as a rough minimum requirement for healthy human brain function, reproduction and civilization. This does not tell us how much oxygen is required in the atmosphere for any self-aware 'observer' life forms to evolve. We could conceive of an intelligent, perhaps sessile, organism that is more efficient than we are at devoting available oxygen to brain function. But taking about 15% oxygen in the atmosphere as a working estimate of what is required to allow the evolution of intelligence, we can ask; when did oxygen exceed this level? If it has exceeded it for most (or all) of the Phanerozoic then clearly it is not lack of oxygen that has been holding back the appearance of intelligent life.

15.2 Charcoal and the lower limit on oxygen

The best constraint we have on atmospheric oxygen levels during the Phanerozoic is provided by evidence of fires in the form of charcoal in the fossil record. With their atmosphere below 15% oxygen, had the 'biospherians' been inclined to light a wood or peat fire and have a sing song to cheer themselves up, they would have been unable to do so, because plant based fuels will not stay alight in an atmosphere containing 15% oxygen. The critical threshold for sustained combustion of plant material is actually around 17% oxygen in the atmosphere. This number comes from combustion experiments that Andrew undertook for his PhD (*6*, *7*). These showed that even completely dry paper will not ignite in 17% oxygen (Fig. 15.2). Recent experiments have confirmed that wood, peat and other plant material does not remain alight when oxygen is below 17% (*8*).

There have always been lightning strikes that could start a fire on land, so when we find charcoal in the fossil record we can be sure there was at least 17% oxygen the atmosphere. Charcoal first appears in the fossil record around 425 million years ago and, after a gap, is present nearly continuously in rocks from 360 million years ago to the present (*9*, *10*). What is remarkable about the first appearance of charcoal in the fossil record is that as soon as there is something on the land that could conceivably burn, it does burn (*11*). Beforehand there would have been very little plant material. It took the first hundred million years of the Phanerozoic for plants to gradually emerge onto the wetter parts of the land surface, and only by the time of the first charcoal were there free standing plants in somewhat drier areas that had some mechanism to stay wet inside whilst their surroundings dried out. Thus, the appearance of the first fossil charcoal, around 425 million years ago

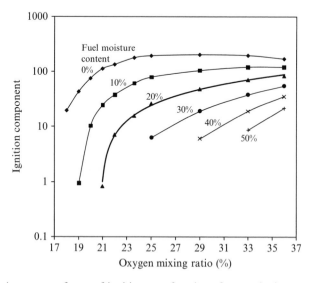

Fig. 15.2 A measure of ease of ignition as a function of atmospheric oxygen mixing ratio from experiments burning paper with different fuel moisture contents (*7*). Note the logarithmic scale on the y-axis

indicates an atmosphere with more than 17% oxygen, and possibly more than the present level of 21%.

Interestingly then, the amount of oxygen required to support combustion and leave a charcoal record, is also sufficient to support intelligent life. Whilst we cannot be sure that there was sufficient oxygen to support intelligent 'observer' life forms in the early part of the Phanerozoic, for at least the last 360 million years, there appears to have been enough. Oxygen might have temporarily dipped below the amount needed to support fire between instances of charcoal in the fossil record, but this is made less likely by the fact that it takes several million years to significantly change the amount of oxygen in the atmosphere and the gaps in the charcoal record are rarely that long (7). Thus, we cannot honestly argue that lack of oxygen held back the evolution of self-aware life. The data refutes it.

15.3 Was oxygen unstable in the early Phanerozoic?

Instead it is possible that oxygen has always been at or near the present level since the start of the Phanerozoic. We have very poor geochemical constraints on the level of oxygen in the atmosphere prior to the charcoal record. The size, mobility and hard shells of the Cambrian organisms can be used to make an estimate of their minimum oxygen requirements. Early work suggested they needed at least 2% oxygen in the atmosphere (or 10% of the present oxygen level) (12), consistent with geochemical arguments that oxygen exceeded this level at the end of the Precambrian (see Chapter 14). But this is only a lower limit based on relatively small shelly fossils. Looking at some of the larger and more mobile predators of the Cambrian that have been pieced together recently, it is hard to escape the conclusion that they probably thrived on more oxygen than this. Our favourite example is the giant swimming predator *Anomalocaris*, which grew up to a metre long. Seeing such giants, we might speculate that the Cambrian explosion was super-charged by high levels of oxygen.

Yet during the early Phanerozoic, anoxic waters were generally more common and widespread in the oceans than they are today. This suggests that, at least some of the time, oxygen was below the present level of 21%. When anoxic waters expand in the ocean, there is a negative feedback that kicks in which tends to stabilize atmospheric oxygen (Fig. 15.3). This feedback hinges on the fact that as anoxic waters expand, the recycling of phosphorus from marine sediments becomes more efficient, tending to increase marine productivity and, in turn, organic carbon burial. In the short term this exacerbates anoxia (a positive feedback), but in the longer term it pushes up atmospheric oxygen (a negative feedback). When we include these feedbacks in a

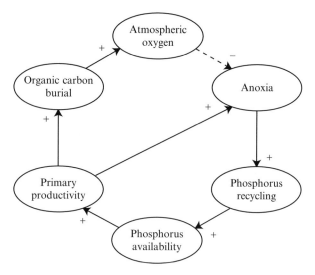

Fig. 15.3 Feedbacks involving enhanced phosphorus recycling under anoxic conditions. This exacerbates anoxia in the short term (positive feedback, inner loop) but buffers declining oxygen in the long-term (negative feedback, outer loop)

model, it predicts 5–10% oxygen in the atmosphere in the early Phanerozoic (*13*).

However, this negative feedback on atmospheric oxygen relies on there being some anoxic water in contact with ocean sediments. In the early Phanerozoic, if some factor had pushed atmospheric oxygen up to the point that there was no anoxic water left in the ocean, then the feedback (Fig. 15.3) would have switched off, leaving nothing to stop oxygen rising unbounded, potentially well above the present level (*7*). So, as far as we can tell, there was nothing to prevent the early Phanerozoic atmosphere becoming very rich in oxygen, and then experiencing potentially large swings in its oxygen content. We are not sure whether this actually happened. An oxygen-rich atmosphere would have been toxic and there are no obvious signs of oxygen toxicity having affecting early Phanerozoic marine life. The end of the Cambrian Period around 488 million years ago was marked by the first major extinction event of the Phanerozoic, but it carries all the signatures of the ocean becoming anoxic and lack of oxygen being the major kill mechanism at the time. Still, we cannot rule out that toxic levels of oxygen sporadically occurred.

The fact that oxygen had the potential to rise to toxic (and highly flammable) levels illustrates a potential problem for the evolution of complex life and especially life on land. Rather than too little oxygen, the evolution of intelligent life could have been threatened by too much oxygen, or unstable variations in oxygen level.

15.4 Oxygen regulation

Since plants colonized the land, has oxygen ever risen much above the present level (and if so, by how much)? There is no completely model-independent upper bound on oxygen during the Phanerozoic (like the charcoal lower bound). However, Andrew's combustion experiments (see Fig. 15.2) show that the relationship between oxygen mixing ratio and ease of ignition is highly non-linear. An increase of just a few percent in oxygen above the present level drastically reduces the energy required to ignite fuel. Furthermore, we know that there have been forests on the land surface continuously since about 370 million years ago. This indicates that oxygen has never been so high that fires prevented the regeneration of trees.

Simple calculations suggest that 25% oxygen might threaten the regeneration of forests (7), but the amount of energy required to cause ignition is also highly sensitive to the moisture content of the fuel (see Fig. 15.2). Wet forests with rapidly reproducing trees would be a lot more tolerant of elevated levels of oxygen than slowly reproducing, dry forests. That said, even rainforests today can be ignited by lightning strikes during spells of drier weather, and considerable fires ensue. Based on these considerations, we can be pretty sure that oxygen has never risen to be, say, twice present levels. In fact our modelling suggests it has never exceeded about 30% of the atmosphere for the last 370 million years, and in times with drier climates the upper limit on oxygen was nearer 25% of the atmosphere.

This is quite remarkable. We have argued that oxygen could have risen essentially unbounded and fluctuated wildly in the earliest part of the Phanerozoic. But since plants and especially forests became established on the land surface, around 370 million years ago, oxygen has remained between about 17% and probably 30% of the atmosphere. This would be unlikely to have happened by chance. All the oxygen in the atmosphere (and ocean) is replaced roughly every 3 million years, yet over a period a hundred times longer, the total amount present has varied by less than a factor of two. This demands some stabilizing negative feedback mechanisms. The question is; what are they? We have already discussed ocean bound mechanisms that can buffer declines in oxygen. The outstanding problem is; what prevented oxygen rising to highly flammable levels? We think the answer must be that vegetation on land became intimately involved in negative feedbacks that counteract rising oxygen.

There are two key ways in which rising oxygen can suppress vegetation, one direct and one indirect. The direct effect is that increasing the ratio of oxygen to carbon dioxide in the air reduces the efficiency of photosynthesis. We described this in Chapter 14; both oxygen and carbon dioxide molecules can react at the active site of Rubisco, but the oxygen reaction (called photorespiration) is useless to the plant and the more of it that occurs the less carbon gets fixed. The indirect effect of rising oxygen is fire, which tends to reduce the biomass of vegetation and shift ecosystems from forest to faster regenerating herbaceous vegetation (in recent time, grasslands). But how can these detrimental effects of oxygen on vegetation suppress its long-term source from carbon burial and hence produce negative feedback? Our suggestion is that phosphorus weathering by vegetation is suppressed, and this in turn limits the total amount of organic carbon that can be buried, and with it the source of oxygen (Fig. 15.4) (*7*). Alternatively, if phosphorus is redirected from the land to the ocean, where less carbon is buried per unit of phosphorus, this would also reduce the source of oxygen (*14*).

We find we have to include a negative feedback on oxygen involving vegetation and Rubisco function for oxygen to be stable *at all* in our models (*7, 13*). Even then it can go to inflammatory high values, reaching 30% of the atmosphere in the Carboniferous and exceeding it in the Cretaceous (*13*). Negative feedbacks involving fire are potentially much stronger because of the highly non-linear relationship between oxygen level and the ease of starting a fire (see Fig. 15.2). In our modelling we varied the strength of the fire feedback on phosphorus weathering and found that potentially it could keep oxygen below 25% of the atmosphere for almost all of the past 350 million years (Fig. 15.5)

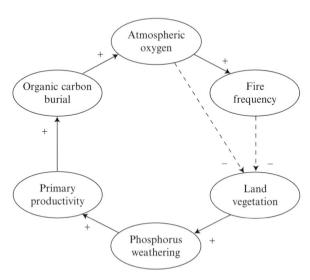

Fig. 15.4 Proposed negative feedbacks on atmospheric oxygen involving vegetation and phosphorus weathering. Oxygen suppresses vegetation directly (a weak effect) and via fires (a strong effect). Vegetation drives phosphorus weathering, which in turn ultimately limits organic carbon burial, which is the long-term source of atmospheric oxygen

(*13*). Thus, we argue that the advent of vegetation on land has been instrumental in stabilizing atmospheric oxygen close to the present level.

Relatively stable, high levels of oxygen have in turn been important for animal evolution. So, we can make a case that the colonization of the land by complex ecosystems played an indirect role in creating a world conducive to the evolution of intelligent life. Our view contrasts with other authors who argue that substantial variations of oxygen have occurred since vegetation colonized the land, and that these fluctuations have in turn played a driving role in animal evolution. So, let us examine in a little more detail why different models give different predictions for oxygen variation, and then consider whether they are consistent with the charcoal data.

15.5 Different oxygen predictions

Different models taking different approaches and input data, not surprisingly, give different predictions of atmospheric oxygen over Phanerozoic time (e.g. Fig. 15.5). However, the first thing to note is that all the models basically agree that oxygen has been surprisingly stable. Over the last 350 million years, all the oxygen in the atmosphere has been replaced around 100 times and yet the amount present has varied by less than a factor of two in either direction. For this to be the case there must be some effective mechanisms for stabilizing the oxygen concentration, both in reality and in the models. Beyond this overarching stability, it seems highly likely that atmospheric oxygen levels have fluctuated since vegetation colonized the land. The current debate is about how much they have fluctuated.

One of the first models of atmospheric oxygen variation over Phanerozoic time, by Bob Berner and Don Canfield, famously predicted up to 35% oxygen in the atmosphere 300 million years ago, as the Carboniferous made way for the Permian period (*15*) (Fig. 15.5a, dotted line). This model used as input data the relative abundances of different rock types over Phanerozoic time, and the concentrations of organic carbon (and pyrite sulphur) in each rock type. In more recent work, Berner has used the isotopic records of carbon and sulphur as input to his models (see Box 15.1). Early efforts to do this resulted in 'oxygen catastrophes' in which oxygen disappeared from the atmosphere or rose to many times the present concentration (*16*). But by making some assumptions about how carbon and sulphur isotope fractionation depend on oxygen concentration, Berner is now able to predict less extreme oxygen variations, which again include a peak of up to 35% of the atmosphere, in the Permian period (*17*).

(a)

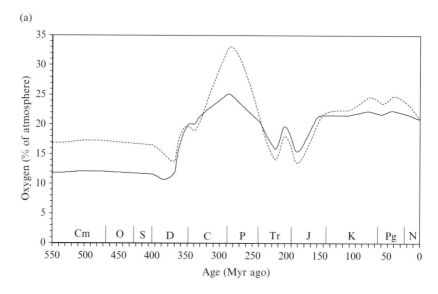

(b)

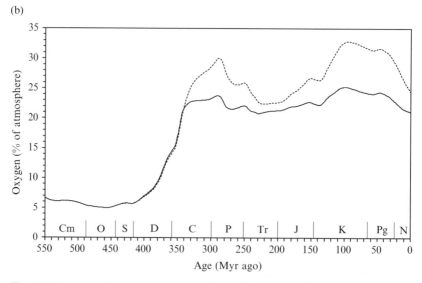

Fig. 15.5 Phanerozoic oxygen predictions from different models, showing the effect of negative feedback involving fires, vegetation and phosphorus weathering. (a) Berner and Canfield's original model (*15*), without this feedback (dotted line) and a variant adding this feedback (*18*) (solid line), (b) our COPSE model with weak fire feedback (dotted line) or strong fire feedback (solid line) (*13*)

Box 15.1 Contrasting approaches to understanding atmospheric oxygen variations

Isotope-driven models

In recent years, the most influential treatment of the atmospheric oxygen problem has been led by Bob Berner at Yale University. His approach has been to constrain the budgets by use of isotope data. A model of the geochemical reservoirs is constructed, and isotope measurements are interpreted to describe changes in the sizes of the carbon and sulphur reservoirs. The fluxes in and out of each are calculated and this is translated into the amount of oxygen evolved. Continue this through time and the oxygen content in the atmosphere can be reconstructed. This school is traceable to the approach of Garrels and Lerman, via classic papers in the 1980s (19). However, it suffers from a problem, in that interpretation of the isotopes is subject to large errors and uncertainties. As noted in Box 10.1, the potential for atmospheric oxygen to change is huge, and unfortunately isotopes are only semi-quantitative guidelines to the fluxes in and out of the reservoirs. Isotopes can be misleading, firstly because isotope data from different locations can scatter widely, so it is quite difficult to obtain appropriate averages for the Earth as whole. Secondly, isotope fractionation is affected by many different processes. Individual species fractionate differently for example; the fractionation varies with changes in the environment and may well change through time as evolution proceeds.

Naïvely applied, the approach can lead to an atmosphere having ten times the amount of oxygen as we see in it today, or conversely, negative oxygen concentrations (16)! Clearly, the former has never occurred (it would be lethal to most life) and the latter is physically impossible. Simply turning the handle on this approach would easily lead us astray, and in practice we need to bring additional insight to bear to get sensible answers out of the models. There is plenty of such 'additional insight' in existing models which predict atmospheric oxygen, but it is often not fully agreed upon. The conclusions of these models are therefore open to challenge and debate. However, many of those not fully familiar with the ins and outs of the subject tend to forget these caveats and treat the reconstructions of atmospheric oxygen as if they were factual, rather than hypothetical. The isotope-led approach is nevertheless very powerful, especially as we come closer to the present when the rock record is more complete, and the assumption that the processes involved are similar to today is more likely to be right.

Mechanism-driven models

Our approach has started from a different premise. We assume that the oxygen concentration in the atmosphere is tightly controlled by negative feedbacks. We know that theoretically, the capacity exists to change oxygen hugely, over geologically short times. Yet, in fact, concentrations have rarely varied wildly. Furthermore, the concentration of oxygen very directly affects the mode of life on the planet, and in turn it is principally affected by life, whether by burying carbon, or by promoting hydrogen loss to space (as discussed in earlier chapters). Add to this the observation that there is plenty of 'gain' available in the system—known processes can change the concentration radically and quickly—and we have the classic conditions (see Chapter 7) in which we expect feedback to dominate its behaviour.

Thus, our approach to the 'oxygen problem' is to make hypotheses about which feedback mechanisms are important, and to test these in models. (We discuss what the important feedback loops might be in the text.) The models can also incorporate what is known about the processes which change isotopes, so we can predict what should happen to the isotopes for any given set of hypotheses about feedbacks. The predictions of the isotopes can be used as a check on the model, and if the predicted isotopic record is incompatible with what is seen, then the model is no good. However, the isotope record is not the starting point for the process, because by itself it does not contain enough information to fully constrain the system.

Berner uses a different method to stabilize oxygen in his models, which he calls 'rapid recycling'. It has nothing to do with recycling in ecosystems; instead it refers to the recycling of recently deposited sediments. It is assumed that organic carbon and pyrite sulphur that have recently been deposited go into a small reservoir of 'young' rocks that are returned to the Earth's surface within a few million years, in contrast to the bulk of the crust, which is recycled over roughly 150 million years. The logic behind this is that sediments deposited in shallow seas or in land basins can be re-exposed and weathered when the sea level falls or the crust is uplifted. Although oxygen is added to the atmosphere when organic carbon (or pyrite sulphur) are deposited, if the same material is weathered again within a few million years the corresponding oxygen is mopped up, and this helps prevent large swings of oxygen in the models. Berner finds he has to include rapid recycling to get sensible results for atmospheric oxygen in his models. But we exclude it in our model and show instead that feedbacks within the ocean-atmosphere-land system can ensure relatively stable values for atmospheric oxygen (*13*).

However, the magnitude of predicted oxygen variations depends strongly on the amount of negative feedback assumed in the models. For example, by including the negative feedback involving fires and phosphorus weathering (see Fig. 15.4) in Berner and Canfield's original model their simulated oxygen peak can be brought down from about 35% to about 25% of the atmosphere (Fig. 15.5a, solid line) (*18*). In our own modelling work, led by our former student Noam Bergman, a similar story transpires (*13*): With weak negative feedbacks, oxygen reaches 30% of the atmosphere in the early Permian and approaches 35% in the Cretaceous (Fig. 15.5b, dotted line). However, with strong fire feedback, oxygen never exceeds 25% of the atmosphere (Fig. 15.5b, solid line). So, the strength of negative feedback in the models clearly affects their oxygen predictions.

The other major determinant of predicted oxygen levels is the choice of input data used to force the models. Both Berner's recent models and our own are driven with reconstructions of geological and biological drivers including continental uplift, volcanic degassing and plants colonizing the continents. Where we differ is over how to use the isotopic records of carbon and sulphur. In Box 15.1 we describe the two contrasting modelling approaches in more detail.

We can understand how the different models come to different oxygen predictions. But can we falsify any of the predictions and thus favour one model or mechanism over another? To do so, we must have some independent data to compare the model predictions against. For all of the models, the charcoal record provides an independent constraint.

15.6 Have oxygen fluctuations driven animal evolution?

So, let us turn to some of the predicted variations in oxygen and their postulated links to animal evolution, and see if they are consistent with the charcoal data.

The first limbed vertebrates staggered or shuffled out of the water around 370 million years ago in the late Devonian. The common picture is of air breathing fish (related to the coelacanth) developing into limbed amphibians pottering around on land. But the first amphibians seem to have spent most of their time in freshwater and did not interact with terrestrial ecosystems. They disappeared at the end of the Devonian and there ensued an interval during the early Carboniferous, 360–345 million years ago, with only very rare fossils of limbed vertebrates. It is known as 'Romer's gap' after the paleontologist who first noted it. Little in the way of terrestrial arthropod evolution seems to have occurred in this interval either. This has led to the proposal that this was a hard time for early land animals because of predicted low levels of oxygen in the atmosphere (20). However, there were already large amounts of charcoal in some latest Devonian deposits and charcoal occurrences become more frequent in the early Carboniferous (10). This suggests atmospheric oxygen approached present levels during this time and possibly rose above them, making it unlikely that lack of oxygen was holding back animal evolution.

Moving forward to the late Carboniferous, around 300 million years ago, most models predict a peak in oxygen somewhere in the range 25–35% of the atmosphere (Fig. 15.5). This has led a number of authors to infer a connection to the appearance of giant arthropods around this time, including dragonfly-like insects (called Meganeura) with wingspans of up to 75 cm. The idea is an old one. Fossils of Meganeura were first described in the 1880s and in 1911 it was proposed that elevated levels of atmospheric oxygen enabled them to fly. In this case abundant fossil charcoal supports the model predictions that oxygen probably exceeded the present level.

However, a rather different line of evidence leads us to question whether Meganeura really required an oxygen-rich atmosphere. The original argument was based on the assumption that the insects only 'breathed' passively by allowing oxygen to diffuse through their trachea. Even if this were the case, a doubling of oxygen concentration would only enable a doubling of the rate of transport of oxygen, since diffusion fluxes vary in proportion to concentration. However, Meganeura was *ten* times the size of modern dragonflies, so the oxygen increase would not alone explain the size difference. Moreover, recent work has shown that some modern insects really do 'breathe' with rapid cycles of compressing and expanding their trachea in the head and thorax, a mechanism of inflation and deflation analogous to vertebrate lungs (*21*). If Carboniferous insects also possessed such an active breathing mechanism then there seems little need to demand higher levels of atmospheric oxygen to account for their large size. Instead the idea that they were the top predators of the air at the time might be a better explanation. The Permian *Meganeuropsis permiana* fed on other insects and even small amphibians, and there is no evidence that anything fed on it. It appears to have filled an ecological niche that much later was occupied by flying reptiles and then birds.

Jumping forward in time to the last 200 million years, it has been hypothesized that the evolution of placental mammals, the group that includes us humans, was due to a rise in oxygen (*22*), on the grounds that placental mammals have relatively high oxygen requirements (*23*). In support of this, one model, when driven with new isotope data, predicts oxygen to have risen from about half the present level over the last 200 million years (*22*). However, fossil charcoal evidence indicates that atmospheric oxygen concentrations have been sufficient to support fires throughout the last 200 million years (*8*). Indeed abundant charcoal from the Cretaceous (146–65 million years ago), including charred flowers, suggests oxygen was above, not below, present levels then. There is relatively little charcoal from more recent periods, hence, if anything, oxygen has declined toward the present. (We return to examine what might have contributed to the evolution of placental mammals in Chapter 17.)

15.7 Summary

In this chapter we have considered whether environmental conditions held back the evolution of intelligent life, the best candidate being a lack of oxygen. The answer is a qualified 'no', instead the problem could have been too much oxygen or unstable variations in oxygen, were it not for the emergence of complex ecosystems on land and their involvement in regulating the oxygen content of the atmosphere. Since vegetation colonized the land, atmospheric oxygen has been maintained close to the present concentration, which in turn has provided a relatively (although far from perfectly) stable environmental context for the evolution of animals.

So, did the land have to be colonized by complex life for any 'observer' species to have evolved? The answer to this is probably not. Flammable levels of atmospheric oxygen are not a problem if you live in water, and toxic levels might be avoided simply by living in a part of the ocean with a weaker oxygen supply. It seems at least conceivable that self-aware life could evolve entirely within water. Indeed it may already have done so; octopuses are remarkably intelligent creatures with complex nervous systems, and who knows what they may be thinking? In Chapter 5, we defined an 'observer' species as one able to ask questions about their origins. Octopuses probably don't ask such questions, but that does not rule out the possibility that an entirely ocean based biosphere could in principle have evolved observers.

What is clear is that the land had to be colonized by plants, fungi and animals for our species to have evolved. All mammals originated on land, even though some (whales, dolphins and other cetaceans) now enjoy life in the ocean. We now turn our attention to this flourishing of life on land, to see how it created a world in which we could evolve and, more importantly, because it is the best preserved example of what makes a successful transformation of the Earth system.

References

1. J. Alroy, *Dynamics of origination and extinction in the marine fossil record. Proceedings of the National Academy of Sciences USA* **105**, 11536 (2008).

2. D.-G. Shu *et al.*, *Lower Cambrian vertebrates from south China. Nature* **402**, 42 (1999).

3. M. E. Raichle, D. A. Gusnard, *Appraising the brain's energy budget. Proceedings of the National Academy of Sciences USA* **99**, 10237 (2002).

4. P. Warshall, *Lessons from Biosphere 2: ecodesign, surprises and the humility of Gaian thought. Whole Earth Review* **89**, 22 (1996).

5. P. J. Napier, J. B. West, 'MMA Memo No. 162 Medical and Physiological Considerations for a High-Altitude MMA Site', National Radio Astronomy Observatory http://www.alma.nrao.edu/memos/html-memos/alma162/memo162.html (1996).

6. A. J. Watson, *Consequences for the Biosphere of Forest and Grassland Fires*. Ph.D., University of Reading (1978).

7. T. M. Lenton, A. J. Watson, *Redfield revisited: 2. What regulates the oxygen content of the atmosphere? Global Biogeochemical Cycles* **14**, 249 (2000).

8. C. M. Belcher, J. C. McElwain, *Limits for Combustion in Low O_2 Redefine Paleoatmospheric Predictions for the Mesozoic. Science* **321**, 1197 (2008).

9. A. C. Scott, *The Pre-Quaternary history of fire. Palaeogeography, Palaeoclimatology, Palaeoecology* **164**, 281 (2000).

10. A. C. Scott, I. J. Glasspool, *The diversification of Paleozoic fire systems and fluctuations in atmospheric oxygen concentration. Proceedings of the National Academy of Sciences USA* **103**, 10861 (2006).

11. I. J. Glasspool, D. Edwards, L. Axe, *Charcoal in the Silurian as evidence for the earliest wildfire. Geology* **32**, 381 (2004).

12. H. D. Holland, *The Chemical Evolution of the Atmosphere and Oceans.* (Princeton University Press, Princeton, 1984).

13. N. M. Bergman, T. M. Lenton, A. J. Watson, *COPSE: a new model of biogeochemical cycling over Phanerozoic time. American Journal of Science* **304**, 397 (2004).

14. L. R. Kump, *Terrestrial feedback in atmospheric oxygen regulation by fire and phosphorus. Nature* **335**, 152 (1988).

15. R. A. Berner, D. E. Canfield, *A new model for atmospheric oxygen over Phanerozoic time. American Journal of Science* **289**, 333 (1989).

16. A. C. Lasaga, *A new approach to isotopic modelling of the variation of atmospheric oxygen through the Phanerozoic. American Journal of Science* **289**, 411 (1989).

17. R. A. Berner, *GEOCARBSULF: A combined model for Phanerozoic atmospheric O_2 and CO_2. Geochimica et Cosmochimica Acta*, 70, (2006) 5653–5664.

18. T. M. Lenton, *The role of land plants, phosphorus weathering and fire in the rise and regulation of atmospheric oxygen. Global Change Biology* **7**, 613 (2001).

19. R. A. Berner, A. C. Lasaga, R. M. Garrels, *The Carbonate-Silicate Geochemical Cycle and Its Effect on Atmospheric Carbon-Dioxide over the Past 100 Million Years. Am. J. Sci.* **283**, 641 (1983).

20. P. Ward, C. C. Labandeira, M. Laurin, R. A. Berner, *Confirmation of Romer's Gap as a low oxygen interval constraining the timing of initial arthropod and vertebrate terrestrialization. Proceedings of the National Academy of Sciences USA* **103**, 16818 (2006).

21. M. W. Westneat *et al.*, *Tracheal Respiration in Insects Visualized with Synchotron X-ray Imaging. Science* **299**, 558 (2003).

22. P. G. Falkowski *et al.*, *The Rise of Oxygen over the Past 205 Million Years and the Evolution of Large Placental Mammals. Science* **309**, 2202 (2005).

23. P. L. Else, A. J. Hulbert, *Comparison of the 'mammal machine' and the 'reptile machine': energy production. American Journal of Physiology— Regulatory, Integrative and Comparative Physiology* **240**, R3 (1981).

16
The grand recycling coalition

By the start of the Phanerozoic, multi-cellular animals, fungi and algae had already evolved, but most complex life was still to be found in the oceans and freshwaters. Some of the late Proterozoic and Cambrian land surface was probably coated by microbial and algal mats and what we have called 'proto-lichens' (Chapter 14), forming multi-coloured carpets on an otherwise barren regolith. These photosynthesizers left their mark in the carbon isotope record, which reveals isotopically-light material being washed from the land into the shallow seas during the Cambrian (*1*), but no fossil evidence of them has been found. Animals appear to have occasionally strayed into this strange terrestrial world. Arthropod tracks have been found preserved in the sand near an ancient early Ordovician shoreline (*2*). Perhaps they left the water to graze on microbial mats. But if they did they would have found relatively little to eat: mats of cyanobacteria and green algae achieve rates of photosynthesis per unit area that are at least an order of magnitude less than those of plant leaves (*3*). The low productivity of these early land ecosystems could not have supported many energy hungry animals, and food webs would have been shorter and simpler than they were in the ocean.

The major new arrivals of the first part of the Phanerozoic, called the Paleozoic, were the land plants. In greening the land surface they doubled the amount of free energy captured by the biosphere, and transformed the face of the Earth. The productivity per unit area on land today is, on average, more than double what it is in the ocean. To achieve this required the emergence of the most remarkable systems for acquiring and recycling water and nutrients to have graced the planet thus far. Probably the best example is the nutrient phosphorus. Plants and fungi selectively extract phosphorus from rocks and it is then recycled around fifty times in modern ecosystems (*4*). The supply of other elements such as potassium, iron, calcium and magnesium is also enhanced by biological rock weathering and recycling within ecosystems. Nitrogen and carbon are fixed from the atmosphere, and they too are internally recycled. Recycling is essential to maintain the high productivity that has enabled the land to support the greatest flowering of biological diversity the Earth has yet seen.

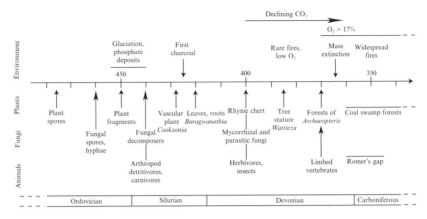

Fig. 16.1 Timeline of the colonization of the land by plants, fungi and animals

The evolution of land plants was a complex and multifaceted process, Fig. 16.1 summarizes the major events. The colonization of the land could be considered the last act in the long revolution that was begun by the evolution of eukaryotes and included the origin of multi-cellularity. Land colonization was a staggering, planet shaping achievement of life. It was not a sudden change, but rather a stepwise process that involved not just land plants, but fungi, animals and bacteria as well, evolving in concert to create entirely new ecosystems and niches, and in the process transforming the global environment.

Plants faced formidable obstacles when first they moved onto the land. Their ancestors' reproductive cycle was strongly tied to water, and on land they also had to find all the materials they required from the air, rainfall, rocks or soil. The 'solution' that evolution came up with hinged on a special symbiosis with fungi, similar to the lichen symbiosis we've already discussed. This symbiosis produced a new unit of organization and enabled large-scale environmental alteration. Animals also became involved, forming a grand coalition that holds to this day. As colonization proceeded, it kept disrupting the environment in ways that could have led to disaster for the responsible organisms. However it eventually succeeded (and disaster was averted) because of the establishment of efficiently recycling ecosystems and the emergence of regulatory feedbacks that ultimately limited the land biosphere.

16.1 Coping with desiccation

The first plants to stray onto the land were faced with a hostile environment. Perhaps the most fundamental problem they had to overcome was lack of water. The ancestors of all land plants today came from a group of multi-cellular green algae called the 'charophyta' that live in freshwaters. Their closest living relatives today are the stoneworts which look like underwater plants, with a branching structure of fronds. However, they lack the ability to cope with drying out, or to erect a free standing structure in the air, and they don't possess true leaves or roots. Conceivably the ancestors of the first land plants inhabited ponds that occasionally dried up, providing a selective force for the evolution of resistance to desiccation, particularly of their reproductive spores.

The first land plants we know of are recognized only from their fossilized spores, which first appear around 470 million years ago in the Ordovician period (5). There are some earlier plant-like fossil spores from the Cambrian and even the Neoproterozoic, but they are not widely accepted as having belonged to land plants. By the end of the Ordovician, the telltale sign which marks out fossil spores as having belonged to a plant living in the air, rather than a freshwater alga, is a characteristic 'Y' shaped scar, which gives them their name of 'trilete' spores. The spores come from clusters of four (known as 'tetrads') and the 'Y' shape is the mark left on each spore at the join with its three neighbours. For a mark to be left, the spore must have a relatively sturdy outer wall, indicating resistance to desiccation.

The oldest plant spores resemble those of modern liverworts, which are members of the bryophytes, along with hornworts and mosses. Molecular analyses support the idea that bryophytes are the most ancient land plants. However, in the tree of life, the branches leading to liverworts, hornworts and mosses each come separately from the stem leading to all other plants. The molecular analyses suggest that liverworts were the first to branch off, and are therefore the oldest (6,7).

Fossils of the first plants themselves have proved extremely hard to find. This is not surprising because the plants that made the Ordovician spores would not have been readily preserved, whereas the spores are hardier and were transported to shallow water environments where new sediments were being formed. Some early 'enigmatic' microfossils resemble tissues of modern liverworts and mosses that are resistant to degradation (8,9). However, the first convincing fossils come from a 450 million year old floodplain, where fragments of a land plant, including parts still holding spores were preserved (10). The fragments resemble parts of modern liverworts, as do the spores they have kept hold of.

These first colonizers had some adaptations to occasional drying out, but they were still tied to relatively wet environments. Bryophtyes have an aquatic stage to the life cycle in which their sperm swim through the environment. Also, if they dry out, they have to put their metabolism on hold and wait for rewetting. They lack the ability to maintain internal water pressure. Most modern plants, in contrast can maintain this water pressure and thus survive varying degrees of dryness in their environments. One of the keys to achieving this is the use of internal tubes with rigid cell walls that take advantage of capillary action to draw water up against gravity into the plant. Continual flow is fuelled by the release of water (transpiration) through holes that can be opened (stomata) at the end of the capillaries. The special tubing and the associated tissues gives the 'vascular' plants their name.

The acquisition of a simple plumbing system was the next key phase of land plant evolution, which allowed the organisms possessing it to stand further above the ground and to colonize drier land. This was achieved by the middle of the Silurian period around 425 million years ago. The first fossils of whole plants are preserved from this time, including various members of the genus *Cooksonia*, which vary in height from a centimetre to a metre tall. They were quite unlike plants as we know them today. The fossils look like branching stems, inside of which are the conducting tubes which identify their owners as the first vascular plants. In some specimens, the stems each have a cone-like spore head on the end of them. They had no leaves and it is unclear whether photosynthesis occurred in the stems or spore heads. They also had no proper roots.

Very soon after this, around 420 million years ago, vascular plants with leaves and proper roots had evolved. They belonged to a genus called *Baragwanathia*, which are related to modern clubmosses. They grew up to several metres in length. Their tiny leaves had a single midvein and are known as 'microphylls' in the jargon, to distinguish them from familiar, large leaves, known as 'megaphylls'.

As plants evolved and more complex ecosystems developed, they made the land surface a much damper place. They did this by creating organic rich soils that hold more water, and by tapping into progressively deeper water reserves in the ground. By actively pumping up water and returning it to the atmosphere, and by reducing runoff from the land surface, plants increased the recycling of water from the land to the atmosphere. This has increased precipitation over land, which in turn has increased the amount of vegetation biomass the Earth can support today by an estimated factor of three (*11*).

16.2 Acquiring and recycling nutrients

Plants are often portrayed as leading the charge in the invasion of the land by complex life, providing the primary production on which animals and fungi depend. But as well as a shortage of water, the first plants were faced with a profound shortage of nutrients, because many essential elements were locked up in rocks. Without proper roots for their first 50 million years, how did plants access the water and nutrients they need to grow?

The answer appears to be that the first plants formed a special symbiotic relationship with fungi, which is still widespread today (*12–14*). In this symbiosis, particular types of fungi play a role we commonly associate with plant roots, forming fungal roots, called 'mycorrhizae', they provide nutrients and water to the plant in exchange for sugars. The partnership is based on the same principles as lichens (described in Chapter 14). The photosynthesizing partner is a multicellular plant rather than a single-celled alga or cyanobacterium, and rather than the plant being contained within the fungus, the fungus invades some of the cells of the plant. Roots that belong to the plant are a later innovation, and they appear to have co-evolved with mycorrhizae (*15*). The great majority (around 80%) of modern plant species still have mycorrhizal fungi.

All the major branches of fungi had diverged by the time the first plants evolved (and probably long before). They include the phylum Glomeromycota, members of which form the most ancient mycorrhizal associations with plants (*16,17*). Some 460 million year old fossils have been likened to this type of fungi (*18*), suggesting that they were present at the time of the first land plants (although others question whether these fossils are definitely fungi (*19*), or whether they might be later contamination). Fungi have not been caught fossilized in the act of symbiosis with the earliest land plants, but as soon as there is a well preserved ecosystem the symbiosis is found (see below).

Symbiotic associations with fungi are certainly common among bryophytes today (*20*). Liverworts from the most ancient lineages, which grow in New Zealand today, have an extraordinary symbiosis with mycorrhizae involving relatively complex cell differentiation on the part of the fungus where it infects the plant (*21*). Fungi are crucial to both the acquisition and the recycling of nutrients in almost all modern terrestrial ecosystems (*22–24*). Mycorrhizae significantly accelerate the weathering of rocks (*25, 26*) but some also extract nutrients from organic matter in soil (*27*). This latter role as decomposers (or 'saprotrophs'), is also fulfilled by free living fungi that break down organic material, recycling the nutrients it contains (especially phosphorus and nitrogen) into a form that the plants can take up again.

Ecosystems including fungi and animals were probably established relatively soon after the first plants colonized the land. By the end of the Ordovician period there were well developed soils with evidence of weathering and with animals (probably related to millipedes) burrowing within them (*28*). By 440 million years ago, when the Ordovician had given way to the Silurian period, a couple of unusual sites now in North America preserve plants, fungi and animals forming recycling terrestrial ecosystems. The animals included detritivores which fed on dead plant material, and predators that fed on the detritivores. Fungi were present as decomposers, recycling plant and animal waste material into a form that could be used again by the plants. By late Silurian times, some remarkable fossils appear which suggest that fungi were having a field day on land. They are giant 'trunks' up to a metre in diameter and many metres long of an organism called *Prototaxites*, which is found from 420 to 370 million years ago, and has long puzzled paleontologists (see Box 16.1).

The first really well preserved terrestrial ecosystem comes from the early part of the Devonian period, around 400 million years ago. It is found today in a small section of rock near the village of Rhynie in Aberdeenshire, Scotland, and is known as the Rhynie chert. The ecosystem thrived near a volcanic hot spring and met its demise when silica-rich water rose rapidly from the spring, petrifying it in its entirety and preserving it in three dimensions. Whilst this was bad news for the organisms concerned, it was of great benefit to future scientists. The Rhynie chert preserves plants, animals, fungi, lichens, cyanobacteria, plant litter, and some of the relationships between them in exquisite detail. It shows that recycling ecosystems with considerable biodiversity were well established by 400 million years ago.

The Rhynie chert ecosystem was fuelled by both bryophytes and early vascular plants. Among them, *Aglaophyton* had fungal mycorrhizal symbionts for which it formed a home (arbuscules) at the bot-

---

Box 16.1 Giant fungi? The mystery of *Prototaxites*

What *Prototaxites* is has been debated for well over a century (*29*). Recent evidence points to it being a giant fungus, possibly in symbiosis with algae, i.e. a giant lichen (*30*). If it were a plant it would have been the first to have achieved tree stature, but the varied carbon isotopic composition of fossils strongly suggests it is not (*31*). Plants have a characteristic and fairly uniform carbon isotopic composition, fungal matter on the other hand 'picks up' the varied isotopic compositions of whatever material it feeds on. The organism is thought to have stood like dead tree trunks rising metres above the ground, although all the preserved specimens are close to horizontal (i.e. are inferred to have fallen over). One possibility, suggested to us by Nick Butterfield, is that they really are dead tree trunks that have been infected and consumed by fungi from the inside out. As Nick notes, the problem with this interpretation is that there is no evidence of plants having reached tree stature by this time. So, what evolutionary advantage would there have been to a fungus attaining tree stature? If its surface contained symbiotic algae or cyanobacteria then these would have had good access to light above the small plants below. But more likely the main advantage to a fungus of sticking metres above the surface would have been to disperse its spores further. Dispersal is naturally selected for because it aids the colonization of fresh habitat, and even when the spores arrive in a habitat that is already colonized, they are less likely to be competing with their near relatives (*32*).

tom of its stems (*33*). There was also a fungus that entered into symbiosis with cyanobacteria—a lichen (*34*). There were fungal decomposers that broke down dead plant material (saprotrophs) and fungi that were parasites on living plants and algae. Fungal infection is known to promote speciation in modern plants, which might help explain the diversity of plants that had appeared by this time. The animals of the Rhynie chert ecosystem were all arthropods and include the earliest known insect, resembling a modern springtail. There were also harvestmen, mites, scorpion-like creatures, and crustaceans. The shape of their mouthparts, and evidence of piercing and boring into plant material, suggest that some of the animals were herbivores. Others were detritivores, feeding on dead plant material. Fossil turds (politely known as 'faecal pellets' or in fossilized form as 'copro-lites') are present and their contents give clues to diet—some are densely packed with spores suggesting they were a major part of some diets. There were also carnivores that fed on the herbivores and detritivores.

The animals contributed fundamentally to the recycling. Herbivores process plant material, returning it to the soil in their excrement, thus providing food for decomposers. Detritivores (such as millipedes and earthworms) act as decomposers feeding on dead organic material. Predators feed on detritivores and herbivores, aiding the processing of organic material. The cycling is completed with the aid of the symbi-otic bacteria and archaea in the animals' guts, and with bacteria free living in the soil, that return the nutrients to a form that plants and fungi can take up. From the perspective of a bacterium, animals are a wonderful invention, which enables them to access more food. Herbivores, for instance, are mobile sacks of bacteria, complete with mechanisms to seek out, chew up and deliver the organic material fixed by photosynthesizers to their gut symbionts. Today therefore, a mature land ecosystem is a grand coalition that involves bacteria, archaea, plants, animals and fungi all working in concert to find, fix and recycle nutrients and carbon. The Rhynie Chert shows that the basics of this system were in place 400 million years ago. Since that time, there has been some elaboration on the theme—some bigger plants and animals have evolved for instance—but the principles have remained the same.

Internal nutrient recycling is essential to the productivity of modern terrestrial ecosystems. If we assume that without recycling, the (biolog-ically-enhanced) phosphorus supply from rocks would set the ultimate cap on the productivity of the land biosphere, then it would be restricted to only about 2% of its present value (*4*). Thus, whilst the evolution of land plants was clearly essential to the successful colonization of the land by complex life, they could not do it alone, and a whole ecosystem view of the process is a more accurate one.

16.3 Trees and their environmental consequences

All the parts that make up most familiar land plants finally came together during the Devonian period. The first big leaves (megaphylls) appeared in the early Devonian. As the period progressed, land plants diversified and colonized lowland regions. By the mid Devonian, plants had reached tree stature, going from 3 metres around 392 million years ago to over 8 metres by 385 million years ago. The first groves of trees are preserved in a long known collection of stumps from Gilboa, New York. Recently these stumps have been reunited with the crown parts of the plant, which lacked proper leaves, but had branch-like appendages that photosynthesized and were occasionally shed (*35,36*). In some ways these first trees (called *Wattieza*) would have resembled unrelated modern tree ferns, with an unbranched trunk. They had limited roots all of a similar size. Their evolutionary strategy appears to have been to rapidly put up a mechanically stable structure to maximize dispersal of their offspring.

The first thing we might recognize as a 'proper' tree, made out of wood, with large branches dividing into smaller branches with leaves, appeared later in the Devonian (*36*). It is called *Archaeopteris*, and individuals were up to 40 metres high, looking something like a modern conifer in structure, although they had fern-like leaves. Their structure would have made them relatively efficient at harvesting light. Underground they had the first substantial, deep root systems and created the first well drained soils. To grow and support their substantial biomass, these trees must have used their deep roots to enhance rock weathering and the extraction of nutrients. *Archaeopteris* were spectacularly successful, forming the first forests that cloaked large areas of the late Devonian land surface. Their enhancement of rock weathering in turn caused global environmental changes.

Although weathered elements may initially enter plants via their fungal mycorrhizae and then be recycled many times within an ecosystem, ultimately they will be washed to the ocean. As we discussed in Chapter 14 there are two key, biologically-enhanced weathering processes; silicate weathering and phosphorus weathering. Let us start with the consequences of silicate weathering: Calcium and magnesium ions liberated from silicate rocks are washed to the ocean where they are deposited in carbonate rocks, removing carbon dioxide from the atmosphere, and cooling the planet.

The magnitude of carbon dioxide removal and planetary cooling would have depended on how much the first forests enhanced weathering relative to what was on the land surface before them. A large amount of work has been done to try and establish the impact of modern trees and their associated mycorrhizal fungi on silicate weathering, using field studies and experiments (summarized in Table 16.1) (*37–42*). The most comprehensive quantification comes from a

Table 16.1 Amplification of weathering by vascular plants

Vegetation	Location	Contrasted with	Element	Amplification factor	Reference(s)
Red pine	Hubbard Brook, New Hampshire	Moss/lichen	Ca^{2+}	10	(38)
			Mg^{2+}	18	
Birch/evergreen trees	West Iceland	Moss/lichen (no soil)	Ca^{2+}, Na^+, HCO_3^-, Si	2–3*	(39)
			Mg^{2+}	3–5*	
			K^+	110–150*	
Deciduous forest	Southern Swiss Alps	Rock (un-vegetated)	HCO_3^-, Si	8**	(39,40)
Spruce and fir forest (6% of watershed)	Colorado Rocky Mountains	Whole watershed (mostly un-vegetated)	$Ca^{2+} + Na^+ + Mg^{2+} + K^+$	3.5**	(41)
Higher plants	Central Iceland	Lichen	HCO_3^-	2–3**	(42)

Taken from Lenton (37).

Results are expressed as an amplification factor for weathering rate relative to 'bare' rock surfaces, which are often covered in lichen and moss, and provide an analogue for what was coating the land surface beforehand.

* Does not account for accumulation in soils hence an underestimate.

** Does not account for accumulation in vegetation or soils hence a greater underestimate.

remarkable experiment with a living forest at Hubbard Brook in New Hampshire, USA (*38*). Giant funnels were buried in the sand in order to catch all the products of weathering leaching from above. On some of the resulting 'sand boxes', red pine trees were planted and allowed to grow, whereas others were kept clear for comparison. The accumulation of weathered elements in vegetation and soil was measured as well as that leached out through the funnels. The growth of trees increased weathering fluxes by tenfold for calcium (Ca^{2+}) and more than that for magnesium (Mg^{2+}).

On the basis of this and other studies (Table 16.1), modelers of the Phanerozoic atmosphere and climate (including ourselves) have assumed that the establishment of vascular plants increased the 'weatherability' of silicate rocks many times, relative to the bryophyte and lichen cover that was present beforehand. To rebalance the carbon cycle, carbon dioxide and temperature must have declined until the removal flux of carbon dioxide again balanced the input from volcanoes. The models predict a roughly tenfold decline in atmospheric carbon dioxide starting in the early Silurian (around 440 million years ago), continuing throughout the Devonian, and ending in the middle or late Carboniferous (roughly 330 to 300 million years ago) (*43,44*). In this case, there is some independent proxy evidence that broadly agrees with the predictions (Fig. 16.2). Box 16.2 describes this evidence in more detail.

The fall of carbon dioxide with the rise of land plants is widely thought to be responsible for an interval of planetary cooling, which included brief intervals of glaciation in the Late Devonian and culminated in the most severe ice ages of the Phanerozoic; the 'Permo-Carboniferous glaciations' that straddle the boundary between the Carboniferous and Permian periods. But we don't think this was the only global environmental consequence of the rise of plants, thanks to the other biological weathering process; the selective extraction of phosphorus from rocks. If this phosphorus reaches the ocean as biologically available phosphate it can fuel productivity there, increasing the tendency for anoxic waters to be created beneath. Ultimately all the biologically available phosphorus entering the ocean is buried, often with organic carbon, providing a long-term source of atmospheric oxygen. So, are there any signs of these environmental consequences with the rise of plants?

Intriguingly, the *Archaeopteris* forests of the late Devonian were accompanied by phosphorite deposits in shallow coastal seas, indicating enhanced phosphorus input from the land (*54*). These pulses of phosphorus due to weathering on land may have been responsible for anoxic events in the late Devonian ocean. The anoxic events have in turn been linked to episodes of extinction, the largest of which, around 364 million years ago, killed off 70–82% of extant species. Shallow warm water species were most affected, and tropical coral reef communities

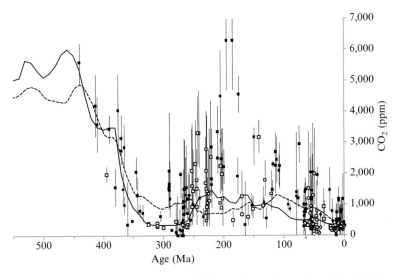

Fig. 16.2 Carbon dioxide over Phanerozoic time. Comparison of model predictions, from GEOCARB II (dotted line) (*43*) and COPSE (dashed line) (*44*), with a compilation of proxy data by Dana Royer *et al.* (*45*) from different sources: $\delta^{13}C$ of paleosols (filled squares), stomatal indices (open squares), $\delta^{13}C$ of phytoplankton (filled circles), and $\delta^{11}B$ of marine calcium carbonate (open circles.)

Box 16.2 Paleozoic proxies for carbon dioxide concentrations

Indirect 'proxy' evidence for CO_2 levels over Phanerozoic time has been compiled by Royer *et al.* (*45, 46*) (based on around 40 studies). Here we are interested in that part of the Paleozoic Era during which vascular plants colonized the land surface, roughly 430 to 300 million years ago. In that interval there are two sources of data for reconstructing CO_2 levels. The primary one comes from the carbon isotopic composition of minerals formed in ancient soils (filled squares in Fig. 16.2) (*45, 46*). There is only one data point in the Silurian, giving an estimate of 5600 ppm for atmospheric CO_2 (20 times the preindustrial level of 280 ppm), 440 million years ago (*47*). By the time of the next data points, early in the Devonian roughly 410 million years ago, CO_2 is estimated to have dropped to around 4000 ppm (*45, 48*). Most studies then find a decline during the Devonian, dropping to roughly 1000 ppm by the end of the period 360 million years ago (*45, 48, 49*). Whether the decline continues into the early Carboniferous is unclear from the scatter in the data. The few estimates from the mid to late Carboniferous 340 to 300 million years ago are in the range 500–1000 ppm (*45, 48*).

Fossil plants also provide a potential record of atmospheric CO_2 in the density of the pores (stomata) on their leaves (open squares in Fig. 16.2). Plants open their stomata to let in the CO_2 they use in photosynthesis, but this also causes them to lose water as vapour. When CO_2 is abundant they tend to have fewer stomata in order to minimize water loss, but when it is scarce they need more stomata to get CO_2 into the leaves. For a given species, in recent times, there is a well established inverse relation between atmospheric CO_2 concentration and the density of fossil stomata (50). The problem with applying such relationships several hundred million years ago is that the same species are rarely present over long intervals. Consequently paleobotanists have searched for ancient relatives of modern plants. The first such study estimated around 4000 ppm CO_2 in the early Devonian, 395 million years ago (51), which was later revised downward to about 2000 ppm (46). The next data points are not until the mid Carboniferous 325 million years ago, when CO_2 is estimated to have dropped to roughly the present level, about 380 ppm (45, 52). Values then hover in the range 250 to 400 ppm during the Permo-Carboniferous glaciations (52).

Not all attempts to reconstruct CO_2 levels in the Paleozoic give early high values and a strong downward trend with the rise of plants. One method gives relatively low values of CO_2 throughout (53), but has arguably the greatest uncertainties associated with it (46) (and is not plotted in Fig. 16.2). Overall, the balance of evidence supports an order of magnitude (tenfold) decline in atmospheric CO_2 with the rise of vascular plants. However, there is considerable uncertainty over the precise timing of the decline and whether it was continuous, as can be seen in the scattered data points in Fig. 16.2.

were decimated, consistent with increased nutrient input from the tropical land surface causing anoxia. At the global scale, the carbon isotope record indicates enhanced carbon burial in the ocean, which should have tended to increase atmospheric oxygen. Sure enough, charcoal is scarce in the middle Devonian but large amounts of it appear in some latest Devonian deposits, consistent with rising oxygen increasing the frequency and extent of fires.

16.4 A failure of carbon recycling

As the Devonian came to a close around 360 million years ago, the lowlands became cloaked in rather different forests of trees related to the clubmosses (or more closely to the quillworts). They grew up to 50 metres tall and their evolutionary strategy seems to have been one of rapid vertical growth for spore dispersal. Unusually large, wet low-lying areas formed due to relatively low sea level, producing perfect conditions to preserve this biomass on land, which ultimately formed coal. This gave the name to the next geological Period; the Carboniferous.

The great coal swamps of the Carboniferous represent ecosystems in which the recycling of carbon was weak or broke down. Large amounts of carbon were preserved because trees of the time had unusually high ratios of bark to wood, and bark is rich in lignin, which is difficult to biodegrade. The trees may have evolved so much bark as protection—a suit of armor—against insect herbivores. Insects that consume roots, leaves, wood and seeds evolved in the Carboniferous in a delayed response to the origin of these plant organs and tissues in the Devonian. In the Carboniferous, these insect herbivores had relatively few predators—insectivores. Furthermore, the types of fungi (Basidomycetes) that thrive on breaking down lignin today, failed to consume much of it (55). Their problem was probably that once dead material had sunk into the swamps it was soon in anoxic conditions and these fungi need at least 5% oxygen to degrade lignin.

Lignin compounds are both rich in carbon and poor in phosphorus, which meant that more organic carbon was buried for the same input of phosphorus from weathering. Thus, the rise of forests and the breakdown of recycling increased the amount of organic carbon being buried globally. This in turn pushed atmospheric oxygen up, probably above the present level (as we discussed in Chapter 15). In the coal swamps of the Carboniferous, fires became widespread, as recorded by abundant fossilized charcoal that is still with us today.

16.5 Re-establishing recycling and regulation

By the time the Earth entered the Permian period around 300 million years ago, coal was still being deposited in equatorial rainforests and temperate forests, and there was a large ice cap in the Southern Hemisphere. The rise of forests had created global environmental changes that were detrimental to them. Proxies suggest that by the end of the Carboniferous, atmospheric carbon dioxide was lowered to levels that became limiting for photosynthesis. The resultant planetary cooling caused continental ice sheets to approach 30° latitude and if sea-ice also extended this far it would have been close to the threshold for triggering a snowball Earth. Meanwhile atmospheric oxygen was pushed up to the point where fires regularly disrupted life on land. So, what stopped this turning into an environmental catastrophe? We don't know for sure, but we can identify at least three important factors.

Firstly, the inputs of nutrients from weathering would have slowed down as plants drained their supply out of rocks. A similar scenario plays out on the much shorter timescale of ecological succession on newly formed lava flows among the islands of Hawaii (56). There, different volcanic islands of different ages capture ecosystems in different stages of development spanning 4 million years. The first colonizers of a fresh lava flow mine their basalt substrate (a relatively easy to weather rock) and gain an ample supply of phosphorus. Typically they become nitrogen limited, having to rely on microbial nitrogen fixers to supply nitrogen from the atmosphere. However, several million years later, the rocks have been ground down and stripped of their nutrients to the point that the ecosystem becomes chronically limited by lack of phosphorus. Then the only input of phosphorus comes in windblown dust (and in Hawaii, thousands of miles from major continents there is very little of this). In the Carboniferous, thick forest cover in shallow basins may also have acted to prevent new rocks being exposed, thus suppressing weathering rates and phosphorus supply.

Secondly, biological evolution produced organisms that learnt to recycle the novel waste products being produced by the first forests. Jennifer Robinson first suggested that the 'white rot' fungi (Basidomycetes) that can degrade lignin first got on top of the job during the Permian, increasing the efficiency of recycling in terrestrial ecosystems (55). If these fungi spread during the Permian this could help explain a drop off in organic carbon burial that occurred at the time. Also, the diverse insect herbivores of the Carboniferous gave way in the Permian to a micro-world dominated (about 90% of insect species) by cockroach-like insects. This may have reduced the evolutionary selection for excessive amounts of bark on trees, and hence reduced the production of lignin. A reduction in organic carbon burial in turn weakened the oxygen source.

Thirdly, the environmental changes triggered by the spread of forests became self-limiting; they generated negative feedbacks that stabilized the Earth system. The combination of high oxygen and low carbon dioxide would have suppressed the activity of Rubisco, in turn suppressing weathering rates (see Fig. 14.5). Fires too probably restricted forests and weathering, generating a negative feedback on oxygen that we described in Chapter 15 (see Fig. 15.4). Of course there may also have been a dose of luck in the Permian Earth system stabilizing before the climate entered a snowball, or the oxygen rose so high that fires became devastating. But in the aftermath, oxygen, carbon dioxide and climate all appear to have been more tightly regulated. At least that is what our modelling predicts; a system with strong biological amplification of weathering responds more rapidly to any perturbations of oxygen, carbon dioxide or temperature (57).

There certainly was a recovery from the brink of catastrophe as the Permian period progressed. Proxies suggest carbon dioxide increased (Fig. 16.2), perhaps sharply (49), and the Earth began to warm from the great Permo-Carboniferous ice age. Nearly all the continents were clustered in the great super-continent Pangaea, and its interior dried out with deserts becoming widespread. The deserts gradually replaced the coal swamps that had girdled the equator. The swamp loving trees that dominated the Carboniferous and early Permian were replaced by better adapted conifers as the climate dried. According to the carbon isotope record, global organic carbon burial declined, and models predict that oxygen levels began to fall (see Fig. 15.5).

16.6 Summary: strengthening regulation?

In summary, the remarkably productive ecosystems we see on land today were formed through symbiosis and are supported by recycling. Land colonization happened in phases and in each phase, plants (and their fungal symbionts) evolved to occupy new habitats, to drill deeper into the ground, or to create new structural materials. As they did so, they kept draining their resource base and screwing up their environment, filling it with their waste products and shifting the composition of the atmosphere and the climate. In order to persist, they relied on the evolution of organisms that recycled the nutrients they needed, producing ecosystems that were closer to being materially closed. They also depended on becoming part of planetary scale regulatory feedbacks that restricted their spread and stabilized the Earth system.

All this should be reminiscent of 'playing Gaia' in Chapter 7. Over Earth history, organisms with increasingly efficient rock weathering capabilities have evolved, and this has tended to lower the steady state level of atmospheric CO_2 and the temperature (thus counteracting the brightening Sun). The first prokaryote colonizers of the Earth's land surface were probably desert crusts like the ones you introduced to planet Ocean in Chapter 7. They are thought to be least effective at amplifying weathering (58). The fungi and lichens, which we argued in Chapter 14 colonized next, clearly amplify weathering rates significantly (59), as do bryophytes. But none of these are as effective as the more recently evolved vascular plants, especially trees. Growing forests can accelerate weathering rates by a factor of ten relative to adjacent lichen and moss covered rocks (38). A system with stronger biotic amplification of weathering has stronger negative feedback. Hence it will respond more rapidly to perturbations, and its state will shift less for a given change in forcing (57). A world with complex plant life entangled in its feedback loops should be one with tighter regulation of the key environmental variables, CO_2 and temperature. Thus, we think the evolution of plants on the Earth's land surface increased the long-term stability of the climate, making it more amenable to complex life—including plant, fungi and animals.

However, there was nothing preordained about the success of land colonization, and the resulting 'complex Gaia' system was far from perfectly stable. The Earth still had an 'Achilles heel' residing in the ocean, which was sensitive to being pushed back into an anoxic state. Evolution continued to drive environmental change, as did perturbations arising from within the Earth or arriving from outer space. We now want to turn to look at some of these instabilities and the creative role they have played in our evolution.

References

1. C. A. Cowan, D. L. Fox, A. C. Runkel, M. R. Saltzman, *Terrestrial-marine carbon cycle coupling in ~500-m.y.-old phosphatic brachiopods. Geology* **33**, 661 (2005).

2. R. B. MacNaughton *et al.*, *First steps on land: Arthropod trackways in Cambrian-Ordovician eolian sandstone, southeastern Ontario, Canada. Geology* **30**, 391 (2002).

3. J. A. Raven, *The early evoluton of land plants: aquatic ancestors and atmospheric interactions. Botanical Journal of Scotland* **47**, 151 (1995).

4. T. Volk, *Gaia's Body—Toward a Physiology of the Earth.* (Copernicus, New York, 1998).

5. C. H. Wellman, J. Gray, *The microfossil record of early land plants. Philosophical Transactions of the Royal Society of London: Biological Sciences* **355**, 717 (2000).

6. J. D. Palmer, D. E. Soltis, M. W. Chase, *The plant tree of life: an overview and some points of view. American Journal of Botany* **91**, 1437 (2004).

7. Y.-L. Qiu, *et al.*, *The deepest divergences in land plants inferred from phylogenomic evidence. Proceedings of the National Academy of Sciences USA* **103**, 15511 (2006).

8. S. B. Kroken, L. E. Graham, M. E. Cook, *Occurrence and evolutionary significance of resistant cell walls in charophytes and bryophytes. American Journal of Botany* **83**, 1241 (1996).

9. L. E. Graham, L. W. Wilcox, M. E. Cook, P. G. Gensel, *Resistant tissues of modern marchantoid liverworts resemble enigmatic Early Paleozoic microfossils. Proceedings of the National Academy of Sciences USA* **101**, 11025 (2004).

10. C. H. Wellman, P. L. Osterloff, U. Mohiuddin, *Fragments of the earliest land plants. Nature* **425**, 282 (2003).

11. R. A. Betts, *Self-beneficial effects of vegetation on climate in an Ocean-Atmosphere General Circulation Model. Geophysical Research Letters* **26**, 1457 (1999).

12. K. A. Pirozynski, D. W. Malloch, *The origin of land plants: A matter of mycotrophism. BioSystems* **6**, 153 (1975).

13. M.-A. Selosse, F. Le Tacon, *The land flora: a phototroph-fungus partnership? Trends in Ecology and Evolution* **13**, 15 (1998).

14. M. Blackwell, *Terrestrial Life - Fungal from the Start? Science* **289**, 1884 (2000).

15. M. C. Brundrett, *Coevolution of roots and mycorrhizas of land plants. New Phytologist* **154**, 275 (2002).

16. L. Simon, J. Bousquet, R. C. Levesque, M. Lalonde, *Origin and diversification of endomycorrhizal fungi and coincidence with vascular land plants. Nature* **363**, 67 (1993).

17. F. Lutzoni *et al.*, *Assembling the fungal tree of life: progress, classification, and evolution of subcellular traits. American Journal of Botany* **91**, 1446 (2004).

18. D. Redecker, R. Kodner, L. E. Graham, *Glomalean Fungi from the Ordovician. Science* **289**, 1920 (2000).

19. N. J. Butterfield, *Probable Proterozoic fungi. Paleobiology* **31**, 165 (2005).

20. D. J. Read, J. G. Duckett, R. Francis, R. Ligrone, A. Russell, *Symbiotic fungal associations in 'lower' land plants. Philosophical Transactions of the Royal Society of London, Series B* **355**, 815 (2000).

21. J. G. Duckett, A. Carafa, R. Ligrone, *A highly differentiated Glomermycotean association with the mucilage-secreting, primitve antipodean liverwort Treubia (Treubiaceae): Clues to the origins of mycorrhizas. American Journal of Botany* **93**, 797 (2006).

22. P. W. Price, *An overview of organismal interactions in ecosystems in evolutionary and ecological time. Agriculture, Ecosystems and Environment* **24**, 369 (1988).

23. T. N. Taylor, J. M. Osborn, *The importance of fungi in shaping the paleoecosystem. Review of Paleobotany and Palynology* **90**, 249 (1996).

24. R. D. Finlay, *Ecoogical aspects of mycorrhizal symbiosis: with special emphasis on the functional diversity of interactions involving the extraradical mycelium. Journal of Experimental Botany* **59**, 1115 (2008).

25. A. G. Jongmans *et al.*, *Rock-eating fungi. Nature* **389**, 682 (1997).

26. N. Van Breemen *et al.*, *Mycorrhizal weathering: A true case of mineral plant nutrition? Biogeochemistry* **49**, 53 (2000).

27. D. J. Read, J. Perez-Moreno, *Mycorrhizas and nutrient cycling in ecosystems—a journey towards relevance? New Phytologist* **157**, 475 (2003).

28. G. J. Retallack, C. R. Feakes, *Trace Fossil Evidence for Late Ordovician Animals on Land. Science* **235**, 61 (1987).

29. F. M. Hueber, *Rotted wood–alga–fungus: the history and life of Prototaxites Dawson 1859. Review of Paleobotany and Palynology* **116**, 123 (2001).

30. M.-A. Selosse, *Prototaxites: A 400 Myr Old Giant Fossil, A Saprophytic Holobasidiomycete, Or A Lichen? Mycological Research* **106**, 641 (2002).

31. C. K. Boyce *et al.*, *Devonian landscape heterogeneity recorded by a giant fungus. Geology* **35**, 399 (2007).

32. W. D. Hamilton, R. M. May, *Dispersal in stable habitats. Nature* **269**, 578 (1977).

33. W. Remy, T. N. Taylor, H. Hass, H. Kerp, *Four hundred-million-year-old vesicular arbuscular mycorrhizae. Proceedings of the National Academy of Sciences USA* **91**, 11841 (1994).

34. T. N. Taylor, H. Hass, W. Remy, H. Kerp, *The oldest fossil lichen. Nature* **378**, 244 (1995).

35. W. E. Stein, F. Mannolini, L. VanAller Hernick, E. Landing, C. M. Berry, *Giant cladoxylopsid trees resolve the enigma of the Earth's earliest forest stumps at Gilboa. Nature* **446**, 904 (2007).

36. B. Meyer-Berthaud, S. E. Scheckler, J. Wendt, *Archaeopteris is the earliest known modern tree. Nature* **398**, 700 (1999).

37. T. M. Lenton, *The role of land plants, phosphorus weathering and fire in the rise and regulation of atmospheric oxygen. Global Change Biology* **7**, 613 (2001).

38. B. T. Bormann *et al.*, *Rapid, plant-induced weathering in an aggrading experimental ecosystem. Biogeochemistry* **43**, 129 (1998).

39. J. I. Drever, J. Zobrist, *Chemical weathering of silicate rocks as a function of elevation in the southern Swiss Alps. Geochimica et Cosmochimica Acta* **56**, 3209 (1992).

40. K. Moulton, R. A. Berner, *Quantification of the effect of plants on weathering: Studies in Iceland. Geology* **26**, 895 (1998).

41. M. A. Arthur, T. J. Fahey, *Controls on soil solution chemistry in a subalpine forest in north-central Colorado. Soil Science Society of America Journal* **57**, 1123 (1993).

42. J. L. Cawley, R. C. Burruss, H. D. Holland, *Chemical Weathering in Central Iceland: An Analog of Pre-Silurian Weathering. Science* **165**, 391 (1969).

43. R. A. Berner, *The carbon cycle and CO_2 over Phanerozoic time: the role of land plants. Philosophical Transactions of the Royal Society B-Biological Sciences* **353**, 75 (1998).

44. N. M. Bergman, T. M. Lenton, A. J. Watson, *COPSE: a new model of biogeochemical cycling over Phanerozoic time. American Journal of Science* **304**, 397 (2004).

45. D. L. Royer, R. A. Berner, I. P. Montañez, N. J. Tabor, D. J. Beerling, CO_2 *as a primary driver of Phanerozoic climate. GSA Today* **14**, 4 (2004).

46. D. L. Royer, R. A. Berner, D. J. Beerling, *Phanerozoic atmospheric CO_2 change: evaluating geochemical and paleobiological approaches. Earth Science Reviews* **54**, 349 (2001).

47. C. J. Yapp, H. Poths, *Ancient atmospheric CO_2 pressures inferred from natural goethites. Nature* **355**, 342 (1992).

48. C. I. Mora, S. G. Driese, L. A. Colarusso, *Middle to Late Paleozoic Atmospheric CO_2 Levels from Soil Carbonate and Organic Matter. Science* **271**, 1105 (1996).

49. D. D. Ekart, T. E. Cerling, I. P. Montanez, N. J. Tabor, *A 400 million year carbon isotope record of pedogenic carbonate: Implications for paleoatmospheric carbon dioxide. American Journal of Science* **299**, 805 (1999).

50. F. I. Woodward, *Stomatal numbers are sensitive to increases in CO_2 from pre-industrial levels. Nature* **327**, 617 (1987).

51. J. C. McElwain, W. G. Chaloner, *Stomatal density and index of fossil plants track atmospheric carbon dioxide in the Paleozoic. Annals of Botany* **76**, 389 (1995).

52. D. J. Beerling, *Low atmospheric CO_2 levels during the Permo-Carboniferous glaciation inferred from fossil lycopsids. Proceedings of the National Academy of Sciences USA* **99**, 12567 (2002).

53. D. H. Rothman, *Atmospheric carbon dioxide levels for the last 500 million years. Proceedings of the National Academy of Sciences USA* **99**, 4167 (2002).

54. T. J. Algeo, S. E. Scheckler, *Terrestrial-marine teleconnections in the Devonian: links between the evolution of land plants, weathering processes, and marine anoxic events. Philosophical Transactions of the Royal Society B-Biological Sciences* **353**, 113 (1998).

55. J. M. Robinson, *Lignin, land plants, and fungi: Biological evolution affecting Phanerozoic oxygen balance. Geology* **15**, 607 (1990).

56. O. A. Chadwick, L. A. Derry, P. M. Vitousek, B. J. Huebert, L. O. Hedin, *Changing sources of nutrients during four million years of ecosystem development. Nature* **397**, 491 (1999).

57. T. M. Lenton, *Testing Gaia: the effect of life on Earth's habitability and regulation. Climatic Change* **52**, 409 (2002).

58. D. Schwartzman, *Life, temperature and the earth: the self-organizing biosphere.* (Columbia University Press, New York, 1999), pp. 241.

59. T. A. Jackson, W. D. Keller, *A comparitive study of the role of lichens and inorganic processes in the chemical weathering of recent Hawaiian lava flows. American Journal of Science* **269**, 446 (1970).

17
Rolls of the dice

By the end of the Permian Period, 251 million years ago, vertebrates had been on land for over a hundred million years. Land vertebrates included a diversity of groups, among them both the ancestors of mammals and of dinosaurs. As the Permian neared its close, vertebrates began to suffer elevated levels of extinction, and the ancestors of mammals fared rather better than those of dinosaurs (*1*). So, were you gambling at the time, you would probably have put your money on mammals inheriting the Earth. But, unbeknown to them, life was about to suffer a massive extinction event—the End-Permian 'big one' (*2*). In its aftermath, it was the dinosaurs that flourished. Mammals would have to wait another two hundred million years for their heyday.

This illustrates the fundamentally random, stochastic nature of the evolutionary process that has produced us and everything else alive today. In the language of Stephen Jay Gould, the dice probably had to be rolled many times for mammals to come to dominance, or for intelligent life to appear. The big shake-ups and rolls of the dice were the mass extinction events that allowed new types of life, and new ecologies, to flourish. Traditionally five mass extinctions are recognized during the Phanerozoic on the basis of the marine invertebrate fossil record (*3*). Recent work has brought some of these 'big five' events into question (*2*), but the end of the Permian still stands out as the biggest of all. Many, if not all, of the big extinction events probably had a trigger arising from the inner Earth or arriving from outer space. However, for environmental changes to become major killers, at least in some cases, required instabilities and amplifying feedbacks from within the Earth system.

In this chapter we want to look in more detail at the two most important rolls of the dice for our story, the End-Permian and End-Cretaceous extinctions. We also want to trace the evolution of the Earth system, and of intelligent animal life, to the point where the modern world and our immediate ancestors are recognizable.

17.1 The End-Permian extinction

The End-Permian event, around 251 million years ago (*4*), was the greatest setback that complex life has faced. Animals were decimated, both on land and in the sea (*5*). Plants, which generally suffer less in mass extinctions (at least at the family level), lost many species, and the dominant groups changed. They suffered high levels of genetic mutation producing unusual structures, termed 'mutagenesis' (*6*). Fungi flourished on decaying material (*7*). Forests were lost and soils were eroded and washed into the ocean (*8*). Streams and rivers were left to cut a more direct path to the sea, their banks no longer stabilized by plant roots (*9*). In some regions, plant diversity actually increased initially with the removal of dominant trees, before undergoing a widespread decline (*10*). Gymnosperms and seed ferns were largely replaced by members of the oldest division of vascular plants, the lycophytes. Insects suffered their only mass extinction, probably linked to changes in the flora.

What caused such a decimation of complex life? Despite considerable research we still don't know the full answer, but some contributing factors can be identified and others largely dismissed. A huge meteorite impact was hypothesized (*11*), but evidence for extraterrestrial material trapped in End-Permian rocks has not been reproducible by other groups. Remnants of a possible impact crater were identified (*12*), but others interpret them differently (*13*). Meanwhile, the 'smoking gun' of an impact—an excess of the element iridium in strata from the time—has not been found. Consequently, the impact hypothesis remains unsubstantiated. Instead, the leading candidate is an Earthly trigger; the episode of volcanism that produced the Siberian traps—a vast flood of basalt covering roughly 2 million square kilometers of what is now Siberia. It was originally 1–4 million cubic kilometers in volume and deposited in or near the Arctic Circle. The outpouring lasted about 700 000 years and its timing coincides with the interval of greatest extinctions (*4*). However, linking this volcanic event to the extinction is not straightforward. In particular, a complete mechanism must explain why extinction was global in extent and decimated life both on land in the sea.

Carbon dioxide emitted with the basalt outpouring may have doubled the total atmospheric concentration and caused a few degrees of global warming (*14*). But carbon dioxide and temperatures have been higher than this at many times during the Phanerozoic without catastrophic consequences (see Fig. 16.2). Sulphur dioxide emitted with the eruptions, if it reached high enough concentrations, could have caused plant mutagenesis. It would also have oxidized to sulphuric acid, causing acid rain that would have harmed plants, especially forests. However, it would not have had much impact in the ocean, even at the surface. The explosive nature of the eruptions probably created a temporary aerosol haze, which could have been thick enough to block out sunlight and interfere with plant photosynthesis. However, these effects would be expected to be strongest in the far North where the eruptions occurred and much weaker in the Southern Hemisphere. Yet the disruption of land ecosystems was global in extent.

To explain the global mutagenic impact on plants it has been hypothesized that the ozone layer was depleted, increasing the flux of harmful ultraviolet radiation to the surface (6). The flood basalt eruptions might have been responsible if they produced a substantial flux of halogens—gases containing chlorine or bromine that catalyze ozone destruction. Hydrogen chloride would have been emitted directly in the volcanic gases, but probably not in sufficient quantity to disrupt the ozone layer. Perhaps the volcanic outpouring occurred through previously deposited coals and evaporite beds and caused them to release long-lived halogen-containing organic compounds. Recent modeling suggests a combination of these mechanisms could have produced substantial ozone depletion (15). However, it is less clear that this would explain extinction in the ocean.

A key kill agent in the ocean was widespread anoxia (16, 17). A number of factors probably contributed to generating it. Traditional explanations focus on warm conditions reducing oxygen solubility, and weak ocean circulation transferring less oxygen to depth (18), but weak circulation also supplies fewer nutrients to surface waters. A more effective mechanism for creating anoxia is to increase total nutrient levels in the ocean (19), or to increase the efficiency of nutrient utilization (20). Plant driven weathering on land had already created a healthy phosphorus input to the ocean, and the continental configuration of the time appears to have produced an ocean circulation pattern in which nutrients could be efficiently utilized. Then something made conditions worse. Perhaps a decline in vegetation and flourishing of fungi on land, led to a flush of nutrients to the ocean. We think a runaway feedback may then have been triggered: once something starts to generate anoxia, there is a positive feedback in the ocean that tends to further increase it (21). We discussed it briefly in Chapter 15 (see Fig. 15.3): Anoxic conditions cause increased recycling of phosphorus from ocean sediments, which in turn fuels increased productivity, which drives increased anoxia. At the time that marine life was worst hit, anoxia extended into the shallow waters that harbored the greatest biodiversity, and some already anoxic waters became euxinic—rich in hydrogen sulphide (22).

Hydrogen sulphide release from the ocean has in turn been implicated in the mass extinction on land (23). If hydrogen sulphide were belched from the ocean and wafted over the land it could have directly caused plant mutagenesis, but this would have been a somewhat localized phenomenon concentrated nearer the seaside. To explain a more global impact on plants it has been hypothesized that hydrogen sulphide depleted the ozone layer and increased the flux of harmful ultraviolet radiation to the surface. However, to produce the amounts of hydrogen sulphide needed to destroy the ozone layer would have required an order of magnitude more phosphorus in the ocean than today. Furthermore, the estimated hydrogen sulphide flux was based on a simple atmospheric chemistry model, and it turns out not to be enough to cause catastrophic ozone depletion in more realistic models (*24, 25*). Instead we suggest that denitrification of the nitrate in the ocean would have happened before hydrogen sulphide production, producing substantial emissions of nitrous oxide, which is well known to cause ozone depletion. Such ocean-to-land kill mechanisms do not explain what triggered the ocean to go anoxic and then euxinic in the first place, but they may have played a strong amplifying role in the extinction.

17.2 Instability of 'complex Gaia'

Although we remain uncertain about the actual scenario, it is clear that land-ocean-atmosphere interactions and positive feedbacks within the Earth system played an important role in amplifying environmental change and making the End-Permian a truly devastating extinction. In the aftermath, the carbon cycle oscillated between extremes (see Box 17.1). This leads us to question the stability of the 'complex Gaia' system that had been generated in the Paleozoic. Clearly the regulatory mechanisms that were in place for atmospheric oxygen, carbon dioxide and climate were not enough to prevent such an extreme event. The Achilles heel of the system appears to have been the ocean and positive feedbacks that arise within it, in particular, the enhanced recycling of phosphorus that occurs once anoxic conditions set in, further amplifying them (Fig. 15.3). This mechanism appears to have sent the ocean back toward a Proterozoic state, partially reversing the revolution when the ocean became oxygenated in the Neoproterozoic. If the stratospheric ozone layer was also depleted, then there was also a partial reversal of the preceding revolution, when the Great Oxidation first formed an ozone layer.

We are led to an important conclusion: the Phanerozoic Earth system, the one in which we live today (for there have been no fundamental biogeochemical changes since the Permian), with its high oxygen and relatively low CO_2 atmosphere and abundant life on land, is not rockstable against all potential shocks. It can be knocked into a state in which internal feedbacks go positive, leading to instability and catastrophic extinctions.

Box 17.1 The aftermath of the End Permian

The aftermath of the End Permian mass extinction was a time of instability in the global carbon cycle as recorded in the carbon isotope records of carbonate rocks and organic carbon. First there was a pronounced negative shift in the carbon isotope record, followed by a series of swings between positive and negative values (*26*). These oscillations are unique in magnitude for the Phanerozoic and comparable to those seen in the Neoproterozoic (*27*). Most potential explanations for the carbon isotope changes concentrate on the initial negative shift and fall short of explaining its full magnitude (*14*).

A currently in-vogue mechanism is a huge release of methane from frozen reservoirs under ocean (and land) sediments, called 'clathrates' (*14*), potentially started by the flood basalt pouring onto polar shelf seas where clathrates are likely to have been present. Global warming itself could also have triggered methane clathrate destabilization, creating a positive feedback loop whereby methane release to the atmosphere contributed to global warming, triggering further methane release from clathrates. This is often wrongly described as a 'runaway' positive feedback. Instead it is a strong but constrained positive feedback that could have roughly doubled the amount of carbon added to the atmosphere by an initial perturbation (the volcanic outpouring). However, there is a problem in that the swings in the carbon isotope record are both rapid and highly symmetrical, implying unrealistic rates and magnitudes of both methane release and clathrate formation (*26*).

Our alternative explanation is that the ocean may have been oscillating between anoxic intervals in which large amounts of organic carbon accumulated in sediments and oxygenated intervals in which organic carbon was efficiently respired away and relatively little was buried. Such an oscillation can be generated by a modest increase in phosphorus inputs from the land surface, and it can be self-sustaining thanks to a combination of fast positive feedbacks involving phosphorus recycling, and slow negative feedbacks involving changes in atmospheric oxygen (see Fig. 15.3) (*21*). In the anoxic conditions, the ocean may have reverted to a state like that in the Neoproterozoic, with productivity dominated by cyanobacteria (and some anoxygenic photosynthesizers).

The End-Permian event suggests that 'complex Gaia' could conceivably have reverted to an earlier state, devoid of complex animals. This would have delayed or prevented the evolution of 'observer' life forms (us). As it was, ecological recovery was unusually slow compared to other Phanerozoic mass extinction events. Many land and ocean ecosystems remained stuck in an alternative (generally low diversity) state for the roughly 5 million years that comprise the Early Triassic (*28*). Forests remained absent until the Middle Triassic and there was a corresponding gap in coal deposition (*28*). But some complex life, including land vertebrates, did survive the extinction.

For millions of years afterwards, the dominant surviving four legged animals on land were the 'shovel lizards' *Lystrosaurus* munching their way through the herbaceous vegetation. Our ancestors did of course survive the extinction (otherwise we wouldn't be here to write about it), and in the earliest part of the Triassic they appear to have been the dominant predators. The fox-sized *Thrinaxodon* (which had whiskers and probably fur) made burrows, perhaps as shelter from inclement environmental conditions (*29*). The ancestors of dinosaurs were also among the lucky few to survive, but the effect of the extinction on them is unclear because of a shortage of fossils.

Many possible futures might have come from the throw of the dice that was the end Permian extinction. However, as it happened, by the mid-Triassic, the group of reptiles that produced the dinosaurs dominated all the large carnivore and herbivore niches. This 'Triassic takeover' from the mammal-like reptiles may have been aided by the dinosaur ancestors' better ability to cope with the arid climate of the time. It would be nearly another two hundred million years before the End-Cretaceous extinction allowed mammals to come to prominence.

17.3 Macro-evolutionary patterns

Whilst mass extinctions were temporary setbacks for complex life, they also played a creative role in allowing different types of life and ecology to dominate. During Phanerozoic time there have been long periods of 'macro' evolutionary and ecological stability. In these intervals, one type of life (e.g. reptiles) may dominate a particular suite of ecological niches and prevent another type of life (e.g. mammals) from filling them. Sometimes it is the whole structure of ecosystems that appears locked in one state for a long interval, before some internal or external perturbation breaks the deadlock. This is best illustrated by the global marine animal fossil record over Phanerozoic time, which shows at least three distinct and apparently stable ecological states (Fig. 17.1) (*30*).

Transitions in marine ecological state are recognized in terms of the balance of mobile and non-mobile animal genera, and the balance of predator and prey taxa (*30*). The biggest extinction at the End Permian marks perhaps the clearest transition between ecological states (Fig. 17.1). Prior to it marine ecosystems were dominated by stationary, filter and suspension feeding animals such as brachiopods and sea lilies, with relatively fewer predators. After it, mobile animals such as snails, urchins and crabs were dominant, bringing an increase in predation pressure. Bony fish and marine reptiles also flourished in the Mesozoic. The more limited fossil record of life on land has yet to yield an equivalent quantitative picture. But a qualitative case can be made that the Mesozoic world of dinosaurs contained longer food chains with more

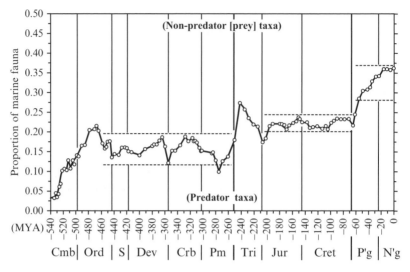

Fig. 17.1 Ecological transitions among marine animals in the proportions of predator and prey taxa. [Taken from Banbach *et al.* (*30*), copyright (2002) National Academy of Sciences, USA.]

levels of predation than Paleozoic terrestrial ecosystems. The next major ecological transition, both in the ocean and on land, did not occur until after the End Cretaceous.

Although particular types of life may hold ecological dominance for long periods, those that ultimately take over from them—after the next roll of the mass extinction dice—are often quietly evolving and diversifying in their shadow. Flowering plants (angiosperms), for example, diversified during the Cretaceous, but remained a minor contributor to plant biomass and ecosystem function until after the End Cretaceous extinction. Mammals were also quietly evolving throughout the Mesozoic.

17.4 Intelligence and societies

After the Triassic takeover of ecological niches by the ancestors of the dinosaurs, the ancestors of mammals were only able to survive as small, mostly nocturnal, insect eaters. However, this selected for improved thermal insulation (including hair), temperature regulation, hearing and smell. In the case of mammals, these heightened senses meant the evolution of a larger brain in proportion to body weight, and hence a greater demand for energy from food and oxygen. Ultimately mammals would evolve a unique outer layer of the brain, the neocortex, which is involved in spatial reasoning, conscious thought, and in humans, language.

One phenomenon that seems to have been intimately tied up with the evolution of intelligence is the evolution of social behaviour. Social interactions require a system of communication—information exchange—between individuals, and the ability to process information is the basis of intelligence. Of course this need not imply a large brain—the adaptive behaviour of insect societies is often described as 'swarm intelligence' because it is not held at the individual level. Instead it is a remarkable emergent property of a lot of relatively simple information exchange between small brained individuals. However, in vertebrates there does seem to be some link between social behaviour and intelligence of individuals. For example, in primates (which of course includes us), the size of social groups is the best predictor of brain size. In short, large mammalian brains may have evolved at least in part as a consequence of social interactions. So, when are the first tentative signs for social behaviour?

Social interactions do not fossilize so geologic evidence for them is indirect. Sociality is usually inferred from mass burials or from fossilized structures that could only have been made by social groups rather than individuals (needless to say, this makes the earliest claims somewhat controversial). Nest building is a common (if not universal) feature of social animals. Remarkably, among the ancestors of mammals (cynodonts), in the Early Triassic, there are fossils preserving multi-chambered burrows containing up to 20 specimens of *Trirachodon*, which probably drowned in a flash flood (*31*). *Trirachodon* were somewhat larger than rabbits. They shared what were clearly communal burrows, with an entrance tunnel wide enough that individuals going in opposite directions could pass one another. Lower down there were branching tunnels and numerous terminating chambers. Such a colonial lifestyle suggests relatively complex social behaviour, which previously was thought to be unique to much more recent mammals.

By the Late Triassic, there is also evidence for social behaviour among insects (*32*); the first and for us most fascinating example, are termites. Termite species today are all social. Molecular phylogeny supports the view that they are basically highly modified, social, wood-eating cockroaches. They may have split from the cockroaches as early as the Permian. The first termite body fossils come from the Cretaceous but some fossilized structures from the Late Triassic are probably termite nests (*33*), and there are more in the Early Jurassic (*34*). These rather large 'trace' fossils have changed remarkably little in the past 225 million years, suggesting insect societies were well established back then (*32*). Modern termite nests are truly remarkable constructions; they are essentially self-regulating structures, which provide air conditioning and regulate the balance of oxygen and carbon dioxide within. Some

are oriented north-south to aid in temperature regulation in the sub-
tropics. How the humble termites 'know' how to create such a structure
is surely one of the most abiding puzzles of swarm intelligence in the
natural world. Some termites also build protective tunnels extending
from their nest, often up trees, to their food source; wood. In order to
digest wood (cellulose), the termites have symbiotic protozoa (meta-
monads) in their guts, and these protozoa in turn have bacteria embed-
ded on their surface producing the necessary digestive enzymes. So, the
termite nest represents at least four nested levels of cooperation; the
bacteria, the protozoa, the termite individual, and the colony. It's those
Russian dolls again.

Ironically by the Late Jurassic, around 150 million years ago, some
mammals were big and 'smart' enough to prey upon termite nests.
Fruitafossor was about the size of a chipmunk and its teeth, forelimbs
and back suggest that it broke open social insect nests and ate the residents (*35*). It is among several examples of early mammals that broke
the stereotype of small, nocturnal insect eaters. By the Early Cretaceous,
130 million years ago, the badger-like, metre long *Repenomamus* was
even preying on young dinosaurs (*36*). By the start of the Late
Cretaceous, 90 to 100 million years ago, both fossils and molecular
clocks agree that some of the main 'crown' mammal lineages had
already diverged (*37, 38*). But the Mesozoic mammals could not challenge the smaller brained dinosaurs for ecological dominance. In the
Mesozoic, intelligence had little bearing on who dominated the
world.

17.5 The End Cretaceous extinction

Land ecosystems and particularly land animals received a fresh roll of
the dice of fate 65 million years ago with the End Cretaceous extinction event, which famously killed off the dinosaurs (with the exception of what we now call birds). The End Cretaceous extinction marks
the transition from the Mesozoic to the Cenozoic Era. Both extraterrestrial and terrestrial causes played a role in the End Cretaceous
extinction. The case for a massive asteroid impact event has been well
publicized (and has entered the popular consciousness). A thin sedimentary layer enriched in the element iridium at the boundary of the
Cretaceous and Paleogene periods provides convincing evidence that
a massive asteroid (or asteroids) hit the planet. The Chicxulub crater
(180 km diameter) on the coast of Yucatan, Mexico is the leading candidate for the spot where it struck, although recent work brings this
into question suggesting it predates the boundary (*39*). More controversial is the larger postulated Shiva crater (600 km by 400 km)
beneath the Indian Ocean, west of Mumbai (*40*). Then there are two
smaller craters (20 km diameter) in Ukraine and the North Sea dating

from approximately the right time. Conceivably they could have been made by broken off fragments of the main impact asteroid. Less well known is that vast volcanic flood basalts—the Deccan traps—erupted over less than a million years leading up to the boundary. The Deccan traps are of a similar magnitude to the Siberian traps implicated in the End Permian extinction. There is also some evidence of ocean anoxia at the End Cretaceous, but it is not recognized as a global oceanic anoxic event.

The End Cretaceous mass extinction—like the End Permian event—impacted land as well as marine life. There were widespread wildfires at the boundary and vegetation was decimated by global deforestation. In the aftermath there was a brief flourishing of fungi feeding on decaying plant material (saprotrophs), then ferns colonized the land (as they did after the 1980 Mount St Helens eruption). Terrestrial ecosystems then recovered to something approximating their pre-extinction state, and they did so much faster than they had after the End Permian extinction, and much faster than marine ecosystems (which took around 3 million years).

Land animals suffered remarkably selective extinctions. Size and the ability to hide underground or fly away from the ravages of environmental change seem to have played a key role in survival. Insect diversity had been high beforehand but was still low nearly 2 million years after, probably because of the temporary loss of plants. Vertebrates that could retreat to the water including amphibians, turtles, snakes, lizards and crocodiles all faired reasonably well. In the air, the last Pterosaurs died out but birds, which had already taken over from them, largely survived. All major mammal lineages survived (including egg laying monotremes, marsupials and placentals) although they suffered losses. Cretaceous mammals were generally small in size, comparable to that of rats, and may have burrowed or lived partly in water. The dinosaurs in contrast, were generally large and this seems to have been a factor in their downfall. In general among land animals, omnivores fared well, whereas pure carnivores or herbivores suffered, and the dinosaurs belonged to the latter categories.

After the End Cretaceous extinction, mammals came to fill many of the ecological niches vacated by the dinosaurs (*41*). They increased in size within 1 million years of the boundary, from an average mass of 150g to 1 kg, suggesting a literal expansion to fill empty niches (*42*). This was accompanied by striking species diversification in the best preserved continental fossil record from North America (*42*). However, this diversification, in the Paleocene Epoch, was mostly among groups of mammals that have subsequently declined or gone extinct (*37*). Molecular techniques suggest that extant (still living) lineages experienced no significant overall change in diversification rates (*37*). This indicates that present day mammals may actually have been inhibited by their Paleocene compatriots.

17.6 The rise of modern mammal orders

We belong to what are called the eutherians—'well developed beasts'—or placental mammals, in which the foetus is nourished during gestation by a placenta. Fossils indicate that eutherian mammal stem groups had already arisen by 125 million years ago (*43*). According to molecular clocks, the crown groups then radiated later in the Cretaceous (*37*). Rodents may have diverged from other placental mammals in the Paleocene (*44*). However, the modern orders of placental mammals did not originate until ten million years after the End Cretaceous (*45*). The three key orders that emerged then were the Primates (to which we belong), the even-toed ungulates (Artiodactyla), and the odd-toed ungulates (Perissodactyla), which taxonomists refer to as the 'APP' orders (see Box 17.2).

So, if it wasn't the End Cretaceous extinction, what was it that triggered modern mammal orders, including our own, to come to prominence? We have already dismissed (in Chapter 15) the idea that there was an increase in atmospheric oxygen that somehow played a role. In truth, we don't know if a changing environment had much to do with it. But there are intriguing signs that all three modern mammal orders appeared in a remarkable time of climatic upheaval called the Paleocene-Eocene Thermal Maximum (or 'PETM' for short).

The PETM was an interval of rapid global warming 55.8 million years ago (Fig. 17.2) (*46*). Global temperatures rose roughly 5°C within 20 000 years and remained high for roughly 100 000 years. There was also a major perturbation of the carbon cycle, which occurred in two steps, each lasting roughly 1000 years and separated by 20 000 years (*47*). The leading explanation is that large amounts of methane were degassed from frozen clathrates, which in turn caused

---

Box 17.2 Members of the APP orders

Today's primates include lemurs, bushbabies, tarsiers, monkeys and apes. They have a variety of social systems. We belong to the family Hominidae along with the other great apes (orangutans, gorillas, chimpanzees). Today's even-toed ungulates (Artiodactyla) include pigs, camels and ruminants. Ruminants include cattle, goats, sheep, antelopes, deer and giraffe. These are foregut digesters, better able to cope with grassland food. One group of Artiodactyla (most closely related to hippos) returned to the sea to become whales, dolphins and porpoises (cetaceans). Odd toed ungulates (Perissodactyla) are hindgut fermenters, digesting plant cellulose in their intestines. They include horses, donkeys, zebras and their allies, as well as tapirs and rhinoceros. Interestingly, many species within both the even-toed and odd-toed ungulate orders were domesticated by humans.

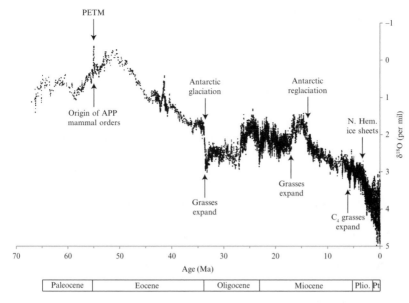

Fig. 17.2 Climate change during the Cenozoic as recorded by the oxygen isotope composition of benthic foraminifera. More positive $\delta^{18}O$ values indicate cooling and/or growth of continental ice sheets. Marked above the data are some climate events, and below it some biological events, discussed in the text. [Redrawn from the data compilation of Zachos et al. (*46*).]

further warming and degassing in a positive feedback loop. The trigger may have been a volcanic intrusion into carbon-rich sediments under the North Atlantic (*48*). Methane released would have been rapidly oxidized to carbon dioxide, contributing to warming and ocean acidification. Acidification in turn dissolved carbonate sediments in the deep ocean and caused the extinction of many calcareous-shelled organisms living there. It took around 100 000 years for the carbon cycle to recover.

During the PETM, the modern (APP) orders of mammals appeared within 10 000 years or so of each other in North America, Europe and Asia (*45*). Primates were the last of the three orders to arrive roughly 10 000 years into the event. The earliest APP species were significantly smaller than their immediate descendants, as were other contemporary mammals. This dwarfism could be due to poor plant food quality in a carbon dioxide rich world. The early part of the PETM was dominated by small leaved plants indicative of a dry climate. Later in the PETM the climate became wetter. The rapidity of appearance of major orders of mammals during the PETM is truly remarkable, and if borne out by the ongoing accumulation of evidence, indicates that evolution can operate surprisingly rapidly.

17.7 Climate cooling and the rise of grasslands

The subsequent evolution of modern mammal orders, including our own, is linked with a profound shift in terrestrial ecosystems; the rise of grasslands. Grasslands provided a habitat in which our ancestors ultimately evolved, as did many of the animals they domesticated (see Box 17.2), whilst grasses provide our major food crops. The first unequivocal fossil parts of grass plants appear 55 million years ago, around the time of the PETM (*49*), but there is an earlier cryptic record, because grasses leave characteristic silicon phytoliths—literally 'plant stones'—in the fossil record. The earliest probable grass phytoliths have been found in fossilized dinosaur dung from the latest Cretaceous, just before 65 million years ago (*50*). Grass pollen then starts to appear from about 60 million years ago. However, grasses remained a sparse component of land ecosystems until long after the PETM. When they finally spread at the expense of woodland, they did so in three phases. This staccato pattern could be explained by global environmental changes triggering local, amplifying positive feedbacks involving increases in fire frequency and grazing herbivores (see Box 17.3).

An overall driver for the spread of grasslands was a cooling and drying of the climate over the last 50 million years (see Fig. 17.2), which in turn was probably (at least partly) driven by declining carbon dioxide concentrations (see Fig. 16.2). The steady trend was interrupted by a

---

Box 17.3 Grassland-woodland bi-stability

Over evolutionary time, grasslands appear to have been 'held back' and then expanded relatively rapidly, which poses an interesting puzzle. It can be related to ecological observations that in large regions of the world—both tropical and temperate—either grassland or woodland can exist under the same global climate conditions. A simple conceptual model is that many regions of low to moderate rainfall can have two simultaneously stable states—covered in woodland or in grassland. Each state acts to maintain itself. Forests tend to recycle water to the atmosphere more efficiently creating wetter conditions with fewer fires, in which they are favoured. They also support less grazing by herbivores. Grasslands, in contrast, create drier conditions with more frequent disturbance by fires and much larger populations of grazing herbivores, in which they are favoured. In a world without fire, exclusion experiments and model simulations suggest that closed forest would double in area (from 27% to 56% of vegetated land), mostly at the expense of C_4 grasslands (*51*). Grazers help maintain the grassland state by indiscriminately munching on young tree saplings as well as grass. The grass can re-grow rapidly, but the saplings cannot. Elephants also play a role in converting tropical woodland disturbed by fire into grass dominated savannah, by disturbing the remaining trees.

relatively rapid cooling and drying of the climate 33.7 million years ago, which marks the boundary between the Eocene and Oligocene Epochs of the Paleogene Period. This occurred when atmospheric carbon dioxide and temperature passed a critical threshold triggering the rapid growth of large ice sheets on the continent of Antarctica. It in turn triggered the first phase of grassland expansion, as the mid-latitudes became cooler and drier and the first desert grasslands replaced earlier woodland in the Great Plains of North America (52). The soils underneath the grasses experienced an increased intensity of silicate weathering, which would have contributed somewhat to declining carbon dioxide and global cooling (52, 53).

The second phase of grassland expansion began around 17 million years ago, early in the Miocene Epoch (52). Short grasses appeared forming a sod, or natural carpet, with more organic-rich soils beneath them. Such grasslands are effective at chocking out the seedlings of woody plants and promoting the spread of fire, and sure enough charcoal becomes more abundant in the rock record at the time. Grazing hoofed mammals (ungulates) expanded with the grasslands, having specially adapted (hypsodont) teeth to grind down the silica-rich grass matter. There was an explosive adaptive radiation of horses and antelope, along with many other mammals adapted to feeding on grasses and living in grasslands. Increased weathering led to increased inputs of phosphorus and silicon to the ocean and increased burial of them in marine sediments, and conceivably contributed to a resumption of climate cooling and Antarctic glaciation (see Fig. 17.2). Also, the presence of grass rather than trees would have increased the reflectivity of the land surface, tending to cool the continental interiors and increase their seasonality.

The third phase of grassland expansion occurred late in the Miocene, beginning around 7 million years ago (52). Tall grasses forming a sod appeared with deep soils under them. They expanded into more humid regions, promoting fire as they displaced forests, again recorded by increased charcoal in the rock record. During this phase, a mechanism of concentrating carbon dioxide at the sites of carbon fixation within the plant, called 'C_4 metabolism', proliferated among the grasses. (C_4 plants were already present by about 15 million years ago, but between 7 and 5 million years ago they underwent a marked expansion in the tropics.) Their carbon concentrating mechanism makes C_4 grasses more efficient under a combination of dry conditions and low atmospheric CO_2. By not having to open their stomata so often, they reduce their water loss. Climate drying, declining atmospheric CO_2 or a mixture of both could have played a role in C_4 grass expansion. Conceivably it was amplified by positive feedbacks as the expansion of C_4 grasses caused another pulse of increased weathering and phosphorus and carbon burial in the ocean.

17.8 The first hominids

In the newly created world of mixed grassland and trees, around 6 million years ago, our own lineage Hominids (which currently includes five genera; *Ardipithecus, Australopithecus, Praeanthropus, Paranthropus* and *Homo*), and the lineage of chimpanzees (*Pan*) diverged from a common ancestor. The first hominid fossils date between 5.8 and 4.4 million years ago and have been assigned to the genus *Ardipithecus*. They may have lived in shady forest. Between about 3.9 and 3 million years ago, members of the genus *Australopithecus* were widespread in Eastern and North Africa. The first footprints of upright hominids walking on two legs are preserved in 3.7 million year old volcanic ash (from Laetoli in Tanzania). As hominids took their first, faltering steps, the long cooling trend of the Earth's climate was triggering the start of a new era of climate instability (which we will explore in the next chapter). The first stone tool use is recorded 2.6 million years ago. Between 2.2 and 1.6 million years ago, *Homo habilis* (probably better reclassified as a member of the *Australopithecus* genus), were using stone flakes as tools, such as for cleaving meat off carrion. But they did not master the use of tools in hunting or defense, and were regularly preyed upon by large carnivorous cats. There is little to suggest that the descendants of these early hominids would one day take over the world.

References

1. R. B. Huey, P. D. Ward, *Hypoxia, Global Warming, and Terrestrial Late Permian Extinctions. Science* **308**, 398 (2005).

2. J. Alroy, *Dynamics of origination and extinction in the marine fossil record. Proceedings of the National Academy of Sciences USA* **105**, 11536 (2008).

3. D. M. Raup, J. J. Sepkoski, *Mass Extinctions in the Marine Fossil Record. Science* **215**, 1501 (1982).

4. S. A. Bowring *et al.*, *U/Pb Zircon Geochronology and Tempo of the End-Permian Mass Extinction. Science* **280**, 1039 (1998).

5. M. J. Benton, R. J. Twitchett, *How to kill (almost) all life: the end-Permian extinction event. Trends in Ecology and Evolution* **18**, 358 (2003).

6. H. Visscher *et al.*, *Environmental mutagenesis during the end-Permian ecological crisis. Proceedings of the National Academy of Sciences USA* **101**, 12952 (2004).

7. H. Visscher *et al.*, *The terminal Paleozoic fungal event: Evidence of terrestrial ecosystem destabilization and collapse. Proceedings of the National Academy of Sciences USA* **93**, 2155 (1996).

8. M. A. Sephton *et al.*, *Catastrophic soil erosion during the end-Permian biotic crisis. Geology* **33**, 941 (2005).

9. P. D. Ward, D. R. Montgomery, R. Smith, *Altered river morphology in South Africa related to the Permian-Triassic extinction. Science* **289**, 1740 (2000).

10. C. V. Looy, R. J. Twitchett, D. L. Dilcher, J. H. A. Van Konijenburg-van Cittert, H. Visscher, *Life in the end-Permian dead zone. Proceedings of the National Academy of Sciences USA* **98**, 7879 (2001).

11. L. Becker, R. J. Poreda, A. G. Hunt, T. E. Bunch, M. R. Rampino, *Impact event at the Permian-Triassic boundary: Evidence from extraterrestrial noble gases in fullerenes. Science* **291**, 1530 (2001).

12. L. Becker *et al.*, *Bedout: A possible end-Permian impact crater offshore of Northwestern Australia. Science* **304**, 1469 (2004).

13. A. Glikson, *Comment on 'Bedout: A Possible End-Permian Impact Crater Offshore of Northwestern Australia'. Science* **306**, 613b (2004).

14. R. A. Berner, *Examination of the hypotheses for the Permo-Triassic boundary extinction by carbon cycle modelling. Proceedings of the National Academy of Sciences USA* **99**, 4172 (2002).

15. D. J. Beerling, M. Harfoot, B. Lomax, J. A. Pyle, *The stability of the stratospheric ozone layer during the end-Permian eruption of the Siberian Traps. Philosophical Transactions of the Royal Society of London A-Physical Sciences* **365**, 1843 (2007).

16. P. B. Wignall, R. J. Twitchett, *Oceanic Anoxia and the End Permian Mass Extinction. Science* **272**, 1155 (1996).

17. Y. Isozaki, *Permo-Triassic Boundary Superanoxia and Stratified Superocean: Records from Lost Deep Sea. Science* **276**, 235 (1997).

18. J. T. Kiehl, C. A. Shields, *Climate simulation of the latest Permian: Implications for mass extinction. Geology* **33**, 757 (2005).

19. T. M. Lenton, A. J. Watson, *Redfield revisited: 1. Regulation of nitrate, phosphate and oxygen in the ocean. Global Biogeochemical Cycles* **14**, 225 (2000).

20. K. M. Meyer, L. R. Kump, *Oceanic Euxinia in Earth History: Causes and Consequences. Annual Review of Earth and Planetary Sciences* **36**, 251 (2008).

21. I. C. Handoh, T. M. Lenton, *Periodic mid-Cretaceous Oceanic Anoxic Events linked by oscillations of the phosphorus and oxygen biogeochemical cycles. Global Biogeochemical Cycles* **17**, 1092 (2003).

22. K. Grice *et al.*, *Photic Zone Euxinia During the Permian-Triassic Superanoxic Event. Science* **307**, 706 (2005).

23. L. R. Kump, A. Pavlov, M. A. Arthur, *Massive release of hydrogen sulfide to the surface ocean and atmosphere during intervals of oceanic anoxia. Geology* **33**, 397 (2005).

24. M. B. Harfoot, J. A. Pyle, D. J. Beerling, *End-Permian ozone shield unaffected by oceanic hydrogen sulphide and methane releases. Nature Geoscience* **1**, 247 (2008).

25. J.-F. Lamarque, J. T. Kiehl, J. J. Orlando, *Role of hydrogen sulfide in a Permian-Triassic boundary ozone collapse. Geophysical Research Letters* **34**, L02801 (2007).

26. J. L. Payne *et al.*, *Large perturbations of the carbon cycle during recovery from the End-Permian extinction. Science* **305**, 506 (2004).

27. A. H. Knoll, R. K. Bambach, D. E. Canfield, J. P. Grotzinger, *Comparative Earth History and Late Permian Mass Extinction. Science* **273**, 452 (1996).

28. C. V. Looy, W. A. Brugman, D. L. Dilcher, H. Visscher, *The delayed resurgence of equatorial forests after the Permian-Triassic ecologic crisis. Proceedings of the National Academy of Sciences USA* **96**, 13857 (1999).

29. R. Damiani, S. Modesto, A. Yates, J. Neveling, *Earliest evidence of cynodont burrowing. Proceedings of the Royal Society B* **270**, 1747 (2003).

30. R. K. Bambach, A. H. Knoll, J. J. Sepkoski, *Anatomical and ecological constraints on Phanerozoic animal diversity in the marine realm. Proceedings of the National Academy of Sciences USA* **99**, 6854 (2002).

31. G. H. Groenewald, J. Welman, J. A. MacEachern, *Vertebrate Burrow Complexes from the Early Triassic Cynognathus Zone (Driekoppen Formation, Beaufort Group) of the Karoo Basin, South Africa. Palaios* **16**, 148 (2001).

32. S. T. Hasiotis, *Complex ichnofossils of solitary and social soil organisms: understanding their evolution and roles in terrestrial paleoecosystems. Palaeogeography, Palaeoclimatology, Palaeoecology* **192**, 259 (2003).

33. S. T. Hasiotis, R. F. Dubiel, *Termite (Insecta: Isoptera) Nest Ichnofossils from the Upper Triassic Chinle Formation, Petrified Forest National Park, Arizona. Ichnos* **4**, 119 (1995).

34. E. M. Bordy, A. J. Bumby, O. Catuneanu, P. G. Eriksson, *Advanced Early Jurassic Termite (Insecta: Isoptera) Nests: Evidence From the Clarens Formation in the Tuli Basin, Southern Africa. Palaios* **19**, 68 (2004).

35. Z.-X. Luo, J. R. Wible, *A Late Jurassic Digging Mammal and Early Mammal Diversification. Science* **308**, 103 (2005).

36. Y. Hu, J. Meng, Y. Wang, C. Li, *Large Mesozoic mammals fed on young dinosaurs. Nature* **433**, 149 (2005).

37. O. R. P. Bininda-Edmonds *et al.*, *The delayed rise of present-day mammals. Nature* **446**, 507 (2007).

38. W. J. Murphy *et al.*, *Resolution of the early placental mammal radiation using Bayesian phylogenetics. Science* **294**, 2348 (2001).

39. G. Keller, T. Adatte, A. P. Juez, J. G. Lopez-Oliva, *New evidence concerning the age and biotic effects of the Chicxulub impact in NE Mexico. Journal of the Geological Society* **166**, 393 (2009).

40. S. Chatterjee, in *Comparative planetology, geological education, history of geology*, W. Hongzhen, D. F. Branagan, O. Ziyuan, W. Xunlian, Eds. (1997), pp. 31–54.

41. M. J. Novacek, *100 million years of land vertebrate evolution: the Cretaceous-Early Tertiary transition. Annals of the Missouri Botanical Garden* **86**, 230 (1999).

42. J. Alroy, The fossil record of North American mammals: Evidence for a Paleocene evolutionary radiation. Systems Biology **48**, 107 (1999).

43. Q. Ji *et al.*, *The earliest known eutherian mammal. Nature* **416**, 816 (2002).

44. R. J. Asher *et al.*, *Stem Lagomorpha and the Antiquity of Glires. Science* **307**, 1091 (2005).

45. P. D. Gingerich, *Environment and evolution through the Paleocene-Eocene thermal maximum. Trends in Ecology and Evolution* **21**, 246 (2006).

46. J. Zachos, M. Pagani, L. Sloan, E. Thomas, K. Billups, *Trends, Rhythms, and Aberrations in Global Climate 65 Ma to Present. Science* **292**, 686 (2001).

47. U. Röhl, T. J. Bralower, R. D. Norris, G. Wefer, *New chronology for the late Paleocene thermal maximum and its environmental implications. Geology* **28**, 927 (2000).

48. H. Svensen *et al.*, *Release of methane from a volcanic basin as a mechanism for initial Eocene global warming. Nature* **429**, 542 (2004).

49. W. L. Crepet, G. D. Feldman, *The earliest remains of grasses in the fossil record. American Journal of Botany* **78**, 1010 (1991).

50. V. Prasad, C. A. E. Stromberg, H. Alimohammadian, A. Sahni, *Dinosaur coprolites and the early evolution of grasses and grazers. Science* **310**, 1177 (2005).

51. W. J. Bond, F. I. Woodward, G. F. Midgley, *The global distribution of ecosystems in a world without fire. New Phytologist* **165**, 525 (2005).

52. G. J. Retallack, *Cenozoic Expansion of Grasslands and Climatic Cooling. The Journal of Geology* **109**, 407 (2001).

53. E. A. Bestland, *Weathering flux and CO_2 consumption determined from palaeosol sequences across the Eocene-Oligocene transition. Palaeogeography, Palaeoclimatology, Palaeoecology* **156**, 301 (2000).

PART VI
A NEW REVOLUTION?

18
Climate wobbles

We have arrived almost at the point of Earth history where our own species makes an entrance. Before describing that event, we first want to look at the Earth system of the last few million years, which provides the context for our evolution. As we come this close to the present, our view becomes ever clearer, as if adjusting the lens of a microscope to bring the image into focus. We can see detail that in more distant ages is completely lost. Looking at the climate record of the last few million years, the detail reveals an Earth system which is complex, puzzling and beautiful.

If you had to choose a time when the Earth was to be subject to shocks and surprises of a revolutionary kind, you'd probably want to avoid periods when it was already disposed to be unstable. Bad luck then, perhaps, that we humans have happened along at this particular point in Earth history. For in the Quaternary—the last two million years or so—all the warning signs are there that the current climate system was entering an unstable phase even before we began experimenting with it. Planetary cooling has culminated in a series of periodic glaciations of the Northern hemisphere—the recent 'ice ages'—which have got progressively longer and deeper. Although they are paced by variations in the Earth's orbit, they are increasingly dominated by internal oscillations and amplifying feedbacks. They illustrate the tightly coupled behaviour of the Earth system and they indicate that the climate system that we have evolved in is unusually sensitive.

18.1 The onset

At these times scales, oscillations of climate that are paced by the orbital wobbles of the Earth known as the Milankovitch cycles, become very apparent. The wobbles occur because the Earth's orbit around the Sun does not repeat exactly each year but is subject to variations, due ultimately to the presence of other bodies in the solar system. It took about a hundred years from the time that the idea that these wobbles might affect the climate was first proposed, by the self-educated

Scottish scientist James Croll (*1*), through its development by the Serbian astronomer Milutin Milankovitch, to its near universal acceptance, with the predicted oscillations being identified in deep sea sediment records (*2*).

The Milankovitch cycles are divided into three categories describing variations of different facets of the orbit (*3*). The highest frequencies, called 'precessional' cycles, have periods near 23 thousand years. These are caused by the regular cycling of the Earth's direction of rotation (like a spinning top that is beginning to wind down) and the orientation of its elliptical orbit. There is also a periodic change of a few degrees in the tilt (or 'obliquity') of the Earth's axis, which is currently 23.5° from the perpendicular to the plane of the orbit. The tilt is responsible for the seasons, and this 41 thousand year cycle has a strong effect on the intensity of summer and winter seasons in both hemispheres. Finally there are longer cycles, at about 100 thousand years and 400 thousand years, which are due to changes in the 'eccentricity' of the orbit—how much it departs from a circle and tends towards an ellipse.

The Milankovitch cycles are comparatively subtle variations. None of them change significantly the total amount of solar radiation received by the planet over a year, but they do redistribute it between the summer and winter hemispheres. For example, at present the difference between the amount of radiation being intercepted when Earth is closest to the Sun (perihelion, which at present occurs on January 3) and furthest distance (aphelion, on July 4) is 6.8%. At times when the Earth's orbit is at its most elliptical, this difference is about 23%, big enough to quite significantly modulate the intensity of the seasons.

Fig. 18.1 shows a composite record of the $^{18}O/^{16}O$ isotope content of carbonate secreted by bottom dwelling foraminifera, over the last five million years (*4*). Like Figure 17.2, it can be read, very roughly, as a thermometer, recording how much ice is on the planet and the temperature of the deep sea. It shows the Earth's response to the Milankovitch oscillations in a wealth of detail. Between 5 million and about 3.5 million years ago, there is little trend on the record, but the rapid variations, too fast to be seen clearly at the scale of this diagram, have a period of 41 thousand years, and are due to the variations in the tilt of the Earth's axis. After 3.5 million years ago a pronounced cooling trend begins, corresponding to the build-up of ice on the planet. As it cools, the amplitude of the oscillations increases: between one and two million years ago, ice is advancing and retreating in large swings, still paced by the obliquity cycle.

At about one million years ago, a slower cycle begins to emerge—the peaks and troughs become more widely spaced, about 100 000 years apart, and the swings become yet larger. This is the modern cycle of ice ages. During this time we know that ice in the northern hemisphere has repeatedly advanced to form major ice sheets on the North American and Eurasian continents, covering most of modern Canada,

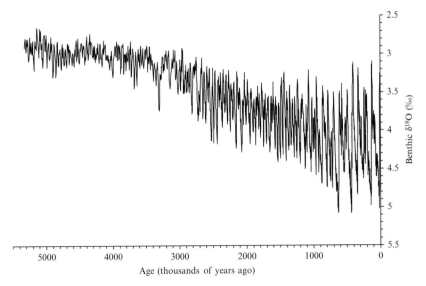

Fig. 18.1 Climate change over the last 5.3 million years—the Lisiecki and Raymo 'benthic stack' of 57 $\delta^{18}O$ records (*4*). More positive $\delta^{18}O$ values indicate greater ice volume and/or lower temperature of the deep ocean hence the axis has been inverted to give an intuitive indication of observed cooling over time. This figure is an enlargement of the most recent section of the curve in Fig. 17.2

Scotland, Scandinavia and Siberia, before melting back to cover only Greenland as it does today (*5*). Meanwhile, in the southern hemisphere, although the Antarctic ice sheet did not grow very much, sea ice coverage waxed and waned, covering twice the extent at glacial maxima that it does today.

It's tempting to look at the climate record of Fig. 18.1 with the eyes of a systems analyst, and see a planet slowly sliding into instability with progressively more and more severe glaciations. In Box 18.1 we discuss this possibility and where the system might be heading.

18.2 The ice core records

For the last million years or so, we have a much more detailed picture of Earth history, derived from long ice cores drilled in Antarctica. Recovering and reading this record represents a stunning achievement by climate scientists, one of the jewels of the last three decades of science.

It was the French glaciologist Claude Lorius who first realized that gas trapped in bubbles in ice cores could provide an archive of old air that might enable us to learn, by direct measurement, what the

Box 18.1 Is the planet heading for another snowball Earth?

Is the Earth heading towards a runaway that might see it locked again in a snowball state? This would see the end of the vast majority of complex multi-cellular life that has developed on Earth since the last snowball, more than 600 million years ago. It would be a disaster of the first magnitude, not just for humans, but for practically all life larger than a pinhead on the planet.

This is a surprising thought: earlier in the book (Chapter 5) we used the long-term warming of the Earth due to the steady increase in luminosity of the Sun to place limits on the time left for which Earth will remain habitable. Yet in the recent past, the planet seems to be courting the opposite fate: the world has been cooling, as indeed it has done over much of the last 50 million years, a trend that might perhaps end sometime in the not-too-distant-future in a catastrophic glaciation.

So will the world end in Fire or Ice? It may be a close-run thing. Over billions of years the steady cooling of the interior of the planet will result in a slowdown in tectonic cycling and loss of volatiles such as CO_2 and water from the surface, these being absorbed back into the mantle. This will, over time, reduce the greenhouse effect and lead to a reduction in surface temperatures. Certainly it is the case that atmospheric CO_2 has been at all-time low concentrations during the recent past. The habitability of the planet on the billion year time scale is threatened potentially both by heating and cooling, the brightening of the Sun and the slow cooling of the interior of the Earth.

However, while our understanding of the Sun's evolution is that it has been and will be slow, smooth and ineluctable, the volatile loss from the surface is a much less steady process. Continental rifting leading to the formation of major sedimentary basins containing large carbonate deposits, such as much of the present day Atlantic for example, may accelerate the removal of CO_2 (see Box 14.2). Conversely, periods of enhanced volcanic activity, such as accompanied the emplacement of large magma outpourings such as the Siberian or Deccan Traps (and many lesser such events), would be accompanied by rapid release of CO_2 and periods of enhanced, if ultimately temporary, warming.

So, the fact that the Sun is getting brighter on a billion year time scale does not mean that the end of complex life on Earth will inevitably come from overheating. In parallel, the loss of volatiles, most particularly CO_2, from the surface is a cooling influence. The Earth seems still to be vulnerable to major glaciations at periods, such as now, when the atmospheric content of greenhouse gases is lowered sufficiently and when the arrangement of continents is conducive to the build-up of major ice sheets.

However, in the depths of the last ice age the CO_2 fell to such a low level that it was limiting for the growth of most vegetation. As plants are intimately tied up in the long term process of silicate weathering that removes CO_2, there exists a negative feedback which should prevent atmospheric CO_2 from being lowered much further (see Fig. 14.5). Furthermore, in the depths of the last ice age, even the winter sea-ice cover fell far short of the threshold for a snowball runaway. So, it seems that some additional cooling agent, such as masses of sunlight scattering aerosol, would be required to trigger another snowball now.

For the immediate future (meaning the next few tens of thousands of years at least), the danger of a runaway icehouse has been averted, by a bizarre and unpredictable biological event: the rise to prominence of humans, with their recently acquired passion for burning fossil fuel. For good or ill, our massive perturbation of the carbon cycle has already raised the CO_2 concentration of the atmosphere sufficiently to cancel the next glaciation. If we stop burning fossil fuels tomorrow, it will take at least 100 000 years to return the carbon cycle to its previous state (*6*).

composition of the atmosphere was in former times. Lorius has described how, as a young postdoctoral worker, he was overwintering in Antarctica with two companions in 1957. The idea came to him when, on a long night when they needed something alcoholic to lift their spirits, they used ice they had recovered from some depth to put in their whisky. It fizzed and crackled, and they realized it contained air bubbles under pressure. By the mid sixties, Lorius was working on the recovery of air from deep Antarctic cores, and also helping to organize the huge logistic efforts required to drill deep into the ice sheet. The first big science payoff came in the early 1980s, when groups in Grenoble and Bern published results that showed that carbon dioxide concentrations in the atmosphere were substantially lower in glacial times than in the warm interglacial that we are now living in (*7, 8*). It had been suspected since the 19th century that changes in greenhouse gases might be implicated in the causes of the ice ages. Lorius and his colleagues provided a proof of this outlandish idea, of the most direct and unequivocal kind.

Fig. 18.2 shows the record from the Vostok ice core, drilled to more than 3 kilometres depth into the Antarctic ice sheet. The Vostok camp is at the 'Pole of inaccessibility'—the summit of the dome of ice covering East Antarctica, and the coldest place on Earth. Though ice core drilling began there in the mid 1970s, the data from the full depth cores was not published until 1999 (*9*) and it thus represents several decades of work by international teams. For climate scientists this data is iconic, because it reveals the Earth's climate to be operating as an orchestrated system. Both the physical and the biological climate systems are shown to be responding to the subtle pacing of the Milankovitch cycles.

The Vostok core reveals that four times over the last 420 000 years, the ice age cycle has repeated. The cycle takes the form of a sawtooth in temperature, with a relatively slow descent into the glaciations, and a rapid warming, called a 'termination'. The cycle takes roughly 100 000 years, but it is not exactly the same length each time and it does not repeat itself identically: it is rather as if, as in a musical symphony, the major themes are repeated with subtle variations each time. The glaciations are cold, dry (the rate of snow accumulation is much less) and

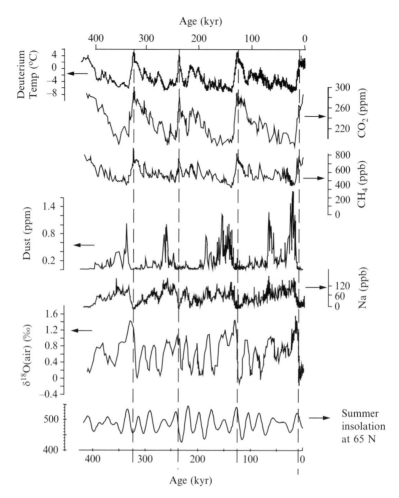

Fig. 18.2 The Vostok ice core record. Shown from the top are: local temperature as recorded by deuterium isotopes, atmospheric CO_2 concentration, atmospheric methane concentration, atmospheric dust, concentrations of sodium from sea salt, which in the ice cores is now thought to be a measure of sea ice coverage, and $\delta^{18}O$ isotope ratio of oxygen in the air, which responds to land vegetation productivity. Also shown at the bottom for comparison is the calculated midsummer solar insolation at 65°N, which responds strongly to the Milankovitch precessional and obliquity cycles. Dotted vertical lines mark the warmest points of the climate cycles, at the ends of the terminations. [Redrawn from Petit *et al.* (*9*).]

dusty, with much higher concentrations of atmospheric dust trapped in the ice cores.

Most strikingly, both carbon dioxide and methane, two important greenhouse gases, were lower in the cold periods and higher in warm periods. The correspondence between CO_2 and temperature in particular, is spectacularly good. It's natural to ask whether the CO_2 is driving the climate change, or vice versa, but the evidence is that this is not an either-or question. Each variable drives the other. The variations are so tightly coupled that it is often quite difficult to tell whether CO_2 changes begin before or after the temperature change. During the rapid changes at the end of the ice ages, the temperature seems to change first, leading the CO_2 by a few hundred years (*10*). However, the great majority of the temperature rise during the deglaciation occurs simultaneously with a CO_2 increase. This behaviour is very much what we expect from a positive feedback system in which the temperature increase is initiated by some other cause (increase in the local solar radiation for instance) but where much of the temperature change is driven by feedbacks, including a CO_2-temperature interaction.

Feedbacks are the key to understanding how such major swings in climate as the recent glaciations arise as a response to the rather modest Milankovitch-driven changes in the distribution in sunlight over the planet—changes which have been occurring throughout Earth's history but have only rarely produced such dramatic oscillations of climate.

18.3 The role of feedbacks

As we discussed in Chapter 7, there are feedback loops within the climate system, both of the positive (destabilizing) and negative (stabilizing) variety. However, during the Quaternary, positive feedbacks have been in the ascendancy, with a number of them combining to constitute a powerful amplifier of small variations. Furthermore, they are nonlinear, so that they don't amplify all the Milankovitch oscillations equally or regularly, but alter the whole pattern of the response.

Table 18.1 lists a number of positive feedbacks on global temperature that are important in the recent climate variations and tend to amplify the change between glacial and interglacial time. We have

Table 18.1 Positive feedbacks of importance in glacial-to-interglacial climate change

Feedback	Strength
Water vapour—temperature	****
Ice/snow albedo—temperature	***
CO_2—temperature	**
Vegetation albedo—temperature	*
Aerosol—temperature	*
Methane and nitrous oxide—temperature	*

attempted a rough ranking of these, with four stars indicating the strongest ones and one star indicating the weakest.

The most enigmatic feedback is the CO_2—temperature one. The reason why increasing CO_2 causes (global) temperature to rise is well established (the greenhouse effect), but why does the opposite also occur? Why does an increase in temperatures (specifically Antarctic temperatures) cause atmospheric carbon dioxide to rise? This turns out to be one of the most difficult questions in Earth system science, and it is one that is not fully answered yet. No single mechanism is responsible, and it is something of a jigsaw puzzle to construct the complete answer. We are quite well advanced on this puzzle, but we are still missing some important pieces. We know that there are several processes acting in the same direction, some biological, some due to the chemistry of CO_2 in the ocean, and some mediated by the changes in the circulation of the Southern Ocean (see Box 18.2).

Box 18.2 Why does atmospheric carbon dioxide increase between glacial and interglacial time?

The relationship between Antarctic temperatures and atmospheric carbon dioxide concentrations as recorded in Antarctic ice cores is spectacularly close. Fig. 18.3 shows a combination of the recent EPICA ice core record with Vostok and others (11). It shows that the relationship stretches back 850 000 years, and includes a major change in the shape of the cycles. Before about 450 000 years ago, the interglacials are longer but cooler, and the record of atmospheric CO_2 exactly matches this change.

Why the exquisitely close match? The fundamental reason is that the Southern Ocean is the place where much of the deep water of the global ocean first leaves the surface, and these waters are the reservoir for practically all the 'labile' carbon on the Earth, the CO_2 that can quickly exchange with the atmosphere. If the concentration of carbon in surface Southern Ocean water is increased by even a modest amount, and this is kept up for a thousand years, the concentration of carbon in the deep sea will be correspondingly increased. Because the total amount of labile carbon is roughly constant, this will mean that the much smaller reservoirs from which the carbon has been taken—especially the atmosphere, will have decreased, and by a much larger amount.

The Southern Ocean is therefore the gatekeeper to the deep sea. Conditions at its surface ultimately determine how much CO_2 from the atmosphere ends up in the ocean, and this is the key variable that determines the partitioning of the labile carbon between the atmosphere and ocean. If it changes, the concentration of carbon dioxide in the atmosphere will be altered (and will only very slowly be restored, over hundreds of thousands of years, by the much slower silicate weathering feedback that we have previously discussed). The North Atlantic is also a place where lots of deep water is formed, so why doesn't this area have a similar influence? It probably does, but today, in the interglacial world, the North Atlantic is already doing an efficient job of sequestering

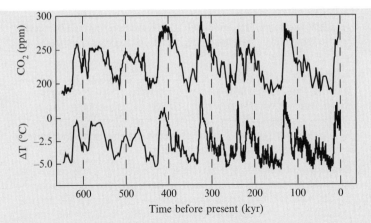

Fig. 18.3 The combined Vostok and EPICA ice core records for deuterium-based temperature and CO_2. Temperature is scaled to be representative of approximate global temperature change. There is a pronounced change in the shape of the cycle between 500 and 400 thousand years ago, with longer and cooler interglacials before that time. The change in the shape of the temperature curve is closely matched by lower CO_2 concentrations in the interglacials. [Redrawn from the data of Siegenthaler *et al.* (*11*).]

atmospheric CO_2, as it did also in the cold phases of the glacial cycles. It didn't change therefore, in this respect at least, whereas the Southern Ocean was much more efficient in the cold periods than it is now.

The Southern Ocean would sequester much more CO_2 if the water at its surface were efficiently stripped of carbon by biological action, allowing more CO_2 to come in from the atmosphere. But the plankton living at the surface of the Southern Ocean today leave unused much of the carbon that they might otherwise remove. This is partly because they lack iron, an essential 'micronutrient', but also because the surface water is deeply mixed, and rapidly renewed by upwelling, so the plankton at the surface just can't keep up with the supply of carbon and other nutrients. Any change in conditions that improves this efficiency, either by increasing the productivity of the plankton or by decreasing the rate at which deep water is supplied to the surface, would allow the ocean to take up more atmospheric CO_2 and should, in time, draw down atmospheric concentrations.

Currently we know of at least four possible mechanisms that might have enabled the Southern Ocean to draw down more atmospheric CO_2 during the glaciations:

1. **Cooling the surface**. CO_2 dissolves better in cold water than in warm water, and a colder Southern Ocean would take up more CO_2 from the atmosphere and supply it to the deep sea, resulting in lower atmospheric CO_2. The surface waters certainly were colder in glacial time—in fact they seem to have been close to the freezing point of sea water, at −1.8°C. However, the deep water being formed today is already rather

cold, so there is a limit to how much CO_2 change one can get from this mechanism. It might have been responsible for a glacial–interglacial change of around 20 ppm.

2. **Adding iron**. Because of the higher supply of iron-rich atmospheric dust in glacial time, the plankton there would have had more iron, and this would have helped draw down the CO_2 from the atmosphere. This is probably the best studied and understood mechanism. However, from what we know at present, it does not seem to be able to produce change in atmospheric CO_2 much greater than 30 ppm (*12*).

3. **Capping with sea ice**. Here the idea is that an impermeable cap of sea ice physically prevents the CO_2 from the deep sea being mixed up to the surface here, breaking the connection between the atmosphere and the deep sea via the surface Southern Ocean. In simple models this can be made to work well, but it's not clear that there was sufficient ice to really make this an efficient mechanism (*13*).

4. **Reducing upwelling rates**. Theory of the Southern Ocean circulation suggests that water may have exchanged more slowly with the deep ocean if there was less wind over the region in glacial time (*14*), or a much colder atmosphere there (*15*). There is much debate over these mechanisms, because there is no generally agreed description for what 'controls' Southern Ocean circulation—it remains a major focus of research.

The first three mechanisms definitely occurred to some extent during glacial time. So, it is apparent that there is no single process responsible for the atmospheric CO_2 variations. However, they don't seem sufficient, either alone or together, to explain the size of the effect that we see—a drawdown in atmospheric CO_2 of more than 80 ppm. Hence we believe that the fourth class of mechanisms may be the most important of all. Currently then, we can't answer this question definitely, but the ice core records have shown us exactly where to look for the explanation (the Southern Ocean), and we have only a short list of possible answers. In the not too distant future, we should have a workable explanation.

While the CO_2—temperature feedback is important, it is actually not the most powerful positive feedback occurring between glacial and interglacial time. More powerful is the ice-albedo feedback that we've already discussed in relation to Snowball Earth. Although this is operating to some extent today, and is amplifying global warming in the Arctic, in glacial time it was stronger. The large ice sheets on the Northern hemisphere, and the much greater sea ice cover in the Southern Ocean, both cooled the planet by reflecting away sunlight, but both were vulnerable to warming.

A third and even stronger positive feedback, which is always present in Earth's climate system, is the response of water vapour, which increases in concentration as the temperature of the oceans increases. This is a 'fast feedback' in the sense that the residence time of water vapour in the atmosphere is only a few days, so it responds instantaneously to changes in surface temperature. Though in detail the response has uncertainties, a good first approximation that we believe captures the most important component of the response is arrived at from the simple physical assumption that relative humidity remains constant.

There are other positive feedbacks too, which are more minor but which all add to the sensitivity of the climate. The atmospheric burden of dust was higher in glacial time, and atmospheric dust and aerosols tend to cool the surface by scattering incoming radiation. Vegetation cover increases from cold to warm climates. Vegetation helps to warm the planet both because it is usually darker than the ground it covers, and because transpiration by forests acts as a source of water vapour to the air, so this too is a positive feedback. Finally, the minor greenhouse gases methane and nitrous oxide both increase between cold and warm phases of the climate, though their concentrations are small enough that they contribute little to the overall warming.

During the ice ages therefore, and in particular during the terminations, when the temperature goes rapidly from full glacial maximum to the warm interglacial in a period of about 6000 years, there are a whole set of positive feedbacks in operation. When feedback is sufficiently positive, it can cause a 'runaway'—meaning that the change in temperature becomes large and self-sustaining, and is no longer related at all to the size of the initial change. This is what seems to have happened at the end of the glaciations. So long as the feedbacks held the climate close to a runaway state, the temperature kept increasing, but eventually, with the melting of much of the Northern hemisphere ice, the strength of the feedbacks lessened sufficiently to allow the climate to stabilize again in a warm state.

18.4 Summary

To summarize, many of the important processes that have driven the recent series of climate changes we can describe reasonably well. But if the test of real understanding is that we know the detailed physics, chemistry and biology of all the relevant processes leading to these climate oscillations, we would certainly fail it. The causes of CO_2 increase in the atmosphere remain particularly obscure (see Box 18.2), but so also does the role that clouds would have played, or many details of ocean or atmosphere circulation in glacial time. However, in a broadbrush sort of way, we can claim to understand the period of dramatic climatic instability during which our own species has developed. Going further, a growing number of researchers are suggesting that our appearance in an interval of unusual climate sensitivity is not entirely accidental.

Instead the striking recent variability of the climate may be causally linked to key events in human evolution. Even the emergence of intelligence itself may be linked to an environment that is changing at the right pace to force the evolution of a mammalian brain.

References

1. J. Croll, *Theory of secular changes in the Earth climate*. (Appleton, New York, 1875).

2. J. D. Hays, J. Imbrie, N. J. Shackleton, *Variations in Earth's orbit - pacemaker of ice ages. Science* **194**, 1121 (1976).

3. A. L. Berger, *Support for astronomical theory of climatic change. Nature* **269**, 44 (1977).

4. L. E. Lisiecki, M. E. Raymo, *A Pliocene-Pleistocene stack of 57 globally distributed benthic d18O records. Paleoceanography* **20**, PA1003 (2005).

5. CLIMAP project, *The surface of the ice age Earth. Science* **191**, 1131 (1976).

6. D. Archer, *The Long Thaw: how humans are changing the next 100,000 years of Earth's climate*. (Princeton University Press, Princeton, NJ, 2008), pp. 192.

7. R. J. Delmas, J.-M. Ascencio, M. Legrand, *Polar ice evidence that atmospheric CO_2 20,000 yr BP was 50% of present. Nature* **284**, 155 (1980).

8. A. Neftel, H. Oeschger, J. Schwander, B. Stauffer, R. Zumbrunn, *Ice core sample measurements give atmospheric CO_2 content during the past 40,000 yr. Nature* **295**, 220 (1982).

9. J. R. Petit *et al.*, *Climate and atmospheric history of the past 420,000 years from the Vostok ice core, Antarctica. Nature* **399**, 429 (1999).

10. L. Loulergue *et al.*, *New constraints on the gas age-ice age difference along the EPICA ice cores, 0–50 kyr. Climate of the Past* **3**, 527 (2007).

11. U. Siegenthaler *et al.*, *Stable carbon cycle-climate relationship during the late Pleistocene. Science* **310**, 1313 (2005).

12. A. J. Watson, D. C. E. Bakker, A. J. Ridgwell, P. W. Boyd, C. S. Law, *Effect of iron supply on Southern Ocean CO_2 uptake and implications for glacial atmospheric CO_2. Nature* **407**, 730 (2000).

13. B. B. Stephens, R. F. Keeling, *The influence of Antarctic sea ice on glacial-interglacial CO_2 variations. Nature* **404**, 171 (2000).

14. J. L. Russell, J. R. Toggweiler, *Shifted westerlies caused low CO_2 during cold glacial periods. Geochimica et Cosmochimica Acta* **68**, A474 (2004).

15. A. J. Watson, A. C. N. Garabato, *The role of Southern Ocean mixing and upwelling in glacial-interglacial atmospheric CO_2 change. Tellus Series B-Chemical and Physical Meteorology* **58**, 73 (2006).

19
The origins of us

What is it that sets us apart from our hominid ancestors and the rest of the living world? Whatever it is, it is to be found in recent Earth history, at least compared to what we have been discussing thus far (Fig. 19.1). Yet despite being relatively close in time, it still proves hard to identify a unique formative event that has given us modern humans our planet changing powers. Our capacities for abstract thinking, advanced planning, behavioural, economic and technological innovation, and cultural and symbolic behaviour all seem to be important. Perhaps the best candidate for an evolutionary innovation behind all of these is the development of 'natural' language with a universal grammar (common word order) and syntax. The evolution of natural language represented a revolution in the transmission of information, which decoupled human social evolution from gene-based evolution giving it a 'genetic material' analogous in many ways to DNA, that could be transmitted both to contemporaries and future generations. This allowed cultural evolution to accelerate, and new levels of social organization to emerge. The result was the rapid development of human imagination and culture. Only after the emergence of language did our ancestors begin to reshape their environment on a large scale. But for environmental change to become truly global in scale relied on human societies increasing their inputs of energy, starting with the transition to agriculture, and culminating (thus far) in the fossil fuel age. Fig. 19.1 summarizes the overall chain of events.

19.1 Our genus

Our genus, *Homo*, originated in the Great Rift Valley of East Africa, roughly 2 million years ago, during the relatively mild climate of the Pliocene Epoch (*1*). *Homo erectus* was the first undisputed member of our genus, with improved long distance walking and running ability, which may have evolved because it enabled them to run down prey to exhaustion (*2*). *Homo erectus* formed the first hunter-gatherer societies. When Northern Hemisphere ice ages began in earnest 1.8 million years

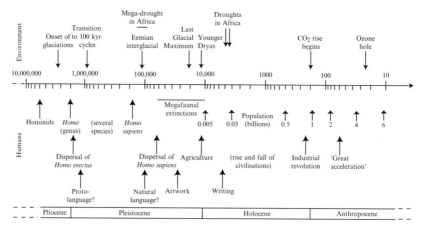

Fig. 19.1 Timeline approaching the present on a logarithmic scale, highlighting major environmental changes and significant events in human evolution. Recent geological time is divided into Epochs; the Pliocene spans 5.3–1.8 million years ago, the Pleistocene 1.8 million years ago to 11 500 years ago, and the Holocene from 11 500 years ago onwards. A new geological Epoch, the Anthropocene has been proposed, to mark out the interval of global scale influence of our species, beginning roughly 200 years ago

ago (marking the transition from the Pliocene to the Pleistocene Epoch), a wet phase of African climate provided a corridor of familiar savannah habitat through what is currently part of the Sahara desert. This allowed *Homo erectus* to make a relatively easy exit from East Africa, and they spread widely, both eastwards to Indonesia (Java) and China, and northwest to North Africa and Europe (Georgia). From these dispersed populations of *Homo erectus*, multiple hominid species appear to have evolved. These include *Homo antecessor* in southern Europe around 800 000 years ago and *Homo heidelbergensis* in central and northern Europe around 500 000 years ago. The heavily built Neanderthals (*Homo neanderthalensis*) may have evolved from *Homo heidelbergensis* in Europe, whilst the origins of tiny *Homo floresiensis* (from Flores, Indonesia) are currently being argued over (they may be a case of island dwarfism, descended from *Homo erectus*, although some of their features are more primitive). Such speciation is not so surprising given periodic geographical isolation of populations caused by the variable climate.

Homo erectus probably weren't capable of producing as complex a range of sounds as in our speech, but they may have developed a form of 'proto-language' as they formed larger social groups and their brain size increased. Examples of proto-languages are the pidgins that have repeatedly developed between modern groups of humans that do not share a common language. They lack syntax (word order is largely arbitrary) and have few grammatical items, but proto-language is still better for communication than no language. You may be wondering how any inferences can be made about the existence or not of language, when clearly it does not fossilize? Well, there is evidence that the area of the brain linked to speech production (Broca's area) appeared in *Homo erectus*. Furthermore, careful study of the distances at which stone tools are found from where the stones were quarried provides some indirect clues (*3*). Early *Homo* transported stone for tool making up to 13 km, which is roughly the home range of an individual. This suggests a planning capacity exceeding that of chimpanzees, but there is no need to invoke language based communication between individuals. However, after 1.2 million years ago, raw materials were transported greater distances of up to 100 km, far beyond the home range of any individual. This suggests the ability to pool information in groups with centralized decision making capabilities—relying on some rudimentary form of language.

Conceivably, it was *Homo erectus* that learnt to control fire, but which *Homo* species mastered this trick, and when, is argued over. The use of fire in hunting and in reshaping ecosystems, encouraging the shift to grasslands, probably pre-dates its domestic use, but human caused and natural fires cannot be readily distinguished from the fossil record. The controlled use of fire may date back as early as 1.5 million years ago, and evidence of its use for domestic purposes such as heating and cooking is scattered through the Pleistocene (*4*). Recent work suggests that by 800 000 years ago, hominids had learnt to create fire (*5*). Stone tool technologies also became more elaborate between about 400 000 and 250 000 years ago and brain size increased.

19.2 Our species

Homo sapiens originated in East Africa around 200 000 years ago, when the high Northern latitudes were in the grip of the penultimate ice age and Africa was particularly arid. The first fossils of anatomically modern humans come from Ethiopia (*6*). 'Mitochondrial Eve', the common female ancestor of all the mitochondria in all the cells of all the humans alive today, lived around 160 000 years ago. However, despite a familiar anatomy, including a large cranial capacity, there is little clue from the first *Homo sapiens* that our species would ultimately take over the world. By around 100 000 years ago *Homo sapiens* had spread throughout Africa and into the Middle East, but other *Homo* species were still much more widespread, and *Homo erectus*, *Homo neanderthalensis* and *Homo floresiensis* all survived until remarkably recently during the last ice age.

Yet something special happened among a group of *Homo sapiens*. The signs left in the archaeological record are relatively subtle;

technology became more sophisticated with the development of bone and antler tools, hunting became more specialized, long distance trade began, pigments were processed and artwork appeared, then our ancestors began ritually burying their dead. This is often portrayed as a revolution in behaviour around 50 000 years ago, although some argue that it occurred progressively over the preceding tens, if not hundreds, of thousands of years (7). At some point during that interval, the ancestors of all of us humans alive today experienced a bottleneck in population of 10 000 or fewer breeding pairs. Members of this founding group have gone on to evolutionary dominance, despite there having been many other individuals from our genus *Homo* (and presumably from our species *Homo sapiens*) alive at the time. It is tempting to assume that they possessed a special trait that gave them a unique advantage.

That special trait may have been a fully developed language capability, including syntax and universal grammar. The theory of universal grammar postulates that there are shared grammatical rules of language innate to all humans alive today, which have some basis in our genes and the structure of the brain. Support comes from observations of what happens when different societies come together and need to devise a new system of communication. Whilst the first contact generation use ungrammatical pidgin to get by, their children typically use the pidgin words to generate their own language—known as a creole—and independent examples of creole languages share common grammatical rules. Even more compelling is the example of a new sign language created among a group of deaf children in Nicaragua—it passed through a pidgin-like stage to a creole-like language which shares various grammatical conventions with other known languages, despite its independent origin. Neurology also provides support; it has been found that Broca's area of the brain is increasingly activated when learning new languages obeying universal grammar, but not when trying to learn arbitrary languages violating universal grammar (8).

The key thing about sharing a universal grammar is that it vastly increases the capacity of those sharing it to exchange complex information. With the emergence of natural language, the accumulation and spread of information was no longer tied to biological reproduction (9). Thus, the first group of humans to acquire universal grammar would have had a formidable advantage over other groups, especially in a changing environment, because they could adapt much faster. Precisely when that event happened we may never know, but there is evidence for some of the necessary precursors. The *FOXP2* gene, which is involved in speech and language, was fixed in the human population sometime during the last 200 000 years (10). Then by 130 000 years ago in Africa, raw materials for stone tool making were being transferred up to 300 km, beyond the range of any one group. This strongly suggests materials were being exchanged between groups, in the beginnings of an economy. This in turn requires some common language, with syntax that can express hierarchical relations (*3*).

The origin of natural language may mark a difficult, perhaps critical, step in evolution (an event so rare that it would only be expected to occur on a tiny fraction of planets that had reached the stage of complex, multi-cellular animal life). However, language can also be viewed as more of a continuum, noting that many species have more primitive forms of communication. Assuming that language is the central innovation that underlies the 'human revolution', the nature of its development and the probability of its occurrence are of great interest to us, and in particular to the usefulness of the 'critical steps' model of Chapter 5. Fully fledged language does appear to have evolved only once thus far on Earth, so fulfils that criterion for a critical step. Also, there have been many social animals during the last few hundred million years of evolution, who did not take this step. However, further analysis to try to establish language as a critical step faces the near-insuperable difficulties discussed in Chapter 5, of distinguishing truly unlikely occurrences in our evolution from moderately unlikely ones.

Our ancestors emerged out of Africa and the Middle East and began to spread around the world around 65 000 years ago. The migration may have been facilitated by one of the periodic wet phases of the Sahara after a mega-drought in Africa from 135 000 to 90 000 years ago (*11*). The exact pattern and timing of dispersal is argued over, but the latest evidence suggests that modern humans first went East along a Southern coastal route through India and onward to Southeast Asia and Australasia (*12*). An early offshoot from this migration led ultimately to the colonization of the Near East and Europe, but was stalled by unfavourable climate conditions in the lands bordering the Eastern shores of the Mediterranean, known as the Levant.

As modern humans arrived in new continents, many large mammal genera (over 40 kg) began to go extinct (*13*). Modern humans arrived in Australia sometime between 72 000 and 44 000 years ago and 14 of 16 genera went extinct thereafter. They reached Europe over 30 000 years ago where 9 of 25 genera went extinct. People reached North America 11 500 years ago where 33 of 48 genera went extinct, and they continued to South America by 10 000 years ago where 50 of 60 genera went extinct. Our ancestors clearly played a part in these 'megafaunal' extinctions, but to what degree continues to be debated. At worst they directly hunted large herbivores to extinction (known as 'overkill'), leaving carnivores and scavengers to suffer. In addition, human use of fire, shifting ecosystems toward grasslands may explain why browsers, mixed feeders and non-ruminants suffered most. Other factors probably contributed, although climate change tends to be over-stressed given that earlier Pleistocene climate changes of the same rate and magnitude had no such extinction effect. Extinction was less severe in Africa (8 of 44 genera), perhaps because existing species were already habituated to and wary of hunting hominids.

19.3 Agriculture and civilization

Human societies have flourished during the Holocene Epoch, the current interglacial (warm interval between ice ages) that started around 11 500 years ago. The rise of human civilizations depended on the origin of agriculture, which allowed high-density, sedentary, stratified societies (*14-16*). But we humans are not the first animals to start farming; there have been several earlier, independent origins of agriculture among social insects (see Box 19.1). This means the transition to agriculture cannot of itself be a difficult evolutionary step, though it seems to be one that is only undergone by highly cooperative, socialized animals. However, what the insect examples show is that it is drastic and possibly irreversible change that constrains subsequent evolution.

Box 19.1 Insect agriculture

Today around 200 species of Attine ants, 330 species of termites, and 3400 species of Ambrosia beetles cultivate fungi as a source of food (*17*). Farming dates back at least 50 million years in ants (*18*). The oldest fossils of termite fungal combs are about 7 million years old, but the activity could be much older (*19*). There have been at least seven independent origins of 'fungicultural' behaviour among the beetles, but only one each in the ants and the termites. The multiple origins indicate that farming is not a really difficult evolutionary transition, but it has only happened in species that are already social. Termites farm the specialized fungus, *Termitomyces*, all the cultivars of which stem from a single domestication event (*20*). The fungus helps the termites break down cellulose, and is fed with the faeces of workers that themselves feed on wood, grass, and leaves. The spores of the fungus survive passage through the termites that eat it, so defecating amounts to 'sowing' a new crop. The fungi farmed by Attine ants stem from three independent domestication events, and their fungal gardens are manured with dead vegetable debris, or in the case of leaf cutter ants, with leaf fragments cut from live plants. To add to the co-evolutionary complexity, bacteria which live on the ants produce a special compound that protects the fungi from a parasite, *Escovopsis*. Those insect societies that have evolved farming—such as leaf cutter ants and *Macrotermes* termites—have often risen to major ecological importance and come to dominate over closely related non-farming species. However, the insect farmers became dependent on their cultivated crops for food, and the transition has apparently never been reversed.

Whilst the transition to agriculture allows a species to rise to ecological prominence, they become so dependent on their cultivars that the farmer and the farmed begin to co-evolve.

Prior to the Holocene, abundant wild cereals were being collected, pounded and ground for food by members of the Natufian culture in the Levant. Then, perhaps in response to the drying effects of the Younger Dryas cooling 12 900 to 11 600 years ago on their regional ecosystem, people in this region domesticated the first cereal crops, triggering the Neolithic revolution (12 000 to 10 000 years ago). Early in the Holocene, around 10 500 years ago, the Sahara re-entered one of its wet and green phases and the region encompassing the Nile, Euphrates and Tigres rivers, connected by the Levant, became the fabled Fertile Crescent. The first farmers and herders there domesticated wild wheat, barley, peas, sheep, goats, cows and pigs, a list which contains what are still two of the most valuable crops and the most valuable livestock. Farming arose independently in a number of other locations during the Holocene. Rice was domesticated in China, as early as 11 500 years ago. Maize (a C_4 grass) was domesticated somewhere in Mesoamerica, sometime before 7 000 years ago. The Andes/Amazonia, eastern North America, Sahel, tropical West Africa, Ethiopia, and New Guinea were also centres of domestication. As a rule these were the places with the most domesticatable wild species, but generally not the best growing conditions (*15*).

Farming spread slowly to other regions, such as Europe (from the Fertile Crescent), indicating a steady migration of farming people, rather than the rapid cultural spread of an advantageous innovation. Although to our civilized eyes, with the benefit of hindsight, farming appears superior to foraging, the one way of life did not simply supplant the other, and in many places they co-existed for long intervals. These were areas rich with wild food where early agriculture required more energy and time investment than foraging. Foragers observing neighbouring farmers working harder than they were, quite logically declined to join them.

However, early agriculture flourished in then fertile regions such as Mesopotamia (the 'land between the rivers'—the Tigris and Euphrates). It permitted more hierarchical societies with division of labour in which not everyone engaged in providing food (as in insect agriculture). Although there was less equality among the members of these societies (i.e. some members were worse off than those in hunter-gatherer bands), the groups as a whole were more successful. Such group selection has been particularly important in the evolution of human societies. The cities of Sumer, beginning around 7300 years ago, were the first to be supported year round by agriculture that produced a surplus of storable food. They had high population density, and the division of labour included the first armies, which undertook aggressive expansion into neighbouring territories.

The need to keep records ultimately led the Sumerians to develop the first writing system around 5500 years ago, and thus they started recorded history (as opposed to pre-history). Their cuneiform characters include a pictogram of a simple plough (ard), an innovation which had spread rapidly around the old world and Europe roughly 5500 years ago. The plough marks the beginning of domesticated animal power replacing human muscle power. The earliest picture of two oxen pulling an ard comes from Babylonia, 5200 years ago.

As people built and expanded agriculturally-fuelled empires, they had an increasing impact on their environments, but they were also subject to natural changes in climate. Although the Holocene used to be portrayed as having an unusually stable climate, and life in the high northern latitudes was certainly less tumultuous than during the last ice age, there were substantial climate changes in the tropics. Sedentary civilizations based on agriculture were (and still are) more sensitive to climate change than mobile foraging societies, and climate change has been implicated in the rise and fall of some civilizations (*21, 22*). Civilizations also often played a role in their own demise by over exploiting their resource base, including deforesting, degrading soil, over-hunting, over-fishing and disrupting their freshwater supplies, all often linked to population growth and/or increased per capita impact of people (*23*).

The meticulous Sumerian records of crop yields tell a salutary tale. At the high points, the Sumerians achieved remarkable yields of 2 tonnes per hectare of wheat and 1.5 tonnes per hectare of barley (which were only surpassed in England around 100 years ago). However, their use of irrigation contributed to salination and siltation of the fields, causing declines in yield and shifts from wheat to the more salt-tolerant barley. Around 5000 years ago, a large scale climate change involving the drying and collapse of savannah ecosystems in the Sahara, brought a 200 year period of drought to the Middle East. This has been implicated in the collapse of Sumerian Late Uruk society (*21*). With the return of wetter conditions, various class based and politically centralized societies flourished in the Old World, including the Akkadian empire in Sumer, but they too collapsed in a further interval of drought around 4200 years ago (*24*).

Early civilizations also began to contribute to climate change, at least at a regional scale. Large scale deforestation and conversion to pasture (e.g. in Europe) would have tended to dry regional climate and increase daily and seasonal extremes. More controversially, it has been suggested that the carbon dioxide released from biomass and soils, and the methane released from domesticated cattle and the creation of paddy fields, began to affect the composition of the atmosphere as early as 8000 years ago (*25*). However, natural factors still dominated large-scale changes in climate. In the New World, the collapse of the Mochica (coastal Peru, 1500 years ago), Classic Maya (Yucatan

Peninsula, 1200 years ago) and Tiwanaku (Bolivia-Peru, 1000 years ago) empires have been linked to sustained (natural) drought intervals (*22*). In more recent human history, correlations have been noted between intervals of cooler climate, poor agricultural production, war and population decline (*26*).

19.4 Population growth

Since the origin of agriculture, human population growth has hinged on both the expansion of agricultural area, and increasing productivity per unit area. Despite regional setbacks, global population grew fairly continuously, going from 5 to 50 million between about 10 000 and 4000 years ago, and from 50 to 500 million by 500 years ago. Early growth was mostly supported by the spread of agriculture throughout the world. New habitats were exploited, for example by terracing hillsides, which has proved successful to this day in Asia and South America, but largely failed in the Mediterranean, where by Plato's time (around 2400 years ago) he was lamenting in the Dialogues that 'What now remains of the formerly rich land is like the skeleton of a sick man...' (*14*). Once the majority of productive land within a region had been turned to agriculture, subsequent population growth depended on increases in agricultural productivity. These have been achieved in a variety of ways, including adding water (irrigation), adding nutrients, and putting more energy in, starting with the replacement of human power with animal power. These inputs have been complemented by improvements in technology, which started with the plough.

As in the great revolutions of Earth history, successful increases in productivity depended not only on adding more nutrients, but also recycling them, in order to maintain or increase soil fertility. When crops are harvested, nutrient-rich material is removed from an agricultural ecosystem, but it can be returned (recycled) in the form of domestic animal dung (or human excrement). Equally, nitrogen can be added by growing plants with nitrogen fixing symbionts (notably the legumes), which may then be dug into the soil as green manure. By Roman times such methods were widely practiced, including crop rotation with legumes growing in alternate years between wheat crops. In Europe in the Middle Ages a three year rotation which boosted production was slowly adopted (first year autumn sown winter wheat or rye, second year spring-sown oats, barley, peas, beans, lentils or chickpeas, third year fallow). Productivity was also increased by the replacement of oxen power with horsepower, after the invention (in China) and widespread adoption of the padded horse collar. Although horses ate a greater fraction of the crop they worked faster to a higher power output, allowing deeper ploughing and overall productivity gains (*16*).

Sustained, rapid population growth began around 500 years ago, especially in Europe and China, with global population doubling to a billion around 200 years ago (in 1825). In Europe this population growth hinged on systematic use of recycling crop rotation systems, introduction of new crops such as potatoes from the New World, the enclosure of the commons, the progressive commercialization of farming and the beginnings of industrialization. This process reached a peak with the agricultural 'revolution' in our home county of Norfolk around the turn of the nineteenth century, with larger farms using a highly synergistic four-crop rotation system. Turnips were planted in the first year as fodder for cattle and the manure returned to the fields, followed by a crop of wheat or barley with the straw used for farmyard manure, followed by a clover crop to restore nitrogen and provide grazing, and finally another crop of barley or wheat.

Whilst this 'revolution' was underway, rapid population growth in England in the 1790s, a series of poor harvests, and escalating food prices caused much concern. They provided the context for Thomas Malthus to publish his famous 'Essay on the Principle of Population' (1798). In it he argued that the growth of population inevitably tends to outstrip the growth of food supply, and cannot be sustained. As Malthus recognized (with an insight of great value to Darwin) there is an inherent positive feedback in population growth; the more individuals there are the more they can beget, and this alone should lead to exponential ('geometric') growth. In contrast, food supply can only increase linearly ('arithmetic' growth). Thus, Malthus argued, food production should ultimately control population, providing negative feedback on its growth. However, developments in the subsequent two centuries were to largely prove Malthus wrong. Human population growth, up to somewhere in excess of 3 billion in the early 1960s, was actually faster than exponential: the annual rate of growth increased, up to a peak of 2.2% per year in 1963. Since then the rate of growth has slowed down and it is now below 1.2% per year.

19.5 The Anthropocene energy revolution

Malthus was (unbeknown to him) writing at the start of a transition from solar powered societies to fossil-fuel powered ones (*16*). Up until this time, human activities had largely been fuelled by solar energy, most of it coming to us, and the animals we domesticated, via plants, as food and biomass. Wood was used in cooking and heating, and converted to charcoal for use in metallurgy. Peat burning was also important and fuelled the Dutch Golden Age in the seventeenth century. The sun's energy was harnessed indirectly by water wheels and windmills. Although coal was used in iron making in ancient China, and coal was being shipped between mainland Europe and England in the Middle

Ages, it was a marginal source of energy. England was the first country to shift from biomass fuels to coal for domestic heating, and some aspects of manufacturing, during the sixteenth and seventeenth centuries. But the transition really took off with James Watt's improvements to the steam engine in 1769, which gave a huge boost to coal extraction, and marked the start of the Industrial Revolution.

With the transition from solar powered to fossil-fuelled civilizations, and the rapid increase in human population that has accompanied it, humans have come to have an unequivocal impact on the global environment. The recognition of this led the Nobel Prize winning chemist Paul Crutzen to propose a new geological Epoch; the Anthropocene (*27*). The Geological Society of London agrees that a case can be made for this on stratigraphic grounds—sediment erosion rates have been increased by more than an order of magnitude by human activities (*28*). In the Anthropocene, for the first time, food and biomass (i.e. solar power) have ceased to be the main source of energy for our societies. Instead the energy contained in annual food production, roughly 50 exajoules (EJ, where an exajoule is 10^{18} joules), has dropped to around a tenth of the total input of energy to human societies, which is now around 500 EJ. This is often misleadingly referred to as total primary energy *production*, but of course we don't really produce energy, instead we *transform* into a form useful to us, with inevitable losses in the transformation process. In terms of a flux of energy per unit time (i.e. power) human societies consume 15 terrawatts (TW). This is equivalent to about a tenth of the energy being captured globally by photosynthesis.

Food production still relies primarily on solar energy, but it has been greatly increased by inputs of fossil-fuel produced fertilizer and fossil-fuelled mechanization (*14*). Population doubled between 1825 and 1927 to 2 billion, doubled again by 1975 to 4 billion, and reached 6 billion by 1998 (Fig. 19.2). The second and third billion were added largely by increasing the area of land under cultivation, but this transition was only linear. The replacement of horse power with tractors increased the yield for human consumption because a significant fraction of the crop had been fed to the horses. Increased irrigation and the introduction of herbicides also helped. The fourth and fifth billion were added mostly by a dramatic increase in fertilizer nutrient inputs (especially nitrogen) to existing land. This was complemented by the introduction (especially to developing nations) of dwarf varieties of wheat and rice, which could thrive on the high nutrient inputs. The sixth billion was added (mostly in China, India and Africa) through increases in crop yield based on the spread of earlier innovations.

Annual fossil energy inputs to agriculture now exceed 10 EJ (estimates ranging over 12.8–18.2 EJ) but the return on this investment (50 EJ) is still a good one. The boost of food production from fossil-fuel produced fertilizer inputs alone currently feeds about 40% of the world's

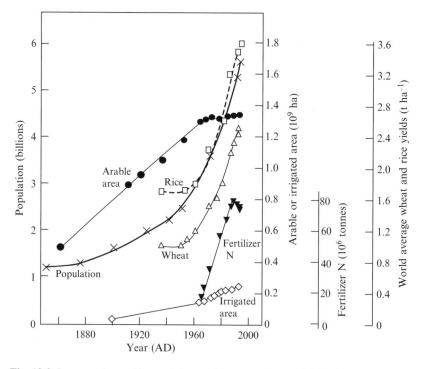

Fig. 19.2 Increases in world population, arable area, average yields of wheat and rice, the amount of nitrogen fertilizer used and the irrigated area of the world. [Redrawn from Evans (*14*)]

population, to which can be added the contribution from fossil-fuelled mechanization. During the twentieth century, agricultural area roughly doubled but the energy harvested in edible crops increased 6-fold. Up to a 3-fold increase in food output was due to what was a roughly 100-fold increase in energy inputs. In other words, up to 4 of the 6 billion people alive at the start of the twentieth century were fed by the fossil-fuelling of agriculture, and without it global population would have stalled somewhere in the range 2–3 billion.

Although fossil fuel inputs have greatly boosted food production, agriculture consumes only about 3% of global primary energy inputs, which totalled around 474 EJ in 2008, 88% from fossil fuels (*29*). Most of the energy inputs to fossil-fuelled societies go elsewhere—into industry (34%, including mining, manufacturing and construction), transportation (20%), residential use (11%), and commercial use (5%). The remainder (27%) is lost in generation and transmission. The 'production' (i.e. extraction) of fossil fuels has increased (generally) exponentially over the last 150 years (*16*). Initially it was dominated by coal, but oil and then gas have become increasingly important. In just the last few years, coal has made a comeback, supplying an increasing share of global primary energy, mostly linked to economic growth in China. In 2008, crude oil supplied 35% of world primary energy, coal supplied 29% and natural gas 24% (*29*). From the non-fossil fuels, hydro-electricity and nuclear power each supplied about 6% of primary energy (*29*). Geothermal, wind, solar and wood combined provided only 0.8% (and are not considered in the preceding totals).

19.6 Waste products (and their environmental consequences)

The exponential growth in human energy consumption has been accompanied by a corresponding increase in material consumption and the accumulation of waste products (*30*). These waste products have generally been dumped (or have ended up) on land, in the atmosphere, or in the ocean. This in turn has led to one environmental problem after another; from air pollution, through silent spring, acid rain, and eutrophication of freshwaters, to the ozone hole, and climate change. These are all symptoms of our prevailing approach to the planet, whilst in a phase of rapid expansion fuelled by an abundant and readily available source of energy. Whilst escalating human numbers are clearly part of these problems, escalating energy and material consumption by a modest fraction of the total population has caused a disproportionate part of these problems thus far. The problems do not come from energy use per se: they are due to the associated material waste products and our use of the environment as a dumping ground for them.

The accumulation of waste products started out as a localized problem for human societies, most keenly felt near the most concentrated sources of waste—often growing cities—especially when natural processes contained the waste nearby. Sewage is the original, pungent example. But as early as 1306, air pollution in London caused Edward I to ban coal fires. Alas, the ban was short-lived and 500 years later with the industrial revolution came the smogs that gave London its nickname 'the big smoke'. These were a consequence of dirty coal burning pouring out masses of smoke and sulphur dioxide (SO_2) into atmospheric conditions prone to trapping the cold pool of smog. December of 1890 marked the low point with Westminster recording no hours of sunshine throughout the whole month (*31*). But it was not until after thousands had died in the great smog of 1952, and several years of debate and denial, that the Clean Air Act (1956) introduced smokeless zones to the city.

It was also in the aftermath of World War II that it began to collectively dawn on humanity that our waste products were accumulating globally, and causing global environmental changes. Some early pioneers had calculated that they could do—before the end of the nineteenth century Svante Arrhenius had painstakingly and accurately

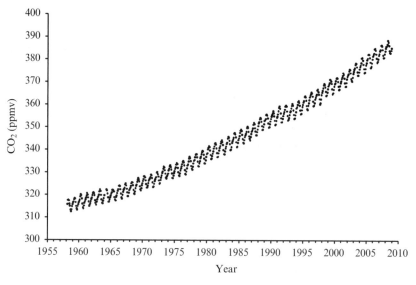

Fig. 19.3 The 'Keeling curve' of rising atmospheric carbon dioxide concentration measured at Mauna Loa, Hawaii (*33*)

calculated the global warming expected if humans managed to double the carbon dioxide content of the atmosphere (*32*)—though this was an event he did not foresee occurring for a thousand years. It was not until the post war era that the first widespread and continuous measurements of globally accumulating waste products were established, including the now famous 'Keeling curve' of rising carbon dioxide, begun by Charles David Keeling in 1958 (Fig. 19.3) (*33*).

The post war era was also the time of what has been dubbed 'the great acceleration' in our impact on the planet. With technological advance came both environmental disruption and our ability to measure it. Jim Lovelock was instrumental (in a literal sense) in this—inventing some of the first exquisitely sensitive measuring devices. Others used them to detect the build-up of pesticides in remote parts of the food chain, which in turn provided key evidence for Rachel Carson's 'Silent Spring'. Then in 1972, Lovelock was the first to measure the accumulation of chlorofluorocarbons (CFCs) in the remote marine atmosphere with his electron capture detector (*34*).

Efforts to tackle the resulting stratospheric ozone depletion are often cited as a success story, but it is our failure to prevent it being created in the first place that is more striking. It was demonstrated in 1907 (and again in 1934) that chlorine containing gases destroy atmospheric ozone (*35*). Two years after Lovelock's first measurements in 1974, it was argued that CFCs would be destroying the ozone layer (*36*). But by the time that Joe Farman and colleagues published their famous detection of the ozone hole from the ground in 1985 (*37*), efforts to reduce CFC use in aerosols had made little headway and emissions from their use in refrigeration were rising. Since the Montreal Protocol in 1987 the ozone hole has continued to get worse because of inherent time delays between CFC emission at the surface and their role in ozone destruction in the stratosphere. Although the ozone layer is just on the point of starting to recover, it will take several decades to return to the state it was even in the 1980s. The European Environment Agency have gathered together more than ten other such case studies, which illustrate that the historical track record of human efforts to anticipate our impacts on the environment has been a salutary one of 'late lessons from early warnings' (*35*).

Our collective activities have now reached the extent that for some elemental cycles we exceed the activities of the rest of the biosphere combined. The synthesis (using fossil fuel energy) and application of nitrogenous fertilizers has roughly doubled the input of available nitrogen to the biosphere. The mining, refining and application of phosphorus as fertilizer has roughly tripled the global mobilization of phosphorus (*38*). Much of the nitrogen and phosphorus we add ends up in freshwaters where it fuels biological productivity (eutrophication), sometimes to the extent that ancient cyanobacteria choke out more recent life forms, with waters turning anoxic and killing fish and other animals. Some of our added nitrogen and phosphorus reaches coastal seas and ultimately the open ocean, where similar consequences ensue.

19.7 Greenhouse gases

Currently our most concerning waste product is the greenhouse gas carbon dioxide (CO_2), which comes directly from fossil fuel combustion and 'land use change'—mostly tropical deforestation. Ice cores show that its concentration was stable at around 280 parts per million (ppm) of the atmosphere for a thousand years before the industrial revolution, but by the start of the 'Keeling curve' (see Fig. 19.3) in 1958 it had risen to 315 ppm and at the time of writing in 2009 it has reached 387 ppm, and is rising at about 2 ppm per year. The ongoing rise is driven primarily by emissions from fossil fuel burning.

Fig. 19.4 visualizes what is going on, in a simplified way (treating the land use change emissions as if they were from fossil fuels). In the pre-industrial state, the massive exchange fluxes of CO_2 between the atmosphere and ocean, and between the atmosphere and land, were in balance (Fig. 19.4a). However, today there is a net input of fossil-fuel-derived CO_2 to the atmosphere, and a net uptake of some of this added 'fossil' CO_2 by the ocean and land 'carbon sinks' (Fig. 19.4b). Thus far, roughly half of the fossil CO_2 that has been released remains in the

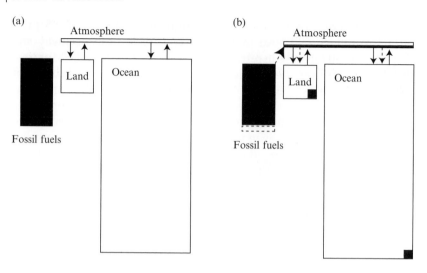

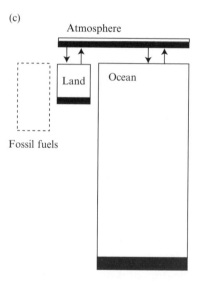

Fig. 19.4 Schematic of the apportioning of carbon between different reservoirs within the Earth system. Box sizes are roughly to scale: (a) Pre-industrial state—solid arrows indicate major exchange fluxes of CO_2 that are in balance; (b) Present state—dotted arrows indicate net fluxes of fossil CO_2, black segments show the accumulation of fossil CO_2 in the atmosphere, ocean and land; (c) Future state assuming all fossil fuels are combusted, showing the new steady state that will be reached after roughly 1000 years, based on (*39, 40*)

atmosphere, and about a quarter has entered each of the ocean and land. Currently our collective activities are increasing emissions at an accelerating rate, and not surprisingly this means that the CO_2 concentration in the atmosphere is going up more rapidly than it has been (for more details of the current carbon balance see Box 19.2).

Box 19.2 Sources and sinks of carbon dioxide

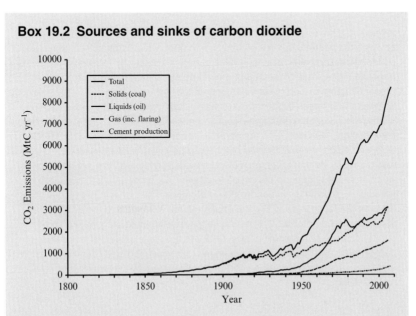

Fig. 19.5 Fossil emissions of carbon dioxide in million metric tons of carbon per year. [Data for all sources up to 2006, and 2007–2008 preliminary estimates for total, cement production and gas (*42*)]

Fossil fuel CO_2 emissions have tracked the growth in human energy consumption, both increasing at an average of 1.9% per year over the last 25 years (Fig. 19.5) (*42*). In the 1990s, there were signs of deceleration as CO_2 emissions grew more slowly at 1.3% per year, due to a decline in energy use per unit of gross domestic product (lower energy intensity). However, CO_2 emissions accelerated at over 3% per year during 2000–2008, linked to a switch toward coal as a source of energy, which generates more CO_2 emissions per unit of energy (higher carbon intensity) (*43, 44*). Much of the historical growth in CO_2 emissions, including the recent acceleration, has been decoupled from population growth (which has been decelerating since the early 1960s). Instead there is a massive inequity in energy use and CO_2 emissions per capita. The developing economies (particularly China) are responsible for much of the recent accelerated growth in emissions, but they still account for less than half of the total annual CO_2 emissions, and less than a quarter of cumulative CO_2 emissions since the eighteenth century. Ultimately, it is the total amount of CO_2 emitted to the atmosphere (the cumulative emission) that determines its atmospheric concentration. The amounts involved are measured in petagrams of carbon (PgC), where a petagram is 10^{15} grams (1 000 000 000 000 000 g) or equivalently 10^9 metric tonnes, called a gigatonne (GtC). Only the weight of the carbon atoms is counted, which is convenient given that it changes chemical form when it is burned. Between 1800 and 2000, 282 PgC were emitted from fossil fuel burning (*45*) and roughly 140 PgC from land use change (*46*). Only about half (213 PgC) of this cumulative emission of carbon has remained in the atmosphere, thanks to net uptakes of carbon by the ocean and land sinks. The ocean sink is the dissolution of CO_2 into the surface water, and the subsequent removal of a small fraction of that into deeper water. It is limited by the slow rate at which the ocean overturns vertically, with deep water forming from the surface. The 'land sink' is a net uptake of carbon, mostly by forests, and it exists because those areas of forest that are not being cleared for agriculture are at present adding carbon mass every year to the live vegetation and to the soil. Both land and ocean sinks can be looked on as negative feedback mechanisms, which kept the atmospheric CO_2 concentration close to 280 ppm before the industrial revolution, and are currently tending to resist its rise. The situation would be much worse if it were not for these negative feedbacks, and CO_2 would already be much higher in the atmosphere. However, the 'set point' to which they are tending to restore atmospheric CO_2 is not invariant but is itself responsive to climate—it clearly changed between glacial and interglacial time for example (see Chapter 18). We expect global warming and the addition of fossil fuel carbon dioxide to affect these feedbacks, most likely weakening them over the course of the coming century. There are some signs that natural carbon sinks, particularly in the ocean may be becoming less effective (*47*) and there is a substantial research effort presently underway to more accurately monitor them.

Once our emissions cease, over a millennial timescale, the added fossil CO_2 in the system will be apportioned between the ocean, atmosphere and land, until a new steady state is reached (Fig. 19.4c) (*39*). Crucially, a fraction of the CO_2 that we add to the atmosphere remains there. Currently this long-lived fraction is about 20%, but the fraction goes up the more CO_2 we add, because the ocean and land become less effective at storing CO_2 (in Fig. 19.4c it is assumed that 4000 billion tonnes of fossil carbon will eventually be released, and in this case about 30% will remain in the atmosphere, 10% will enter the land, and 60% will end up in the ocean) (*40*). Over even longer timescales of tens of thousands to millions of years, processes of weathering and exchange with ocean sediments will gradually return the fossil CO_2 to the Earth's crust, via the ocean. However, the processes take so long that our fossil CO_2 waste is actually more persistent than nuclear waste (*41*).

We worry particularly about carbon dioxide because it is making the largest human addition to the greenhouse effect, it is acidifying the ocean, and it is an especially long-lived pollutant. However, some of our other waste products that we inadvertently dump in the atmosphere are also potent greenhouse gases, notably methane (CH_4), nitrous oxide (N_2O) and CFCs. Increased methane emissions to the atmosphere have come from food production (mostly ruminant livestock and paddy fields), fossil fuel use (mostly leaks of natural gas during its extraction, transport and use), and from landfills, fires and waste treatment works. These have caused methane concentration in the atmosphere to more than double from around 800 parts per billion (ppb) in 1800, to around 1800 ppb today. Meanwhile, a fraction of the nitrogen we have synthesized and added to agricultural soils has been converted to the long-lived greenhouse gas (and ozone-depleter) nitrous oxide, by the ancient microbiological processes of nitrification and denitrification. This in turn has been largely responsible for increasing the atmospheric concentration of nitrous oxide from 272 ppb to 310 ppb.

19.8 Climate change

With such a mass of our waste products accumulating in the atmosphere, we should not be surprised that we are changing the climate. Global temperature increased by about 0.75°C over the last century, but the most striking rise of around 0.5°C has occurred during the last 40 years and is most clearly linked to growth in greenhouse gas concentrations (Fig. 19.6) (*48*). There are clearly other factors besides rising CO_2 that influence the global temperature and mean that the temperature rise does not exactly mirror that of CO_2. However, to our minds the combination of the well developed theory for the greenhouse effect, and the observed temperature increase, makes the case for this causal connection unequivocal. There is an unprecedented level of scientific consensus about it, and there should be no need to debate the matter further, but for some powerful vested interests seeking to promote the appearance of uncertainty and lack of consensus (*49*). The climate is actually changing faster than many existing model projections, which have used 1990 as a start year. The warming since 1990 is at the upper bound of the projections. This recent temperature trend could be partly enhanced by natural variability of the climate system, but equally it could indicate a relatively high sensitivity of the climate system to increasing greenhouse gas concentrations.

One factor that has unintentionally masked some of the warming effect of increasing greenhouse gases is our emissions of tiny aerosol particles that tend to scatter sunlight. Sulphate aerosol forms from sulphur dioxide, emitted by fossil fuel burning especially combustion of sulphurous (brown) coals. However, when it enters solution in water,

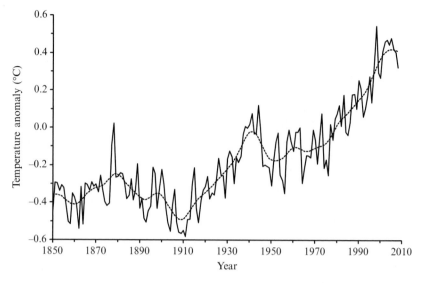

Fig. 19.6 The global temperature record 1850–2008 (HadCRUT3). [Compiled by our colleagues in the Climatic Research Unit, and at the Hadley Centre (*48*).]

sulphate forms sulphuric acid and hence acid rain. Successful efforts have been made to scrub sulphur dioxide out of power station flue gases in order to curb acid rain, but this in turn has unmasked the increasing greenhouse effect. A second key aerosol waste product is soot (or 'black carbon') produced from burning biomass and fossil fuels. It tends to absorb sunlight (by virtue of its dark colour), thus warming the atmosphere but cooling the surface underneath, unless it is deposited on snow and ice where it contributes significantly to melting. Soot emissions have been increasing of late, especially in China and India, and inhaling the particles poses a direct health threat. Further up in the atmosphere, they combine with sulphate to form an 'atmospheric brown cloud' haze that is particularly prominent over Southern Asia. Whilst this is cooling the surface somewhat, it is also disrupting the seasonal monsoons that are critical for food production (*50*).

Currently our net effects on the global climate are far weaker than some of those caused by past biological innovations we have discussed earlier in the book. However, we are changing the composition of the atmosphere much faster than anything seen in the ice core record, or the recent geological record. So, we should expect climate change to accelerate, and we should prepare for much larger climate changes in the long term. Indeed, scientists have been stunned by the rate of recent changes in the climate system, including the shrinking area of summer sea-ice in the Arctic, accelerating shrinkage of the Greenland ice sheet, and loss of mass from the West Antarctic ice sheet (*51*).

19.9 Outlook

There is no convincing evidence that we humans are about to collectively take our foot off the gas of increasing energy and material consumption. Although population looks set to stabilize at 8-10 billion mid-century, resource consumption is already profoundly decoupled from basic human requirements for food and water and thus from population trends. The inhabitants of less developed countries are understandably aspiring to a better life, and currently this equates to higher per capita energy and material consumption. Under the current global economic system this means increased consumption of fossil fuels and many other materials mined out of the Earth's crust. Whilst environmentalists have been arguing for some time that the most profligate over-consumers should rein in their wasteful habits, there is little sign that we are actually doing so. Despite many fine words and good (though weak) intentions, there has been no sign of even a slowing in the rate of increase of global carbon dioxide emissions, instead they have accelerated. The current direction and rate of change is actually worse than 'business as usual' and might be better termed 'burn baby burn'. It is close to even the most pessimistic scenarios of the Intergovernmental Panel on Climate Change for emissions growth developed in the late 1990s. It should not be surprising then that the world is warming and the sea level is rising faster than was forecast. The question is: where is such profligate squandering of resources taking us and the Earth system?

References

1. I. Tattersall, J. H. Schwartz, *Evolution of the Genus Homo. Annual Review of Earth and Planetary Sciences* **37**, 67 (2009).

2. D. M. Bramble, D. E. Liberman, *Endurance running and the evolution of Homo. Nature* **432**, 345 (2004).

3. B. Marwick, *Pleistocene Exchange Networks as Evidence for the Evolution of Language. Cambridge Archaeological Journal* **13**, 67 (2003).

4. S. R. James, *Hominid Use of Fire in the Lower and Middle Pleistocene. Current Anthropology* **30**, 1 (1989).

5. N. Alperson-Afil, *Continual fire-making by Hominins at Gesher Benot Yaaqov, Israel. Quaternary Science Reviews* **27**, 1733 (2008).

6. I. McDougall, F. H. Brown, J. G. Fleagle, *Stratigraphic placement and age of modern humans from Kibish, Ethiopia. Nature* **433**, 733 (2005).

7. S. McBrearty, A. S. Brooks, *The revolution that wasn't: a new interpretation of the origin of modern human behavior. Journal of Human Evolution* **39**, 453 (2000).

8. M. Musso *et al.*, *Broca's area and the language instinct. Nature Neuroscience* **6**, 774 (2003).

9. J. Maynard Smith, E. Szathmáry, *The Major Transitions in Evolution*. (Oxford University Press, Oxford, 1995).

10. W. Enard *et al.*, *Molecular evolution of FOXP2, a gene involved in speech and language*. Nature **418**, 869 (2002).

11. C. A. Scholz *et al.*, *East African megadroughts between 135 and 75 thousand years ago and bearing on early-modern human origins*. Proceedings of the National Academy of Sciences USA **104**, 16416 (2007).

12. V. Macaulay *et al.*, *Single, Rapid Coastal Settlement of Asia Revealed by Analysis of Complete Mitochondrial Genomes*. Science **308**, 1034 (2005).

13. A. D. Barnosky, P. L. Koch, R. S. Feranec, S. L. Wing, A. B. Shabel, *Assessing the Causes of Late Pleistocene Extinctions on the Continents*. Science **306**, 70 (2004).

14. L. T. Evans, *Feeding the Ten Billion: Plants and population growth*. (Cambridge University Press, Cambridge, 1998).

15. J. Diamond, *Evolution, consequences and future of plant and animal domestication*. Nature **418**, 700 (2002).

16. V. Smil, *Energy in nature and society: general energetics of complex systems*. (MIT Press, Cambridge, 2008).

17. U. G. Mueller, N. Gerardo, *Fungus-farming insects: Multiple origins and diverse evolutionary histories*. Proceedings of the National Academy of Sciences USA **99**, 15247 (2002).

18. T. R. Schultz, S. G. Brady, *Major evolutionary transitions in ant agriculture*. Proceedings of the National Academy of Sciences USA **105**, 5435 (2008).

19. P. Duringer *et al.*, *The first fossil gardens of Isoptera: oldest evidence of symbiotic termite fungiculture (Miocene, Chad basin)*. Naturwissenschaften **93**, 610 (2006).

20. D. K. Aanen *et al.*, *The evolution of fungus-growing termites and their mutualistic fungal symbionts*. Proceedings of the National Academy of Sciences USA **99**, 14887 (2002).

21. H. Weiss, R. S. Bradley, *What Drives Societal Collapse?* Science **291**, 609 (2001).

22. P. B. deMenocal, *Cultural Responses to Climate Change During the Late Holocene*. Science **292**, 667 (2001).

23. J. Diamond, *Collapse: How Societies Choose to Fail or Succeed*. (Viking Press, Northampton, 2005).

24. H. M. Cullen *et al.*, *Climate change and the collapse of the Akkadian empire: Evidence from the deep sea*. Geology **28**, 379 (2000).

25. W. F. Ruddiman, *The Anthropogenic Greenhouse Era Began Thousands of Years Ago*. Climatic Change **61**, 261 (2003).

26. D. D. Zhang, P. Brecke, H. F. Lee, Y.-Q. He, J. Zhang, *Global climate change, war, and population decline in recent human history*. Proceedings of the National Academy of Sciences USA **104**, 19214 (2007).

27. P. J. Crutzen, E. F. Stoermer, The Anthropocene. *Global Change Newsletter*. (2000), 41, pp. 17–18.

28. J. Zalasiewicz *et al.*, *Are we now living in the Anthropocene? GSA Today* **18**, 4 (2008).

29. BP, 'BP Statistical Review of World Energy June 2009' http://www.bp.com/liveassets/bp_internet/globalbp/globalbp_uk_english/reports_and_publications/statistical_energy_review_2008/STAGING/local_assets/2009_downloads/statistical_review_of_world_energy_full_reports_2009.pdf (2009).

30. W. Steffen *et al.*, *Global Change and the Earth System—A Planet Under Pressure*. Global Change—The IGBP Series (Springer-Verlag, Berlin, 2004), pp. 336.

31. A. Woodward, R. Penn, *The Wrong Kind of Snow*. (Hodder & Stoughton, London, 2007).

32. S. Arrhenius, *On the Influence of Carbonic Acid in the Air upon the Temperature of the Ground. Philosophical Magazine and Journal of Science* **41**, 237 (1896).

33. R. F. Keeling, S. C. Piper, A. F. Bollenbacher, J. S. Walker, in *Trends: A Compendium of Data on Global Change*. (Carbon Dioxide Information Analysis Center, Oak Ridge National Laboratory, U.S. Department of Energy, Oak Ridge, Tenn., U.S.A., 2009).

34. J. E. Lovelock, R. J. Maggs, R. J. Wade, *Halogenated Hydrocarbons in and over the Atlantic. Nature* **241**, 194 (1973).

35. EEA, 'Late lessons from early warnings: the precautionary principle 1896–2000' (European Environment Agency, Copenhagen, 2001).

36. M. J. Molina, F. S. Rowland, *Stratospheric Sink for Chlorofluoromethanes: Chlorine Atom-Catalysed Destruction of Ozone. Nature* **249**, 810 (1974).

37. J. C. Farman, B. G. Gardiner, J. D. Shanklin, *Large losses of total ozone in Antarctica reveal seasonal ClOx/NOx interaction. Nature* **315**, 207 (1985).

38. V. Smil, *Phosphorus in the Environment: Natural Flows and Human Interferences. Annual Review of Energy and Environment* **25**, 53 (2000).

39. T. M. Lenton, *Land and ocean carbon cycle feedback effects on global warming in a simple Earth system model. Tellus* **52B**, 1159 (2000).

40. T. M. Lenton, *Climate Change to the end of the Millennium. Climatic Change* **76**, 7 (2006).

41. D. Archer *et al.*, *Atmospheric Lifetime of Fossil Fuel Carbon Dioxide. Annual Reviews of Earth and Planetary Sciences* **37**, 117 (2009).

42. Boden, T.A., G. Marland, R.J. Andres. Global, Regional, and National Fossil-Fuel CO_2 Emissions. Carbon Dioxide Information Analysis Center, Oak Ridge National Laboratory, U.S. Department of Energy, Oak Ridge, Tenn., U.S.A. doi 10.3334/CDIAC/00001(2009)

43. C. Le Quere, M. R. Raupach, J. G. Canadell, G. Marland, *et al.*, *Trends in the sources and sinks of carbon dioxide. Nature Geoscience* **2**, 831 (2009).

44. M. R. Raupach *et al.*, *Global and regional drivers of accelerating CO_2 emissions. Proceedings of the National Academy of Sciences USA* **104**, 10288 (2007).

45. G. Marland, R. J. Andres, T. A. Boden, in *Trends: A Compendium of Data on Global Change*. (Carbon Dioxide Information Analysis Center, Oak Ridge National Laboratory, Oak Ridge, Tenn., U.S.A., 2008).

46. R. A. Houghton, in *TRENDS: A Compendium of Data on Global Change*. (Carbon Dioxide Information Analysis Center, Oak Ridge National Laboratory, U.S. Department of Energy, Oak Ridge, Tennessee, U.S.A., 2008).

47. C. LeQuéré *et al.*, *Saturation of the Southern Ocean CO_2 Sink Due to Recent Climate Change. Science* **316**, 1735 (2007).

48. P. Brohan, J. J. Kennedy, I. Harris, S. F. B. Tett, P. D. Jones, *Uncertainty estimates in regional and global observed temperature changes: A new data set from 1850. Journal of Geophysical Research* **111**, D12106 (2006).

49. O. Houck, *Tales from a Troubled Marriage: Science and Law in Environmental Policy. Science* **302**, 1926 (2003).

50. V. Ramanathan *et al.*, *Atmospheric brown clouds: Impacts on South Asian climate and hydrological cycle. Proceedings of the National Academy of Sciences USA* **102**, 5326 (2005).

51. T. M. Lenton *et al.*, *Tipping Elements in the Earth's Climate System. Proceedings of the National Academy of Sciences USA* **105**, 1786 (2008).

20
Review

Our story has reached the cusp between the past and the future, with the planet in the midst of what might be a new revolution. A single species is now in the process of transforming the planet. But for the first time in the history of the Earth the agents of planetary change have a dawning collective awareness that they are changing the world. We can't be sure if this will come to rank alongside the great revolutions that made the present Earth, not least because it is very much still underway. So let us pause briefly to review the similarities between what is happening now and the past revolutions.

20.1 Common features of revolutions

There are several common features to the great revolutions that have made the present Earth (Fig. 20.1). These can be broadly categorized as the main characteristics of change, and the features that must emerge for it to be 'successful'—in the sense that permanent change persists and includes a thriving biosphere.

Two out of the three past revolutions have been underlain by a step change in information transmission between living organisms. An increase in the amount of information that can be passed on allows life to build more complex structures. Complex structures can in turn improve the information transmission mechanisms, allowing more information to be passed on, so these two processes mutually reinforce one another. The original example was the origin of the genetic code supporting prokaryote cells—and of cells supporting the mechanics of genes. The evolution of the eukaryote genome, simultaneously with the new structure of the eukaryote cell, was another. These revolutions in information and organization do not of themselves lead to changes in their planetary environment. Rather, they light a fuse that can be very slow burning.

Changes of the planetary environment also require revolutions in energy and matter flow through the biosphere. Virtually all the important metabolisms of life evolved among prokaryotes, on the early

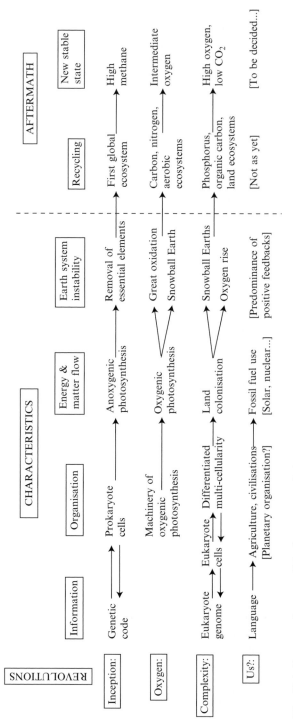

Fig. 20.1 Common features of Earth system revolutions

Earth. The origin of DNA and prokaryote cells was the fuse for them. The most important and difficult-to-evolve example was the origin of oxygenic photosynthesis, an essential precondition for the Great Oxidation. The subsequent revolution eventually involved eukaryotes taking this already existing metabolism onto land and evolving ways to mine rocks for nutrients, with the consequence of boosting productivity both on land and in the ocean. Still these innovations in metabolism did not necessarily nor immediately lead to planetary scale disruption.

The revolutions did however cause such disruption at the global scale once they began to interact with fundamental feedbacks in the Earth system. Strong positive feedbacks exist in the system, separating multiple stable states for planetary variables such as atmospheric oxygen, ice cover and temperature, which are locally stabilized by negative feedbacks. The Earth system can be knocked out of one stable state and into another by a change in forcing conditions or a significant perturbation. In the case of the oxygen revolution, the origin of oxygenic photosynthesis eventually knocked the system from a very low oxygen state into a higher one in the Great Oxidation. In the Neoproterozoic, it was pushed into a yet higher oxygen state. In both cases, changes in atmospheric composition triggered the strong, non-biological ice-albedo feedback, causing climate instability lasting tens of millions of years. Once these positive feedbacks took hold, major changes of the environment became unstoppable.

Whilst the combination of information revolution, energy revolution and instability in the system can start a planetary scale revolution, there is no guarantee it will end well. Of course all the past revolutions had to have succeeded for us to be here to remark upon them. But we think there always lurked the possibility of failure—the system reverting back to an earlier state—or even of disaster—life wiping itself out. In each case, it required a revolution in recycling for the system to recover stability in a new state that supported a thriving biosphere. Recycling is so important because the Earth system is almost closed to mass and so all the elemental cycles must be very nearly closed. The biological innovations at the source of revolutionary changes in the environment involve the accumulation of a novel waste product—be it oxygen in the atmosphere, or the flushing of weathered elements to the ocean. To stop these waste products accumulating indefinitely, and maintain high productivity, it is necessary that new types of life evolve to form closed recycling loops.

For the Earth system to find a new stable state also involves the reassertion of negative feedbacks, which can involve the planetary effects of new metabolisms becoming self-limiting to their carriers. For example, we argue that the ending of snowball Earth events in the Neoproterozoic relied on the high oxygen and low carbon dioxide atmosphere created by land colonizers becoming limiting to them.

20.2 Does the present fit the pattern?

Comparing what is happening to the planet now with the start of previous revolutions the main preconditions appear to be at least partly in place.

The origin of language represents a revolution in the transmission of information, which has largely decoupled information flow from reproduction, and allowed a phenomenal acceleration of social evolution within our species. Natural selection still operates, but (unusually) it often does so more strongly at the level of human social groups than of individuals or genes. The information revolution has supported the emergence of complex social structures including hierarchical societies supported by agriculture, as well as the literal structures of the built environment. There are precursors of all these features among the social insects, most remarkably some fungus-farming termites that build towering air-conditioned nests. These insects achieve their communication using pheromones and touch. We humans on the other hand have evolved language, an entirely new system of communication, which has opened new horizons for the type of information we can share. We have taken society and industry to new levels, including building information processing methods and machines. These started with writing and clay tablets, accelerated with the invention of the printing press, and recently went into exponential increase with the invention of computers and the internet. Most of the information being transmitted appears to be junk (as it does in eukaryote genomes), and we are still a long way from creating a truly self-replicating machine. But optimists can at least see the potential for the emergence of yet higher levels of organization, whilst pessimists warn that our machine creations will usurp us.

An energy revolution is also underway. Once we realized we had a plentiful source of energy—in the form of fossil fuels—and the ability to make use of it, we set about exploiting it, thus following the same evolutionary imperative that photosynthesizers and land colonizers followed before us. This time the ability to make use of fossil energy was not a metabolic innovation but a suite of technological ones. But just as those organisms gifted with oxygenic photosynthesis came to dominate the planet, those societies with the technology to make the swiftest and greatest use of fossil fuels have become the dominant ones. The flow of fossil energy through human civilization is now about a tenth of the size of the flow of solar energy through the biosphere as a whole. Only a small fraction of this energy is going into supplying our biological needs for food. Much of it is being used for heating, cooling, transportation, construction, and information processing. Of course a fundamental problem is that fossil fuel reserves are finite. Our fossil-fuelled societies are like respiring organisms arriving in a world full of food, but without any primary producers—soon enough they are doomed to a meagre existence unless they can evolve the equivalent of photosynthesis. If our dependence on fossil fuels continues then this energy revolution will sooner or later stall in its tracks. However, diverse alternative sources of energy are actively being developed and used. Energy demand continues to increase in a roughly exponential manner despite population growth slowing down, and we see no fundamental reason why humans will not continue to meet that rising energy demand.

A third similarity with previous revolutions is also in place; the Earth system is subject to strong feedbacks and is prone to instability. Instability is abundantly clear in the recent geological record of climate change, with the planet oscillating between glacial and interglacial conditions. The oscillation has recently got larger in magnitude and longer in period. In the last million years, its shape has changed to a saw-tooth in which ice ages end in abrupt warming dominated by positive feedback, temporarily in a runaway state. Although all our civilizations have developed in a 10 000 year interglacial interval of apparent stability, underlying positive feedbacks are still there. The PETM event 55 million years ago shows that the Earth system can propel itself into a much warmer state. On the flip side, the snowball Earth state might still be available. It's not just the climate either, the ocean is still potentially vulnerable to being pushed into an anoxic state, and we have already collectively demonstrated that the ozone layer can be disrupted. Whichever direction we push the Earth system, the key point is that if we push it too far there are positive feedbacks that can act to continue carrying the system in that direction, toward a new stable state.

20.3 Conditions for success

So, the main preconditions for an Earth system revolution appear to be in place, but what will it take for this to be a successful revolution? The past revolutions of the Earth had to be successful in order for us to be here at all, but that does not mean it was inevitable or even likely that they would succeed. It could be that success is rare and failure (reversion to an earlier state) or even disaster (the wiping out of life) is the more likely outcome. It is just that we could only have evolved on the rare, fortunate planet that, after the inception of life, got through at least two more dangerous (and potentially fatal) revolutions. If humans represent the start of another revolution, there is certainly no guarantee it will be successful, and at present the omens don't seem very promising that it will be.

The one obvious thing that is missing, which was needed to make previous revolutions successful, is recycling. As our collective energy use has rocketed up, so too has the dumping of waste materials in our surroundings. The past revolutions were also driven by new sources of energy and materials tapped by the biosphere, but ultimately they were only successful because the system 'learned' to recycle the materials needed by the innovators from their waste products. Environmental regulatory feedbacks also kicked in which stabilized the system. Typically it took tens of millions of years for these solutions to arise. During that time there were major swings of climate that could have destroyed the emerging system. Fortunately, we are equipped with intelligence and imagination, and we ought to be able to foresee and understand these dangers. Thereby we could avoid them, and also the millions of years of climatic upheaval that previously had to be endured before the 'blind watchmaker' of evolution, and the reorganization of feedback loops, restored a smoothly working global system.

21
Where next?

Now we want to look ahead, equipped with the knowledge of what makes a successful revolution of the Earth, and what can cause it to fail. There is no guarantee that what we are doing now as a species will become revolutionary for the planet. If it does, there is certainly no guarantee that the outcome will be a success, for the planet or for us. But the great thing about the future is that it is not preordained: by our actions we can influence the path taken. So, rather than offer one vision of the future for humans and our planetary home, we want to explore three. The first two should be fairly familiar to you. But the third we hope is something of a pleasant surprise. There is a way that we can make this a successful revolution of the Earth.

21.1 Apocalypse?

Let us start with the worst possible future: that we tip the planet into an uninhabitable (or barely habitable) state, eliminating ourselves in the process, and worse still, destroying the jewel of an Earth system that has developed over the past four billion years. With it would go not just everything we cherish now, but the potential for things far grander than we can even imagine. This literally would be the end of the world, and not just as we know it. This is an improbable scenario. It is hopefully just hubris to think we have the power to sterilize the Earth, but we cannot completely rule it out as impossible. Risk is likelihood multiplied by impact. So, although the likelihood may be tiny, because the impact is essentially infinite, it is a risk we must consider.

The ultimate danger, slim though it may be, comes from the runaway greenhouse effect, in which the Earth's atmospheric blanket becomes so thick with water vapour and other heat trapping gases that the planet cannot shed heat at the rate it is coming in from the Sun. Then the oceans boil, the atmosphere becomes a steam pressure cooker, and ultimately water is lost from the planet, taking it to a Venus-like state. Existing assessments generally put the threshold for runaway far from the present climate state. Yet, remarkably, its precise location is not well tied down (see Box 21.1). In particular, the degree to which human increases in greenhouse gases can bring the Earth closer to the runaway threshold has not been thoroughly assessed. All the same, though our current climate-altering behaviour is irresponsible, most climatologists think it very unlikely that the Earth is already so close to runaway that we could bring on this catastrophe.

Setting aside full runaway to a Venus-like state, the next-worst-case scenario for us and the planet would be if we could flip it into a much

---

Box 21.1 How close is the Earth to a runaway greenhouse?

The runaway greenhouse delineates the inner edge of the 'habitable zone', discussed in Chapters 5 and 7. As the Sun slowly heats up and the habitable zone moves steadily outwards, the runaway state comes closer and may ultimately be the fate of the Earth. If it turned out that the planet was already close to this limit, is it conceivable that the actions of humanity in rapidly increasing greenhouse gas concentrations could push it over the edge?

So extreme are the consequences for life on Earth that you might think that the runaway greenhouse would have been extensively researched and we would know exactly the circumstances under which the Earth would suffer it, including whether we are likely to precipitate such a fate. You'd be wrong. The runaway greenhouse was introduced in an elegant but very idealized theory, more than 40 years ago (1). The paper showed that there is a critical value of absorbed solar radiation above which a planet with a water ocean experiences the runaway. Subsequent work (2-4) suggests this critical flux may be around 20% higher than the amount presently absorbed by the Earth. However, the precise value of the critical flux is difficult to define, because in the approach to runaway, an atmosphere with a great deal of water vapour will have radiative properties that are very different from the present climate. The heat trapping properties of water vapour at high concentrations are not very accurately known, and neither are the properties of clouds in such a steamy atmosphere. So while we know there is a cliff edge out there, that the Earth is moving very slowly towards, we aren't sure just how far away it is.

It might seem logical that, since the Earth has so far survived 4.5 billion years of turbulent history, it is unlikely that mere humans will bring it to an end tomorrow. We should be wary of this argument however, since as we've seen (Chapter 5), we expect the remaining lifetime of the Earth to be much shorter than the period it has lasted up to now. Furthermore, whereas it *had* to have a long history for us to exist at all, nothing suggests it must have a similarly long future. It is just possible that the Earth is closer to runaway than we think. If so, the climate will be very sensitive to any perturbation, and we will find that our actions in forcing up atmospheric CO_2 will cause a much larger rise in temperature than most models predict. Over the next century we are going to find out what the 'climate sensitivity' to increased CO_2 is, by direct experiment. In the unlikely event that doubling CO_2 increases temperatures by much more than the 3°C that is expected, we should be very alarmed.

higher carbon dioxide, hotter steady state that was uninhabitable to pretty well all complex multi-cellular life (but still amenable to many prokaryotes and simple eukaryotes). Here we are talking about something in the region of 50°C average temperature with orders of magnitude more carbon dioxide in the atmosphere. Such a state might be indefinitely stable if it eliminated all plants, fungi and lichens that accelerate silicate weathering on today's world. Then they would be unable to bring the carbon dioxide and temperature down again. Instead it would be the heat and high CO_2 that maintained high weathering rates and kept the carbon cycle in balance. There is a model (discussed at the end of Chapter 7) in which this alternative stable state exists under the present day Sun (5). The good news is that to push the model system into this alternative steady state would require the addition of more than 50 000 billion tonnes of carbon to the atmosphere. This is at least ten times more carbon than estimated in available fossil fuel reserves, and more than all the carbon stored in the oceans. Additional positive feedbacks in the Earth system, missing from the model, might bring the threshold closer. But again, it looks unlikely that we could trigger such a fate.

So let us consider a less extreme but still apocalyptic vision, for humans and many other life forms we cherish, if not for the Earth as a whole system. This is the all-too-familiar, depressing hellhole of dangerous climate change, destroyed ecosystems, war, famine and pestilence that we may be heading for if 'burn baby burn' continues. We are losing count of the number of recent popular science books and articles devoted to this vision. Even the normally cheerful and creative Jim Lovelock argues that we are already doomed, and nothing we can do now will stop the Earth system being carried by its own internal dynamics into a different and inhospitable state for us. If so, all we can do is try to adapt. We disagree—in our view the game is not yet up. As far as we can see no one has yet made a convincing scientific case that we are close to a *global* tipping point for 'runaway' climate change (see Box 21.2). We have just described two extreme examples of such a threshold but both of these, we think, are far away and likely inaccessible by current human activities. We cannot rule out that an as-yet-unidentified threshold for runaway climate change could be reached, perhaps even this century, but there is no evidence that we have reached one yet.

Yet even without truly 'runaway' change, the combination of unmitigated fossil fuel burning and positive feedbacks from within the Earth system could still produce an apocalyptic climate for humanity. We could raise global temperature by up to 6°C this century, with more to come next century. On the way there, many parts of the Earth system could pass their own thresholds and undergo profound changes in state. These are what Tim and colleagues have christened the 'tipping elements' in the climate system (6). They warrant a book in themselves, so

Box 21.2 What 'runaway' climate change actually means

In the popular media and the blogosphere there is much loose talk about 'runaway' climate change. A supposed example is that methane out-gassing from the unfreezing Arctic permafrost will lead to runaway. We use it to highlight a more general mistake. The argument starts on safe ground noting that warming is triggering extra methane release, which it is, and that methane contributes to global warming, which it does. The extra warming will in turn trigger a little bit more methane release—a positive feedback that will amplify the initial change. But then a fatal error of reasoning creeps in; it is assumed that this feedback can enter a runaway state of escalating change. What this requires is that any perturbation, such as a rise in temperature, propagated around the feedback loop will give rise to an additional nudge—rise in temperature—that is larger than the initial one. Then that goes through the feedback system and produces a yet larger response, and so on. The result is that the system runs away into a new state, carried by its own dynamics. This is what happens in the true runaway greenhouse, as discussed earlier, and it also apparently occurred in the transition from glacial to interglacial climate, though in that case the slow disintegration of the ice sheets meant that it took thousands of years (see Chapter 18). However, in most scenarios, including the methane release one above, the feedback is not powerful enough to produce runaway. In this case, any perturbation propagated around the feedback loop gives rise to an additional nudge that is smaller than the initial one. Further iterations give rise to yet smaller changes. The crucial point is that the system converges—it comes to rest—in a warmer version of the same state. In the case of methane release from the Arctic tundra the best scientific estimates of the strength of the feedback are that it is two orders of magnitude weaker than the runaway condition. To put this in temperature terms it can turn what would have been 1°C of global warming into 1.01°C of global warming. In our view it is dangerously misleading to talk wrongly of 'runaway' change because it is likely to produce a psychological response of despair and nihilism, when in fact the climate is not out of control and there is still a great deal to play for.

we will just touch on them briefly here. The tipping elements include the great ice sheets covering Greenland and West Antarctica that are already losing mass and adding to sea level rise. In the tropics, there are already changes in atmospheric circulation, and in the pattern of El Niño events. The Amazon rainforest suffered severe drought in 2005 and might in future face a climate drying-triggered dieback, destroying biodiversity and adding carbon to the atmosphere. Over India, an atmospheric brown cloud of pollution is already disrupting the summer monsoon, threatening food security. The monsoon in West Africa could be seriously disrupted as the neighbouring ocean warms up. The boreal forests that cloak the northern high latitudes are threatened by warming, forest fires and insect infestation. The list goes on. The key point is that the Earth's climate, being a complex feedback system, is unlikely to respond in an entirely smooth and proportional way to significant changes in its energy balance being caused by human activities. Tipping points are among the most difficult responses to capture in climate models, which tend to show smooth responses and not sudden changes in state, so model predictions are of limited help here.

The total amount of fossil fuel carbon we emit to the atmosphere will determine the long-term extent of climate change and ocean acidification (unless humans start actively removing CO_2 from the atmosphere). Emissions to date mean we have already made a commitment to 330 ppm of CO_2 and a roughly 0.75°C warmer world for the coming millennia. The 'airborne fraction' of added CO_2 that remains for the long-term actually increases the more we add, because the capacity of the ocean to mop it up decreases (7). So, taking a popular estimate that there are around 4000 billion tonnes of carbon in known fossil fuel resources, if humans eventually emit all of this to the atmosphere then the long-term commitment will be around 1000 ppm of CO_2 and about 6°C of global warming. This is more than the temperature change between the last glacial maximum and the present climate. The closest analogue from the past is the Paleocene-Eocene thermal maximum (PETM) event 55 million years ago. Then there was a comparable release of carbon to the atmosphere, and the resulting spike in global temperatures lasted at least 100 000 years. The sea level rise that would accompany a human caused release of carbon of this magnitude is likely to eventually reach several tens of metres. The next ice age would be prevented, and probably several more after that. Possibly the Earth could be snapped out of ice age cycling altogether.

Although most current discussions focus on carbon dioxide pollution and climate change, it is possible that one of our other waste products will turn out to be more dangerous, and that there could be genuine 'runaway' change elsewhere in the Earth system. For example, we have documented how the ocean has been prone to switching to an anoxic and sometimes sulphidic state repeatedly in the past. There is clear evidence that today, the low-oxygen zones of the oceans are expanding, as a result of warming, stratification and nutrient inputs (8). It should be carefully considered whether escalating human fertilizer inputs, especially of phosphorus, could trigger runaway positive feedbacks that carry the ocean into an anoxic state. Tim has examined this in a simple model that can reproduce oceanic anoxic events of the past (9). Already phosphorus input to the ocean has been increased nearly ten-fold by human activities. If this is sustained for a thousand years then anoxic regions double in extent and feedbacks within the ocean start to kick-in and carry the system towards wholesale anoxia. The saving grace for the ocean (though potentially disastrous for us), may be that phosphorus-rich rock reserves will run out before an anoxic event can be triggered. The latest estimate is that it will take the order of 100 years to deplete the known reservoir of phosphate rocks on which our food production depends (*10*). However, the warmer future climate will in turn support higher natural weathering rates of phosphorus that may be enough to eventually trigger an anoxic event.

Apocalyptic visions, in the modern sense of 'destruction of the Earth as we know it' are, it seems, strangely seductive. But they are becoming so familiar now that they defy the original meaning of the word apocalypse—which was a 'lifting of the veil' or 'revelation'. There seems little revelation left in yet another pronouncement of climate change doom and gloom. And one has to wonder what the motivations are for some of the heralds of the apocalypse. It would appear, in many cases that they are trying to motivate a change of action away from business-as-usual. That is a laudable aim, but research shows that fear is not a constructive motivator (*11*). Apocalyptic visions can have quite the opposite effect on behaviour, breeding depression and inaction, hedonism or nihilism. It is not enough to know what we shouldn't be doing, if we are going to change, we need a constructive vision to turn towards. But which vision?

21.2 Retreat?

A popular answer to apocalyptic visions of the future is *retreat*, into a lower energy, lower material consumption, and ultimately lower population world. In this future world the objective is to minimize human effects on the Earth system and allow Gaia to reassert herself, with more room for natural ecosystems and minimal intervention in global cycles. The noble aim is long-term sustainability for people as well as the planet. There are some good and useful things we can take from such visions of the future, especially in helping to wean ourselves off fossil fuels, achieve greater energy efficiency, promote recycling, and redefine what we mean by quality of life. However, we think that visions of retreat are hopelessly at odds with current trends, and with the very nature of what drives revolutionary changes of the Earth. They lack pragmatism and ultimately they lack ambition. Moreover, a retreat sufficient to forestall the problems outlined above might be just as bad as the problems it sought to avoid.

Many environmentalists argue that we should all adopt low energy and low material consumption lifestyles. Of course the majority of people on the planet still have such a lifestyle. But very few of us enjoying (or acquiring) high-consumption lifestyles have thus far been willing voluntarily to make the switch to low consumption. The trend is currently heading strongly in the opposite direction of increasing per capita consumption. We cannot rule out a collective tipping point in human behaviour, in rich nations, toward much lower energy and material consumption. Indeed we might wish for one. But many commentators seem to think it will take a large-scale disaster (the Apocalypse scenario) to trigger it. Furthermore, it is the product of population and per capita consumption that determines total effect on the planet.

Around nine billion of us are projected to be on the planet in the middle of this century, and at least 10 billion by the end of the century, which will make our collective impact significant whatever our average consumption level. Defenders of the vision of retreat will sometimes counter that high population numbers simply cannot be reached because a catastrophic cull will befall us on the way there (again, the Apocalypse scenario), or that they should not be reached because the use of land they imply will be too much for the Earth and its natural ecosystems to bear. Some advocate long-term targets of less than a billion people on the planet. But the same individuals rarely map out how we get there. Scratch the surface of many environmentalists and you will find a benign dictator underneath, who will equitably cap per capita consumption at the same level for everyone, and then presumably set limits on numbers of offspring. The problem with such visions has always been enforcing them without violating fundamental human rights.

The argument that there is not enough land to sustain 10 billion people within a healthy Earth system actually depends strongly on the nature of our future food production and distribution systems. A popular vision for 'sustainable' agriculture is that food production should be localized and food miles minimized, for the sake of reducing CO_2 emissions. In fact, if everyone in the world tries to satisfy their food needs from nearby land, and to eat what is considered an adequate diet in the developed world, this would put huge pressure on heavily populated and relatively unproductive regions such as India. To some degree this is what has already happened globally. Some of the most productive land for agriculture is either underused or has cities rapidly expanding over it, whilst lots of unproductive land is overused. The result is that the overall area needed to feed the world population is greatly increased, leaving less room for natural ecosystems. Modelling suggests that feeding 12 billion through localized food production (on a 1995 diet) would demand around twice the area currently used by agriculture (over 30 million km^2) (*12*). This certainly would be a disaster for natural ecosystems.

There are useful elements that we should adopt from visions of retreat. It is abundantly clear that rising CO_2 concentrations are already causing both undesirable climate change and ocean acidification. These are both (on their own) compelling reasons to try and stop CO_2 rising, stabilize its concentration, and then potentially bring it down, to whatever we can collectively decide is a safe level for ecosystems and human systems. In visions of retreat, CO_2 is stabilized by reducing emissions. Whilst the current talk is about targets for mid-century of 50% or greater cuts in CO_2 emissions, this alone will not be enough to stabilize CO_2 concentration. The only way to do that is for sinks to match sources. So in the retreat scenario, emissions have to continue to decline towards zero: even then, atmospheric CO_2 concentrations will remain above pre-industrial levels for thousands of years. If the only policy lever is reducing emissions, it actually makes more sense to be negotiating the total amount of fossil carbon that we can collectively emit to the atmosphere from now on (and this total allowance can then be translated into an annual global cap on CO_2 emissions that is reduced year on year). Current estimates put the future emissions allowance, in order to limit global warming to no more than 2°C, at less than a trillion tonnes of fossil fuel carbon (*13*). This is significantly less than the fossil fuels known to exist, implying that a large amount of fossil fuel must be left in the ground, or, if this fuel is combusted, that the resulting CO_2 must be captured (either at the point of emission or from the free atmosphere).

The future under retreat, whether it came about through the dawning of a new Aquarian age of global ecological awareness and mutual restraint, or was imposed by a global dictatorship, would be sustainable. We could ensure our survival on a planet that remained indefinitely habitable for ourselves and the other multi-cellular life forms on Earth. With the human population reducing considerably from present day levels, and those that remain treading lightly on the planet's life support systems, our future would be assured. Biogeochemically, it would be sound, but not spectacular. Looking back from a future of retreat, the last two centuries and the current one would appear as a striking 'blip' in the rock record, but not as the start of a further revolution in the Earth system. The Earth would drift gradually but inexorably into another ice age, around fifty thousand years hence, and the Quaternary climate cycles would resume.

From a human point of view however, this future doesn't seem very convincing or satisfactory. Firstly there is the practical consideration that it will be nearly impossible to achieve by consent. The prospects of global retreat happening by mutual, peaceful, agreement seem extremely remote at present, when we are heading as fast as we can in the exact opposite direction. Though we may voluntarily cut our energy use so long as it does not affect our standard of living, reducing it to the degree necessary to balance the global carbon budget will almost certainly involve some real sacrifice, and is unlikely to be a vote winner either in the developed or the developing world. Retreat will involve a species-wide limitation of personal freedoms that those of us in the developed world currently enjoy and to which many more aspire. (No, you *can't* travel long distance, put more coal on the fire, eat meat…) We will need to limit our ambitions both as individuals and as a species. Yet it is appealing to think that we have not yet exhausted our full potential as a species; that our art, architecture, medicine and science will continue to advance, that we will build better cities and infrastructure, cure more diseases, solve the secrets of the cosmos with our telescopes and space missions, and discover the fundamental nature of matter with our particle accelerators. These can all at times be very energy-intensive activities, and it's hard to see how one could justify pursuing them actively if the population as a whole has to save every watt of power. Humanity in retreat may last forever, but it may also be a boring species compared to us, the *Homo sapiens industrialis* from which it evolved.

21.3 Revolution?

Our alternative vision of the future is of a *revolution*, into a high energy, high recycling world that can support billions of people as part of a thriving and sustainable biosphere. The key to reaching this vision of the future is to learn the lessons from past revolutions: future civilizations must be fuelled from sustainable energy sources, and they must undertake a greatly enhanced recycling of materials. Two ideas are central to this vision. Firstly, there is an awful lot of energy potentially available for human use and there is nothing wrong or immoral about increasing human energy use per se. Secondly, the Earth's surface and near-surface has only a finite amount of materials on it and the input of new materials is, on human time scales, infinitely slow, so we must recycle all the materials we want to use with nearly 100% efficiency, much more than we currently do.

Humans are in an analogous position to those key organisms that went before us and sparked earlier revolutions. Consider for example the cyanobacterium that first invented oxygenic photosynthesis. Today, we can see that we owe virtually everything that makes the Earth habitable to that invention, and no one could argue that what the cyanobacterium did was not a good thing. At the time however, it resulted in an upheaval on the Earth lasting hundreds of millions of years of atmospheric pollution and climate instability. At length, by trial and error, the Earth system evolved to recycle the products of photosynthesis and could regain a new stable state. We however have foresight. We don't have to wait for millions of years of trial and error. We can and should intervene and design our own recycling loops to close the biogeochemical cycles that we are already disrupting with our increased energy use. Otherwise we will quickly run out of raw materials, and will end up suffering from the planetary changes wrought by our accumulating waste products.

Before we get into the specifics, let us illustrate our case with a thought experiment: Imagine a contraption that converts whatever material waste products you put into it into new products that you desire. To do this without breaking the laws of thermodynamics, the contraption needs a source of energy. And it is not an alchemist (or a nuclear reactor); the material that comes out has to have the same masses and proportions of elements as what goes in. But beyond that, if there is no limit on energy supply, there is no theoretical limit to its recycling capability. The amount of stuff being recycled can far exceed the inputs of fresh material. This is of direct relevance to human society: what economists call gross domestic product—the things we make, use, buy and sell—is limited by the energy supply, but not by the materials supply, provided that recycling is highly efficient. What this thought experiment implies is that, in principle, as long as humans have an abundant source of energy, we ought to be able to increase our standard of living, and avoid the problems of accumulating waste products, by increasing the efficiency of recycling.

You might have pictured this recycling contraption in quite a metallic and industrial form, but the closest current example is not a human construction at all; it is the Earth system as a whole—whose development we have been charting throughout this book. Our argument is that human civilizations should emulate the biosphere, and design and construct what we will call 'Gaia devices' that recycle our waste products far more efficiently than we are currently doing. Of course there will be many incarnations of such devices and some of them will be more organic than synthetic—for example, highly recycling agricultural ecosystems. But they will all carry an element of human design. In the following, we will explore what some of these Gaia devices might look like. But first we must address the sources of energy that will drive future societies and their Gaia devices.

21.3.1 Sustainable energy

The current energy 'problem' is not a fundamental one of lack of energy. There is plenty of energy available on the planet. Instead it is a combination of problems surrounding how we are sourcing energy, how inefficient we are being with it, and how we deal (or rather don't deal) with the waste products. Our discussion here draws liberally on an excellent recent treatment of the problem, David MacKay's 'Sustainable energy—without the hot air' (*14*).

In our vision of revolution, human energy demand will continue to increase long after population stabilizes. We will need a lot of energy therefore, into the indefinite future. But this need not conflict with sustainability, provided the energy is produced cleanly and the waste products are contained. On the contrary, given that really efficient recycling is energy-intensive, higher energy use may be needed if a large population is to be sustained indefinitely. It should be abundantly clear however that we need to wean ourselves off fossil fuels, because of the CO_2 pollution they cause, and because even if we don't they will run out soon enough anyway, so cannot sustain a high energy future. The reason we are so hooked on fossil fuels is that (at the moment) they are inexpensive compared to other sources, and they have a relatively high energy density—a lot of energy is contained per unit volume of fossil fuel. Sources with a high energy density are necessary for a high-energy future, because low energy-density sources, which include the majority of renewable energy sources such as wind, tide and biomass, take up impractically large amounts of space if used to generate large quantities of energy.

There is one renewable energy source presently available however that could meet high energy demand indefinitely: solar power. There is a second source which, while not renewable, will last a very long time, perhaps a thousand years at current utilization rates: nuclear fission. A third near-inexhaustible source, which may in the future become viable, is nuclear fusion. A future human-driven Earth system revolution will probably be powered by some combination of these.

Among truly renewable energy sources by far the greatest is the primary one—sunshine. The power in sunlight reaching the Earth's land surface is about 2.5×10^{16} W, making current total human power consumption of just over 1.5×10^{13} W look measly in comparison. Even when averaged out, sunlight has a relatively high power density of around 170 Wm^{-2} at the Earth's surface. But only a modest fraction of the available energy can be captured in a useful form, e.g., 20% for expensive solar photovoltaic (PV) panels converting sunlight to electricity. This implies large areas would need to be devoted to solar power to meet even current energy demand; assuming a 10% conversion efficiency (from cheap but inefficient PV panels), about 0.4% of the Earth's land surface. The figure is similar for concentrated solar power in deserts. Given that suitable deserts (without drifting sand dunes) cover about 2% of the Earth's land surface, it is quite possible that a future globalized society could source the majority of its energy from desert solar power. Thus, we can envisage a relatively high energy solar powered future, though probably not using ten or a hundred times more energy than is used by current societies.

Nuclear fission (splitting heavy nuclei into medium-sized nuclei) is already responsible for about 6% of human primary energy production (and 16% of global electricity production), but its potential is considerably greater. It is not sustainable indefinitely because the amount of fissionable material on the Earth is finite, but this energy source could be significantly larger than fossil fuels. The amount of energy available depends crucially on two factors; the size of the fuel reserves (uranium or thorium) and whether 'fast-breeder' (for uranium) or 'energy amplifier' (for thorium) reactors are used. Considerable uncertainty surrounds the amounts of uranium and thorium that might be extractable from the Earth's crust. However there are 4.5 billion tons of uranium in the ocean and a tested method for extracting it. Current 'once-through' reactors use the minority isotope of uranium, ^{235}U, and discard the majority, ^{238}U, limiting the long term energy yield. However, fast-breeder reactors which convert ^{238}U to fissionable plutonium-239 (^{239}Pu) yield 60 times the energy per unit of uranium. If uranium from seawater is used as the fuel for fast-breeder reactors then it could supply everyone on the planet today with about twice the energy consumed by the average Brit for 1000 years (*14*). Whether you approve or disapprove of nuclear energy (for reasons we are not discussing here), the point is that there is an awful lot of it. There is of course a problem of disposing safely of the waste products. But fissionable fuels have an energy density that is millions of times higher than fossil fuels, so by comparison their waste disposal problem is 'beautifully small' (*14*). It may not appear that way today, but that is because we make no effort to dispose of CO_2 in a responsible fashion, but simply dump it into the atmosphere, leaving the consequences for future generations to deal with.

Nuclear fusion is ultimately the source of almost all the energy available on the Earth—it is the process which lights the Sun, and fusion in stars that formed supernovae before the solar system formed produced almost all the elements from which the Earth is made. However, producing useful power from a fusion reactor on Earth is still far from becoming a reality. Infamously this is a 'solution' that has steadfastly remained at least thirty years in the future. Assuming that one day the many problems can be cracked, the energy source would be enormous. For example, for the type of fusion being used in experimental reactors (deuterium fusion with tritium) the ultimate limiting factor would be the supply of the element lithium from which the tritium is made. Current global human energy demand could be met for about a million years providing that lithium could be extracted from sea water. Deuterium fusion with deuterium could produce many orders of magnitude more energy still. For all practical purposes, the source of energy would be infinite.

We have concentrated on energy sources with a high energy density. Renewable sources other than solar have significant but ultimately limited potential. The problems are a combination of low power density and limited areas available. Hydroelectricity can achieve 10 Wm^{-2} but perhaps half of global capacity is already realized, yielding about 6% of primary energy production (and 15% of global electricity generation). Tidal power can capture 3–6 Wm^{-2} and wind power 2–3 Wm^{-2} demanding areas of 167–500 km^2 to produce the equivalent power output of a typical fossil fuel or nuclear power station (1 GW). Thus, we cannot envisage a high energy future powered entirely from non-solar renewable sources.

Some have argued that the production of biomass energy (which is essentially solar power converted into chemical form) should be ramped up such that it displaces fossil fuel energy. But it does not make sense for a high energy future society to be based predominantly on biomass energy because the power density that can be achieved is rather low. Plants are limited in how efficiently they can convert sunlight into chemical energy with the best energy crops currently achieving about 0.5 Wm^{-2} (only about 3% of what can be captured by solar PV or solar thermal in deserts). This in turn demands a massive area be devoted to biomass energy, roughly 12% of the Earth's land surface if it were to meet current global energy demand. This would be sure to drive conflicts with other land uses. Instead the priorities for land use should be maximizing the efficiency of food production and thus maximizing the area left to natural ecosystems.

21.3.2 Feeding the world efficiently

In our vision of revolution we accept that human population will grow substantially before stabilizing. But land area and coastal seas are finite. A fundamental challenge then is how to feed the still growing human population without wiping out yet more natural ecosystems. These ecosystems are essential to the healthy functioning of the biosphere and to the maintenance of biodiversity. So if they must be replaced with 'agro-ecosystems', the challenge becomes to ensure that key biosphere functions are maintained—in short, to make them Gaia devices.

How much Earth would we need to sustain a potential doubling of population, a probable upper limit for 2100? Since by some analyses 40% of global photosynthesis is already 'used, co-opted or foregone' by humans (*15*), the intuitive answer is that we would need to clear for agriculture most of the remaining natural ecosystems. But the real answer depends strongly on how localized or globalized food production is (as well as a host of other factors, including diet). A localized food production system (as in the Retreat vision) aims to satisfy demand within a region by production in that region, which often leads to extensive use of unproductive land. In contrast, a truly globalized system would use only the most productive agricultural land for the most important crops (and so on) regardless of where the food was destined to be consumed. Current agricultural production is partly globalized but still predominantly localized—many nations aim for self-sufficiency. Remarkably, some modelling (*12*) suggests that in a fully globalized system of food production, 12 billion people (eating a 1995 diet) could be fed on as little as a third of the currently used agricultural land! Even with doubled consumption of animal products, less than half of this land would suffice.

This globalized food production scenario is a thought experiment; it ignores a host of trade barriers, subsidy regimes and political constraints that exist at present. However, it suggests that, in principle, the Earth could support a much greater human population without expanding current agricultural land, and we might actually be able to give land back to the rest of the biosphere. The scenario implies high energy inputs especially in the transport of food products, mechanization of agriculture and the production of fertilizers. But energy is not the problem. Current energy inputs to agriculture are only around 3% of total human primary energy consumption, and food production should surely be a priority use for energy. What is a problem are the high inputs of fertilizer nutrients implied and the waste products that would follow.

To make future agricultural systems into Gaia devices we need to shift from piling on fertilizer that is inefficiently utilized and leads to harmful waste products, toward high recycling and high productivity agricultural ecosystems. The efficiency of nutrient and water use by crop plants needs to be maximized, within carefully designed crop rotation systems. Nutrients removed with the crop need to be returned to the land as animal dung and appropriately treated human sewage. The conversion of farm waste to charcoal ('biochar'), which is returned to the soil, can act as a sink for carbon and improve nutrient and water retention, thus reducing required fertilizer loading. Most importantly, for long-term sustainability, the phosphorus cycle needs to be much more efficiently closed, with phosphate extracted from wastewater and recycled for agricultural use. Phosphorus recycling is particularly critical because known rock phosphate reserves are finite, and some recent estimates suggest we may soon pass 'peak phosphate' (*10*). Although nitrogen can be fixed from the atmosphere indefinitely, the generation of waste products including the potent greenhouse gas nitrous oxide, need to be reduced, by reducing fertilizer loading. Methane production associated with agriculture should, where possible, be used as an energy source, for example in anaerobic digesters. The science required to achieve this vision is underway.

21.3.3 Recycling the manufactured environment

Human technology and construction also require massive fluxes of 'nutrients' in the form of metals, which we mine out of the Earth's crust. The global cycles of the most important 'technological nutrients' have only recently been quantified (*16*). By far the largest mass flux is of iron, followed by aluminium, and together they represent more than 95% by mass of all metals mined. Copper and zinc are next, followed by chromium, lead and nickel. In our vision of revolution, human use of these technological nutrients will shift increasingly towards recycling. The Gaia devices in this case will literally be Gaia machines.

Currently, global use of metals is increasing rapidly, especially with industrialization in China. Stockpiles of metals, in the form of manufactured goods (especially in construction, transportation and manufacturing industries) are accumulating. This can only be supported by a net input of metals from the Earth's crust. However, there is evidence that in already industrialized nations, stockpiles of metal in use can stabilize. This means that as global population stabilizes and global industrialization completes, a shift towards closed metal recycling systems will become feasible and (in many cases) energetically favourable. Cities (and their scrap heaps) will become 'the mines of the future' (see Box 21.3) (*17*).

Globally, there is already significant recycling, especially for iron and aluminium. We calculate that current global recycling ratios are 1.58 for iron, 1.58 for aluminium, 1.14 for copper, and 1.06 for zinc, based on the global flux estimates of Rauch and Pacyna (*16*). Here the recycling ratio is the ratio of total material entering production to material recycled from fabrication or discard (scrap). A ratio of 1 means no recycling, 1.58 means production is 58% greater than could be supported by fresh input from mineral resources alone. However, if our

---

Box 21.3 Metal recycling

To illustrate what might be possible in the future and how much work needs still to be done to better recycle our manufactured goods, we borrow from the work of the Center for Industrial Ecology at Yale, and their historical example of iron use in the US (*18,19*). Per capita use stabilized at around 11–12 metric tons around 1980. This equates to about 3.2 billion metric tons (or petagrams, Pg) for the entire present US population. This is similar to the amount remaining in national mineral reserves (4.6 Pg), and the material in use is generally of higher quality than mineral ores. US iron ore grades have decreased in the last 60 years from 50–60% to 25–30% demanding significantly increased water and energy use and iron waste (tailings) production. Currently crude ore extraction and recycling of scrap are of similar magnitude in the US, but with a perfect recycling system for scrap, and current imports, the US could stop domestic mining of iron. This would save energy as steel recycling eliminates the most energy intensive step of converting iron ore to pig iron in blast furnaces, reducing energy consumption by 70–75%.

Global stocks of iron in use (21 Pg) are currently about a tenth of the mineral reserve base (180–230 Pg) but if a future population of 10 billion adopted the use level of current US citizens, iron stocks would rise to 120 Pg and 'urban mines' would make increasing economic sense. Total closure of an industrial iron cycle can never be completely achieved (there will always be some losses and need for ore extraction) but increased recycling will greatly reduce the production of waste products. Although iron is abundant and viewed as environmentally harmless, as the dominant metal cycle it controls the cycles of many other metals (e.g. manganese, nickel, chromium, cobalt, molybdenum, niobium, and vanadium), including toxic ones. By shifting towards recycling for them as well, harmful waste products will be reduced.

metals-based society is to continue into the indefinite future, fresh mineral resources will eventually decline towards zero even for materials as abundant as iron and aluminium, meaning that these ratios will eventually need to increase to values of 10 or 100.

21.3.4 Dealing with carbon dioxide

If our aim is an indefinitely sustainable society, we must have the ambition of recycling the great majority of the material we use. However, so long as we are powered by fossil fuel energy, the first law of thermodynamics will not allow us to recycle the carbon we burn, because the energy needed to return the CO_2 to hydrocarbon would be as great as was released by burning it in the first place (or actually greater, because no energy conversion process is ever 100% efficient). Recycling carbon in the fossil fuel age is therefore impractical. We need to wean ourselves off fossil fuel as an energy source with as little delay as possible, but this is a difficult and slow process, and for the near future we will remain addicted to it.

Besides reducing our burning of fossil fuel, there are other actions we could be taking to slow the rise, and then perhaps begin to reduce, CO_2 concentrations in the atmosphere. We should for example be storing as much as possible of the CO_2 we create in locations away from the atmosphere, and we could also be creating new sinks that actively remove some of the CO_2 we have already released. Some of the advantages of a high-energy lifestyle, such as air travel, may be impossible without burning hydrocarbons and releasing the CO_2 into the air. Even in the long term future therefore, actively removing atmospheric CO_2 may be necessary if we are to become overall CO_2-neutral, but we still want to fly in jet aircraft. The simplest example of a Gaia device that could do this for us is of course a photosynthesizing plant, where if the plant is subsequently turned into biofuel and burned, carbon and oxygen are both recycled. While it is not a good use of agricultural land to replace a major proportion of our present fossil fuel use with biofuels, some limited use of biofuels to offset unavoidable use of hydrocarbon fuels in a sustainable future, makes more sense.

If we drive up atmospheric CO_2 due to fossil fuel burning in the first half of the present century, but have non-fossil fuel energy increasingly available later, it might be sensible to use some of that energy to remove CO_2 from the atmosphere. This would offset climate change and restore the CO_2 to a comfortable level—which might be the pre-industrial 280 ppm, but might also be rather higher if it was considered safe, given that plant productivity increases with CO_2 concentration. Carbon dioxide could be removed using 'air capture' devices that use a chemical scrubbing process to absorb it from the atmosphere (20, 21). These devices use energy, and are therefore costly to operate. The end product is liquid

CO_2 that then must be sequestered in geological formations, or perhaps put into the deep sea—the carbon is not recycled, and we end with a bulky waste product that must be stored long-term. The ultimate capacity of these methods to remove CO_2 will likely be set by their energy use (hence their cost) and by the availability and acceptability of such storage sites. The storage capacity is actually estimated to be quite large and comparable to known fossil fuel reserves. In a variant scenario, plants could be used to get the CO_2 out of the atmosphere for us and their biomass converted to a mixture of biochar (that is put back on soils) and liquid CO_2 (that needs storage), but because of the resulting demands on land-use, this approach can only ever make a limited contribution. So, we envision a short-term future where fossil fuel CO_2 emissions are significantly reduced—through carbon capture at the point of emission as well as switching to other energy sources—but they are not eliminated. Instead dispersed sources, especially from transport (including that of food) are counterbalanced by CO_2 capture and storage from the atmosphere.

21.3.5 Avoiding a tragedy of the global commons

The practical problem with this vision of the future is that it would cost a great deal of money to establish the indefinitely sustainable society that we have described. Up to now, we have deliberately not talked about what would make 'economic sense', for the very good reason that our current economic valuation system does not make Earth system sense. The problem lies not with the economy itself, but with what it fails to value—the goods and services provided by the Earth system—and what it fails to penalize—waste products that disrupt the Earth system. We are not against the notion of economic growth per se. The problem is that at present, growth is coupled to increased waste production.

Why is to so hard for us to collectively reduce our production of waste products? To answer this we need to recognize that we are all seeking to maximize our own well being (if not our reproductive success). The environment, on the other hand, is a resource that we all share. We all affect it a little bit but our individual contributions are tiny, even though our collective impact is profound. Anything we do to make a difference will have only a tiny impact and any benefit will be shared by everyone—including those not making an effort themselves. We have all heard or made this argument. In the UK of late it often takes the form: 'Why should I do anything about climate change if the Chinese don't do anything? And even if we do something, the UK is only 2% of global emissions...' and so on.

This a familiar problem that has been with life since its first stirrings; why should any organism act for the collective good? To an evolutionary biologist it is the problem of 'cheats'. In philosophy it was first expressed by Thucydides and later by Aristotle. Its most familiar modern incarnation was in an essay by Garrett Hardin entitled 'The Tragedy of the Commons' (22). Hardin introduces a hypothetical example of a common pasture shared by local herders. The herders wish to maximize their yield, so will increase their herd size whenever possible. For each additional animal added to the commons the herder receives all the proceeds. The pasture is also slightly degraded, but this negative impact is shared among all the herders. The division of costs and benefits is unequal and the rational course of action for each herder is to add an extra animal to the pasture, and another, and so on. The result is overgrazing and long-term degradation of the pasture*.

Nowadays we have a tragedy of the global commons—the atmosphere and oceans. The tragedy is that as yet we have no effective system for regulating their use, and penalizing activities that pollute them. So our waste products are just accumulating. There are ways out of the tragedy, but all of them hinge on recognizing that the global commons needs to be valued: some of our wealth needs to be directed at its preservation.

As we explored in Chapter 7, if it does not cost anything to improve the environment, because it is just a by-product of some other activities, then it can happen spontaneously (and positive feedback will kick in to amplify the effect). Alas, however, for most options there is currently at least a short-term cost barrier to making the shift. To lower this barrier requires some regulations imposed at the national and international level to make us behave more for the common good and less for individual gain. For the environmental commons, Hardin suggests a polluter pays principle as a way out of the tragedy. Equally, we think that active removal and recycling of existing pollution should be rewarded. In short, we need to alter the system of valuation so that economic growth encourages increased recycling, instead of increased waste, and this needs to be led by governments and international groupings such as the United Nations.

---

* Interestingly, in medieval England the common lands did not suffer tragic over-exploitation because they were *regulated*. Most historical commons were largely reserved for their own commoners (members of the local parish) who had rights of grazing that were inherited. Passing herders had few rights but could pay 'thistle rent' to lease grazing land. If overgrazing occurred, a limit would be placed on the number of animals each commoner was allowed to graze. Thus community management and oversight—deploying regulatory feedbacks where necessary—led to a largely sustainable use of the commons. Hardin sees the transition from unregulated access by all, to regulated access by some, as a first step towards privatization—which he argues is one solution (but not the only one) to the tragedy.

Currently, CO_2 and other greenhouse gases are the headline pollution problem. The UN system of international negotiation that produced the Kyoto protocol, and floundered in Copenhagen in December 2009, is an attempt to regulate greenhouse gas pollution entering the atmosphere. But it is clearly not up to the task. Out of the mishmash of compromise between different and often conflicting vested interests, emerge targets that are too weak, that are not truly global—when they need to be—and that lack a real mechanism for enforcing them anyway. The question is; what could work better?

Nearly everyone agrees that there needs to be a price on CO_2 pollution, whether a tax on emissions, sale of permits for emissions, or sale of permits for fossil fuel extraction (23). These proposals would generate income, which could then be used to further safeguard the assets to be protected—in this case the global atmosphere, oceans and climate. One stimulating suggestion is to establish an Earth Atmospheric Trust (24). The key idea behind this is to deposit the revenues into a trust fund, managed by trustees appointed with long terms and with a mandate to protect the asset. A fraction of the revenues would be returned to everyone on Earth on a per capita basis. This amount will be insignificant to the rich, and much smaller than their per capita contribution to the fund, but for the world's poor it could be enough to lift them out of poverty. The remainder of the revenues would be used to enhance and restore the asset. They could be used to fund research, development and deployment of sustainable energy, payments for carbon captured by Gaia devices, and so forth.

Even if we solve the CO_2 problem, there will surely be other waste products from our expanding societies that will pollute the global commons and need to be dealt with in the future. So an effective regulatory system needs to be broadened out to encompass other pollutants, and other parts of the global commons, especially the ocean. Although much of the land surface is currently in private ownership it might make more Earth system sense for some of it, including the best agricultural land to be treated as part of the global commons. That would be consistent with a truly globalized food production system.

What we are fumbling towards is the need for an effective system of global governance that could help steer the Earth and us through revolution to a successful and sustainable outcome. Such top-down governance, should it emerge, needs to somehow complement and value the bottom-up processes that drive creative change. All the great biological innovations that have revolutionized the Earth up to now started with one individual or with a new type of group. We have every reason to think that potential is still present now, in the thick of revolutionary change. We must dispel the myth that individual and local actions are worthless. They can help change the world—for better or for worse—and in a deep sense they are the only thing that ever has.

21.4 Conclusion

In its long history, the Earth system has undergone several cataclysmic upheavals, sparked by blind biological evolution and innovation, coupled to the geochemical cycles that control the surface environment of the planet. These revolutions resulted in potentially planet-sterilizing climatic disturbances and pollution events. Had there been a committee of organisms overseeing the planet, they would surely have vetoed the proposals of the first cyanobacterium and later, the first land colonizers. Probably too, they would have eliminated early humans when they first began to develop symbolic language, and before they started to get ideas above their station.

There was no such oversight committee, and the planet is in the midst of yet another upheaval. This time we are the irresponsible new organisms, equivalent to the cyanobacteria, about to change the planet forever. The combination of our inventive minds and opposable thumbs has resulted in the awakening of a technically adept but dangerously naïve species. But in contrast to previous Earth system revolutions, we have foresight—we know what harm we may do and how we can prevent it. Rather than retreat in a last ditch effort to minimize our impacts, our best hope now is to embrace revolution and intervene more in the Earth system, to close the recycling loops we have opened, and restore stability.

From within the Earth system has emerged the potential for conscious thought that could help shape the future state of the Earth and steer a course to a stable and sustainable future. However, that consciousness remains embodied within a single species: a newly evolved ape with all its animal emotions and instincts still intact. Are we as yet sufficiently grown up to take responsibility for a whole planet? Over the next century we will find out.

References

1. A. P. Ingersoll, *The Runaway Greenhouse: a history of water on Venus. Journal of the Atmospheric Sciences* **26**, 1191 (1969).
2. J. F. Kasting, *Runaway and moist greenhouse atmospheres and the evolution of Earth and Venus. Icarus* **74**, 472 (1988).
3. S. Nakajima, Y. Y. Hayashi, Y. Abe, *A study on the runaway greenhouse-effect with a one-dimensional radiative convective equilibrium-model. Journal of the Atmospheric Sciences* **49**, 2256 (1992).
4. T. Pujol, G. R. North, *Runaway greenhouse effect in a semigray radiative-convective model. Journal of the Atmospheric Sciences* **59**, 2801 (2002).
5. T. M. Lenton, W. von Bloh, *Biotic feedback extends the life span of the biosphere. Geophysical Research Letters* **28**, 1715 (2001).

6. T. M. Lenton et al., *Tipping Elements in the Earth's Climate System. Proceedings of the National Academy of Sciences USA* **105**, 1786 (2008).

7. T. M. Lenton, *Climate Change to the end of the Millennium. Climatic Change* **76**, 7 (2006).

8. R. F. Keeling, A. Kortzinger, N. Gruber, *Ocean Deoxygenation in a Warming World. Annual Review of Marine Science* **2**, 199 (2010).

9. I. C. Handoh, T. M. Lenton, *Periodic mid-Cretaceous Oceanic Anoxic Events linked by oscillations of the phosphorus and oxygen biogeochemical cycles. Global Biogeochemical Cycles* **17**, 1092 (2003).

10. D. Cordell, J.-O. Drangert, S. White, *The story of phosphorus: Global food security and food for thought. Global Environmental Change* **19**, 292 (2009).

11. S. O'Neill, S. Nicholson-Cole, *'Fear Won't Do It' Promoting Positive Engagement With Climate Change Through Visual and Iconic Representations. Science Communication* **30**, 355 (2009).

12. C. Muller, A. Bondeau, H. Lotze-Campen, W. Cramer, W. Lucht, *Comparative impact of climatic and nonclimatic factors on global terrestrial carbon and water cycles. Global Biogeochemical Cycles* **20**, GB4015 (2006).

13. M. R. Allen et al., *Warming caused by cumulative carbon emissions towards the trillionth tonne. Nature* **458**, 1163 (2009).

14. D. J. C. MacKay, *Sustainable Energy—Without the Hot Air.* (2009).

15. P. M. Vitousek, P. R. Ehrlich, A. H. Ehrlich, P. A. Matson, *Human appropriation of the products of photosynthesis. Biosciences* **36**, 368 (1986).

16. J. N. Rauch, J. M. Pacyna, *Earth's global Ag, Al, Cr, Cu, Fe, Ni, Pb, and Zn cycles. Global Biogeochemical Cycles* **23**, GB2001 (2009).

17. J. Jacobs. *The Economies of Cities* (Random House, New York, 1969).

18. D. B. Muller, T. Wang, B. Duval, T. E. Graedel, *Exploring the engine of anthropogenic iron cycles. Proceedings of the National Academy of Sciences USA* **103**, 16111 (2006).

19. T. Wang, D. B. Muller, T. E. Graedel, *Forging the anthropogenic iron cycle. Environmental Science & Technology* **41**, 5120 (2007).

20. D. W. Keith, *Why Capture CO_2 from the Atmosphere? Science* **325**, 1654 (2009).

21. K. S. Lackner, *Capture of carbon dioxide from ambient air. European Physical Journal-Special Topics* **176**, 93 (2009).

22. G. Hardin, *The Tragedy of the Commons. Science* **162**, 1243 (1968).

23. O. Tickell, *Kyoto2.* (Zed Books, London, 2008).

24. P. Barnes et al., *Creating an Earth Atmospheric Trust. Science* **319**, 724 (2008).

Index